Applied Discrete Structures

Part 2 - Applied Algebra

Applied Discrete Structures

Part 2 - Applied Algebra

Al Doerr
University of Massachusetts Lowell

Ken Levasseur
University of Massachusetts Lowell

May 21, 2023

Edition: 3rd Edition - version 10

Website: discretemath.org

©2023   Al Doerr, Ken Levasseur

Applied Discrete Structures by Alan Doerr and Kenneth Levasseur is licensed under a Creative Commons Attribution-NonCommercial-ShareAlike 3.0 United States License. You are free to Share: copy and redistribute the material in any medium or format; Adapt: remix, transform, and build upon the material. You may not use the material for commercial purposes. The licensor cannot revoke these freedoms as long as you follow the license terms.

To our families

Donna, Christopher, Melissa, and Patrick Doerr

Karen, Joseph, Kathryn, and Matthew Levasseur

Acknowledgements

We would like to acknowledge the following instructors for their helpful comments and suggestions.

- Tibor Beke, UMass Lowell
- Alex DeCourcy, UMass Lowell
- Vince DiChiacchio
- Warren Grieff, UMass Lowell
- Matthew Haner, Mansfield University (PA)
- Dan Klain, UMass Lowell
- Sitansu Mittra, UMass Lowell
- Ravi Montenegro, UMass Lowell
- Tony Penta, UMass Lowell
- Jim Propp, UMass Lowell
- Ivan Temesvari, Oakton College
- Thao Tran, UMass Lowell
- Richard Voss, Florida Atlantic U.

I'd like to particularly single out Jim Propp for his close scrutiny, along with that of his students, many of whom are listed below.

I would like to thank Rob Beezer, David Farmer, Karl-Dieter Crisman and other participants on the pretext-xml-support group for their guidance and work on MathBook XML, which has now been renamed PreTeXt. Thanks to the Pedagogy Subcommittee of the UMass Lowell Transformational Education Committee for their financial assistance in helping getting this project started.

Many students have provided feedback and pointed out typos in several editions of this book. They are listed below. Students with no affiliation listed are from UMass Lowell.

- Ryan Allen
- Rebecca Alves
- Anonymous student from Florida Atlantic U.
- David Arakelian
- Junaid Baig
- Anju Balaji
- Carlos Barrientos
- Raymond Berger, Eckerd College
- Ron Burkey, Independant Contributor
- Chris Berns
- Brianne Bindas
- Nicholas Bishop
- Nathan Blood
- Cameron Bolduc
- Sam Bouchard
- Amber Breslau
- Rachel Bryan
- Nam Bui

- Courtney Caldwell
- Joseph Calles
- Rebecca Campbelli
- AJ Capone
- Eric Carey
- Emily Cashman
- Cora Casteel
- Rachel Chaiser, U. of Puget Sound
- Sam Chambers
- Vanessa Chen
- Hannah Chiodo
- Sofya Chow
- David Connolly
- Sean Cummings
- Alex DeCourcy
- Ryan Delosh
- Hillari Denny
- Matthew Edwards
- John El-Helou
- Adam Espinola
- Josh Everett
- Christian Franco
- Anthony Gaeta
- David Genis
- Lisa Gieng
- Holly Goodreau
- Lilia Heimold
- Kevin Holmes
- Benjamin Houle
- Alexa Hyde
- Michael Ingemi
- Eunji Jang
- Matthew Jarek
- Kyle Joaquim
- Mathew John
- Devin Johnson
- Jeremy Joubert
- William Jozefczyk
- Joel Keaton
- Antony Kellermann
- Yorgo A. Kennos
- Thomas Kiley
- Cody Kingman
- Leant Seu Kim
- Jessica Kramer
- John Kuczynski
- Auris Kveraga
- Justin LaGree
- Daven Lagu
- Kendra Lansing
- Gregory Lawrence
- Pearl Laxague
- Kevin Le
- Thien Tran Le
- Matt LeBlanc
- Maxwell Leduc
- Ariel Leva
- Robert Liana
- Tammy Liu
- Anson Lu
- Laura Lucaciu
- Kelly Ly
- Kevin Mackie, Learning Assistant
- Alexandra Mai
- Andrew Magee
- Matthew Malone
- Logan Mann
- Sam Marquis
- Amy Mazzucotelli
- Colby Mei
- Adam Melle
- Jason McAdam
- Nick McArdle
- Christine McCarthy
- Shelbylynn McCoy
- Conor McNierney
- Albara Mehene
- Joshua Michaud
- Max Mints
- Charles Mirabile
- Timothy Miskell
- Genevieve Moore
- Mike Morley

- Zach Mulcahy
- Tessa Munoz
- Zachary Murphy
- Logan Nadeau
- Carol Nguyen
- Hung Nguyen
- Tam Nguyen
- Shelly Noll
- Steven Oslan, the champion typo finder!
- Harsh Patel
- Beck Peterson
- Donna Petitti
- Paola Pevzner
- Zach Phillips
- Sam Pizette
- Angelo Pocoli
- Samantha Poirier
- Roshan Ravi
- Ian Roberts
- John Raisbeck
- Adelia Reid
- Derek Ross
- Tyler Ross
- Jacob Rothmel
- Zach Rush
- Ryan Saadah
- Steve Sadler, Bellevue College (WA)
- Doug Salvati
- Chita Sano
- Noah Schultz
- Anna Sergienko
- Ben Shipman
- Florens Shosho
- Lorraine Sill
- Jonathan Silva
- Joshua Simard
- Mason Sirois
- Gabriel Shahrouzi
- Sana Shaikh
- Joel Slebodnick
- Greg Smelkov
- Andrew Somerville
- Samuel Stanley
- Alicia Stransky
- Brandon Swanberg
- Joshua Sullivan
- James Tan
- Steven Tang
- Amitha Thalanki
- Bunchhoung Tiv
- Andy Tran
- Tina Tran
- Mary Tsykora
- Joanel Vasquez
- Rolando Vera
- Anh Vo
- Nick Wackowski
- Ryan Wallace
- Uriah Wardlaw
- Phoebe Watkins
- Zach Weaver
- Steve Werren
- Laura Wikoff
- Henry Zhu
- Several students at Luzurne County Community College (PA)

x

Preface

This is part 2 of *Applied Discrete Structures*. In order to maintain uniform numbering and avoid broken links, stubs of the first ten chapters are included and contain the introduction elements that are referenced in chapters 11 through 16. To see all of chapters 1 through 10, visit our web page at http://faculty.uml.edu/klevasseur/ADS2.

Applied Discrete Structures is designed for use in a university course in discrete mathematics spanning up to two semesters. Its original design was for computer science majors to be introduced to the mathematical topics that are useful in computer science. It can also serve the same purpose for mathematics majors, providing a first exposure to many essential topics.

We embarked on this open-source project in 2010, twenty-one years after the publication of the 2nd edition of *Applied Discrete Structures for Computer Science* in 1989. We had signed a contract for the second edition with Science Research Associates in 1988 but by the time the book was ready to print, SRA had been sold to MacMillan. Soon after, the rights had been passed on to Pearson Education, Inc. In 2010, the long-term future of printed textbooks is uncertain. In the meantime, textbook prices (both printed and e-books) had increased and a growing open source textbook movement had started. One of our objectives in revisiting this text is to make it available to our students in an affordable format. In its original form, the text was peer-reviewed and was adopted for use at several universities throughout the country. For this reason, we see *Applied Discrete Structures* as not only an inexpensive alternative, but a high quality alternative.

The current version of *Applied Discrete Structures* has been developed using *PreTeXt*, a lightweight XML application for authors of scientific articles, textbooks and monographs initiated by Rob Beezer, U. of Puget Sound. When the PreTeXt project was launched, it was the natural next step. The features of PreTeXt make it far more readable, with easy production of web, pdf and print formats.

Ken Levasseur
Lowell MA

Contents

Acknowledgements   vii

Preface   xi

1   Set Theory I   1

2   Combinatorics   3

3   Logic   5

4   More on Sets   7

5   Introduction to Matrix Algebra   9

6   Relations   11

7   Functions   13
    Exercises 13

8   Recursion and Recurrence Relations   15

9   Graph Theory   17

10   Trees   19

11   Algebraic Structures   21
    11.1 Operations 21
    11.2 Algebraic Systems 25
    11.3 Some General Properties of Groups 30
    11.4 Greatest Common Divisors and the Integers Modulo n 35

11.5	Subsystems	44
11.6	Direct Products	49
11.7	Isomorphisms	55

12 More Matrix Algebra · 63

12.1	Systems of Linear Equations	63
12.2	Matrix Inversion	72
12.3	An Introduction to Vector Spaces	76
12.4	The Diagonalization Process	83
12.5	Some Applications	91
12.6	Linear Equations over the Integers Mod 2	97

13 Boolean Algebra · 101

13.1	Posets Revisited	103
13.2	Lattices	107
13.3	Boolean Algebras	109
13.4	Atoms of a Boolean Algebra	112
13.5	Finite Boolean Algebras as n-tuples of 0's and 1's	116
13.6	Boolean Expressions	117
13.7	A Brief Introduction to Switching Theory and Logic Design	121

14 Monoids and Automata · 129

14.1	Monoids	129
14.2	Free Monoids and Languages	132
14.3	Automata, Finite-State Machines	138
14.4	The Monoid of a Finite-State Machine	143
14.5	The Machine of a Monoid	146

15 Group Theory and Applications · 149

15.1	Cyclic Groups	149
15.2	Cosets and Factor Groups	155
15.3	Permutation Groups	161
15.4	Normal Subgroups and Group Homomorphisms	170
15.5	Coding Theory, Linear Codes	177

16 An Introduction to Rings and Fields · 187

16.1	Rings, Basic Definitions and Concepts	187
16.2	Fields	195
16.3	Polynomial Rings	199
16.4	Field Extensions	204
16.5	Power Series	208

Appendices

A Algorithms · 213

A.1	An Introduction to Algorithms	213
A.2	The Invariant Relation Theorem	217

B Hints and Solutions to Selected Exercises	**221**
C Notation	**257**
D Glossary	**259**
An Informal Glossary of Terms.	259

Back Matter

References	**263**
Index	**267**

Chapter 1

Set Theory I

Goals for Chapter 1. This is a stub for Part 2 of Applied Discrete Structures. To see the whole chapter, visit our web page at http://faculty.uml.edu/klevasseur/ADS2.

In this chapter we will cover some of the basic set language and notation that will be used throughout the text. Venn diagrams will be introduced in order to give the reader a clear picture of set operations. In addition, we will describe the binary representation of positive integers (Section 1.4) and introduce summation notation and its generalizations (Section 1.5).

Algorithm 1.0.1 Binary Conversion Algorithm. *An algorithm for determining the binary representation of a positive integer.*

Input: a positive integer n.

Output: the binary representation of n in the form of a list of bits, with units bit last, twos bit next to last, etc.

(1) $k := n$ //initialize k

(2) $L := \{\ \}$ //initialize L to an empty list

(3) While $k > 0$ do

    (a) $q := k$ *div* 2 //divide k by 2

    (b) $r := k$ *mod* 2

    (c) $L := $ prepend r to L //add r to the front of L

    (d) $k := q$ //reassign k

Chapter 2

Combinatorics

This is a stub for Part 2 of Applied Discrete Structures. To see the whole chapter, visit our web page at http://faculty.uml.edu/klevasseur/ADS2.

Throughout this book we will be counting things. In this chapter we will outline some of the tools that will help us count.

Counting occurs not only in highly sophisticated applications of mathematics to engineering and computer science but also in many basic applications. Like many other powerful and useful tools in mathematics, the concepts are simple; we only have to recognize when and how they can be applied.

Chapter 3

Logic

This is a stub for Part 2 of Applied Discrete Structures. To see the whole chapter, visit our web page at http://faculty.uml.edu/klevasseur/ADS2.

In this chapter, we will introduce some of the basic concepts of mathematical logic. In order to fully understand some of the later concepts in this book, you must be able to recognize valid logical arguments. Although these arguments will usually be applied to mathematics, they employ the same techniques that are used by a lawyer in a courtroom or a physician examining a patient. An added reason for the importance of this chapter is that the circuits that make up digital computers are designed using the same algebra of propositions that we will be discussing.

Chapter 4

More on Sets

This is a stub for Part 2 of Applied Discrete Structures. To see the whole chapter, visit our web page at http://faculty.uml.edu/klevasseur/ADS2.

In this chapter we shall look more closely at some basic facts about sets. One question we could ask ourselves is: Can we manipulate sets similarly to the way we manipulated expressions in basic algebra, or to the way we manipulated propositions in logic? In basic algebra we are aware that $a \cdot (b+c) = a \cdot b + a \cdot c$ for all real numbers a, b, and c. In logic we verified an analogue of this statement, namely, $p \wedge (q \vee r) \Leftrightarrow (p \wedge q) \vee (p \wedge r))$, where p, q, and r were arbitrary propositions. If A, B, and C are arbitrary sets, is $A \cap (B \cup C) = (A \cap B) \cup (A \cap C)$? How do we convince ourselves of it is truth, or discover that it is false? Let us consider some approaches to this problem, look at their pros and cons, and determine their validity. Later in this chapter, we introduce partitions of sets and minsets.

Definition 4.0.1 Minset. Let $\{B_1, B_2, \ldots, B_n\}$ be a set of subsets of set A. Sets of the form $D_1 \cap D_2 \cap \cdots \cap D_n$, where each D_i may be either B_i or B_i^c, is called a minset generated by $B_1, B_2,\ldots$ and B_n. ◊

Definition 4.0.2 Minset Normal Form. A set is said to be in minset normal form when it is expressed as the union of zero or more distinct nonempty minsets. ◊

Chapter 5

Introduction to Matrix Algebra

This is a stub for Part 2 of Applied Discrete Structures. To see the whole chapter, visit our web page at http://faculty.uml.edu/klevasseur/ADS2.

The purpose of this chapter is to introduce you to matrix algebra, which has many applications. You are already familiar with several algebras: elementary algebra, the algebra of logic, the algebra of sets. We hope that as you studied the algebra of logic and the algebra of sets, you compared them with elementary algebra and noted that the basic laws of each are similar. We will see that matrix algebra is also similar. As in previous discussions, we begin by defining the objects in question and the basic operations.

Chapter 6

Relations

This is a stub for Part 2 of Applied Discrete Structures. To see the whole chapter, visit our web page at http://faculty.uml.edu/klevasseur/ADS2.

One understands a set of objects completely only if the structure of that set is made clear by the interrelationships between its elements. For example, the individuals in a crowd can be compared by height, by age, or through any number of other criteria. In mathematics, such comparisons are called relations. The goal of this chapter is to develop the language, tools, and concepts of relations.

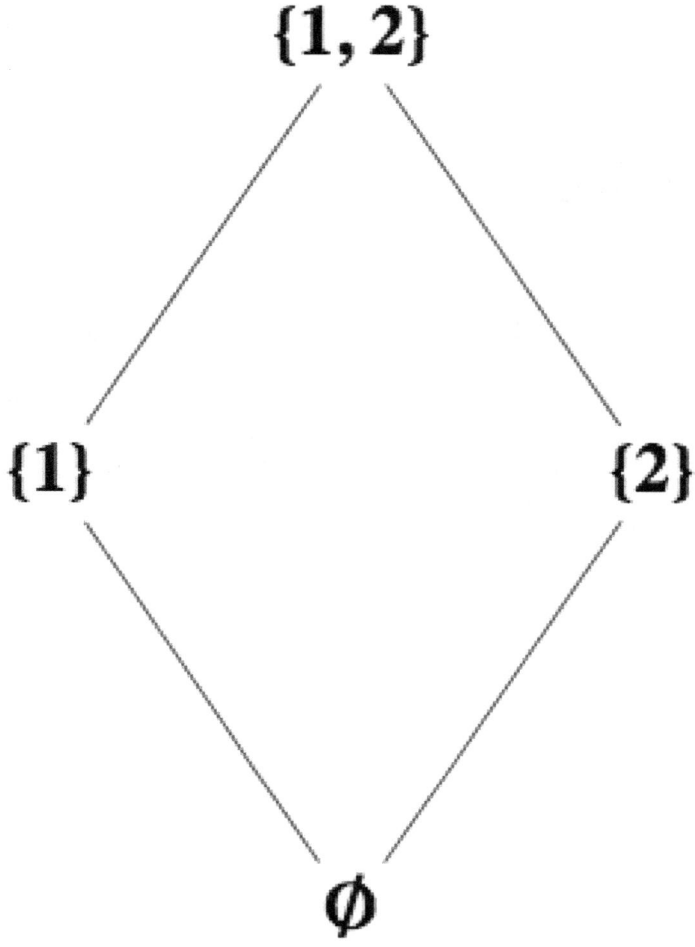

Figure 6.0.1 Hasse diagram for set containment on subsets of $\{1,2\}$

Definition 6.0.2 Congruence Modulo m. Let m be a positive integer, $m \geq 2$. We define **congruence modulo m** to be the relation $\equiv_m$ defined on the integers by

$$a \equiv_m b \Leftrightarrow m \mid (a-b)$$

Chapter 7

Functions

This is a stub for Part 2 of Applied Discrete Structures. To see the whole chapter, visit our web page at http://faculty.uml.edu/klevasseur/ADS2.

In this chapter we will consider some basic concepts of the relations that are called functions. A large variety of mathematical ideas and applications can be more completely understood when expressed through the function concept.

Theorem 7.0.1 The composition of injections is an injection. *If $f : A \to B$ and $g : B \to C$ are injections, then $g \circ f : A \to C$ is an injection.*

Theorem 7.0.2 The composition of surjections is a surjection. *If $f : A \to B$ and $g : B \to C$ are surjections, then $g \circ f : A \to C$ is a surjection.*

Exercises

7. If A and B are finite sets, how many different functions are there from A into B?

Chapter 8

Recursion and Recurrence Relations

This is a stub for Part 2 of Applied Discrete Structures. To see the whole chapter, visit our web page at http://faculty.uml.edu/klevasseur/ADS2.

An essential tool that anyone interested in computer science must master is how to think recursively. The ability to understand definitions, concepts, algorithms, etc., that are presented recursively and the ability to put thoughts into a recursive framework are essential in computer science. One of our goals in this chapter is to help the reader become more comfortable with recursion in its commonly encountered forms.

A second goal is to discuss recurrence relations. We will concentrate on methods of solving recurrence relations, including an introduction to generating functions.

Algorithm 8.0.1 Algorithm for Solving Homogeneous Finite Order Linear Relations.

(a) *Write out the characteristic equation of the relation $S(k) + C_1 S(k-1) + \ldots + C_n S(k-n) = 0$, which is $a^n + C_1 a^{n-1} + \cdots + C_{n-1} a + C_n = 0$.*

(b) *Find all roots of the characteristic equation, the characteristic roots.*

(c) *If there are n distinct characteristic roots, $a_1, a_2, \ldots a_n$, then the general solution of the recurrence relation is $S(k) = b_1 a_1^k + b_2 a_2^k + \cdots + b_n a_n^k$. If there are fewer than n characteristic roots, then at least one root is a multiple root. If a_j is a double root, then the $b_j a_j^k$ term is replaced with $(b_{j0} + b_{j1} k) a_j^k$. In general, if a_j is a root of multiplicity p, then the $b_j a_j^k$ term is replaced with $\left(b_{j0} + b_{j1} k + \cdots + b_{j(p-1)} k^{p-1}\right) a_j^k$.*

(d) *If n initial conditions are given, we get n linear equations in n unknowns (the b_j's from Step 3) by substitution. If possible, solve these equations to determine a final form for $S(k)$.*

Example 8.0.2 Solution of a Third Order Recurrence Relation. Solve $S(k) - 7S(k-2) + 6S(k-3) = 0$, where $S(0) = 8$, $S(1) = 6$, and $S(2) = 22$.

(a) The characteristic equation is $a^3 - 7a + 6 = 0$.

(b) The only rational roots that we can attempt are $\pm 1, \pm 2, \pm 3,$ and ± 6. By checking these, we obtain the three roots 1, 2, and -3.

(c) The general solution is $S(k) = b_1 1^k + b_2 2^k + b_3 (-3)^k$. The first term can simply be written b_1 .

(d) $\left\{\begin{array}{c} S(0)=8 \\ S(1)=6 \\ S(2)=22 \end{array}\right\} \Rightarrow \left\{\begin{array}{c} b_1+b_2+b_3=8 \\ b_1+2b_2-3b_3=6 \\ b_1+4b_2+9b_3=22 \end{array}\right\}$ You can solve this system by elimination to obtain $b_1 = 5$, $b_2 = 2$, and $b_3 = 1$. Therefore, $S(k) = 5 + 2 \cdot 2^k + (-3)^k = 5 + 2^{k+1} + (-3)^k$

□

Chapter 9

Graph Theory

This is a stub for Part 2 of Applied Discrete Structures. To see the whole chapter, visit our web page at http://faculty.uml.edu/klevasseur/ADS2.

This chapter has three principal goals. First, we will identify the basic components of a graph and some of the features that many graphs have. Second, we will discuss some of the questions that are most commonly asked of graphs. Third, we want to make the reader aware of how graphs are used. In Section 9.1, we will discuss these topics in general, and in later sections we will take a closer look at selected topics in graph theory.

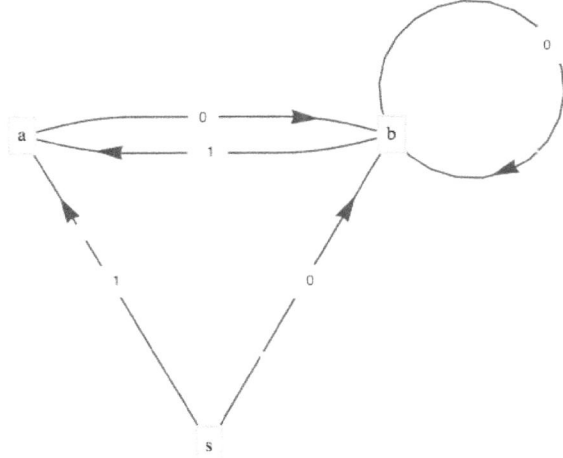

Figure 9.0.1 A directed graph

Mentioned in section 11.7:

Definition 9.0.2 Degree Sequence of a Graph. The degree sequence of an undirected graph is the non-increasing sequence of its vertex degrees. ◊

Mentioned in section 15.5:

Definition 9.0.3 The n-cube. Let $n \geq 1$, and let B^n be the set of strings of 0's and 1's with length n. The n-cube is the undirected graph with a vertex for each string in B^n and an edge connecting each pair of strings that differ in exactly one position. The n-cube is normally denoted Q_n. ◊

Chapter 10

Trees

This is a stub for Part 2 of Applied Discrete Structures. To see the whole chapter, visit our web page at http://faculty.uml.edu/klevasseur/ADS2.

In this chapter we will study the class of graphs called trees. Trees are frequently used in both mathematics and the sciences. Our solution of Example 10.0.1, p. 19 is one simple instance. Since they are often used to illustrate or prove other concepts, a poor understanding of trees can be a serious handicap. For this reason, our ultimate goals are to: (1) define the various common types of trees, (2) identify some basic properties of trees, and (3) discuss some of the common applications of trees.

Example 10.0.1 How many lunches can you have? A snack bar serves five different sandwiches and three different beverages. How many different lunches can a person order? One way of determining the number of possible lunches is by listing or enumerating all the possibilities. One systematic way of doing this is by means of a tree, as in the following figure.

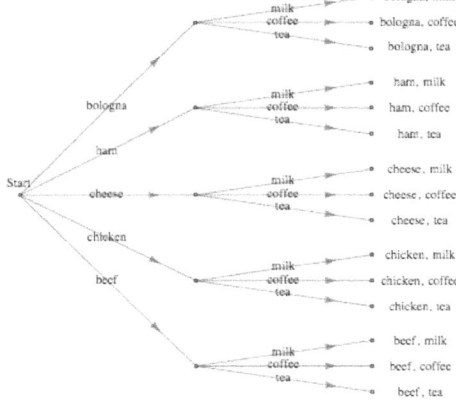

Figure 10.0.2 Tree diagram to enumerate the number of possible lunches.

Every path that begins at the position labeled START and goes to the right can be interpreted as a choice of one of the five sandwiches followed by a choice of one of the three beverages. Note that considerable work is required to arrive at the number fifteen this way; but we also get more than just a number. The result is a complete list of all possible lunches. If we need to answer a question that starts with "How many ... ," enumeration would be done only as a last resort. In a later chapter we will examine more enumeration techniques. □

Chapter 11

Algebraic Structures

Abelian Group

In **Abelian groups**, when computing,
With operands there's no refuting:
The expression bc
Is the same as cb.
Not en route to your job, yet commuting.

Howard Spindel, The Omnificent English Dictionary In Limerick Form

The primary goal of this chapter is to make the reader aware of what an algebraic system is and how algebraic systems can be studied at different levels of abstraction. After describing the concrete, axiomatic, and universal levels, we will introduce one of the most important algebraic systems at the axiomatic level, the group. In this chapter, group theory will be a vehicle for introducing the universal concepts of isomorphism, direct product, subsystem, and generating set. These concepts can be applied to all algebraic systems. The simplicity of group theory will help the reader obtain a good intuitive understanding of these concepts. In Chapter 15, we will introduce some additional concepts and applications of group theory. We will close the chapter with a discussion of how some computer hardware and software systems use the concept of an algebraic system.

11.1 Operations

One of the first mathematical skills that we all learn is how to add a pair of positive integers. A young child soon recognizes that something is wrong if a sum has two values, particularly if his or her sum is different from the teacher's. In addition, it is unlikely that a child would consider assigning a non-positive value to the sum of two positive integers. In other words, at an early age we probably know that the sum of two positive integers is unique and belongs to the set of positive integers. This is what characterizes all binary operations on a set.

11.1.1 What is an Operation?

Definition 11.1.1 Binary Operation. Let S be a nonempty set. A binary operation on S is a rule that assigns to each ordered pair of elements of S a unique element of S. In other words, a binary operation is a function from $S \times S$ into S. ◇

Example 11.1.2 Some common binary operations. Union and intersection are both binary operations on the power set of any universe. Addition and multiplication are binary operators on the natural numbers. Addition and multiplication are binary operations on the set of 2 by 2 real matrices, $M_{2\times 2}(\mathbb{R})$. Division is a binary operation on some sets of numbers, such as the positive reals. But on the integers ($1/2 \notin \mathbb{Z}$) and even on the real numbers ($1/0$ is not defined), division is not a binary operation. □

Note 11.1.3

(a) We stress that the image of each ordered pair must be in S. This requirement disqualifies subtraction on the natural numbers from consideration as a binary operation, since $1 - 2$ is not a natural number. Subtraction is a binary operation on the integers.

(b) On Notation. Despite the fact that a binary operation is a function, symbols, not letters, are used to name them. The most commonly used symbol for a binary operation is an asterisk, $*$. We will also use a diamond, $\diamond$, when a second symbol is needed.

If $*$ is a binary operation on S and $a, b \in S$, there are three common ways of denoting the image of the pair (a, b). They are:

$$*ab \qquad a * b \qquad ab*$$
$$\text{Prefix Form} \quad \text{Infix Form} \quad \text{Postfix Form}$$

We are all familiar with infix form. For example, $2 + 3$ is how everyone is taught to write the sum of 2 and 3. But notice how $2 + 3$ was just described in the previous sentence! The word sum preceded 2 and 3. Orally, prefix form is quite natural to us. The prefix and postfix forms are superior to infix form in some respects. In Chapter 10, we saw that algebraic expressions with more than one operation didn't need parentheses if they were in prefix or postfix form. However, due to our familiarity with infix form, we will use it throughout most of the remainder of this book.

Some operations, such as negation of numbers and complementation of sets, are not binary, but unary operators.

Definition 11.1.4 Unary Operation. Let S be a nonempty set. A unary operator on S is a rule that assigns to each element of S a unique element of S. In other words, a unary operator is a function from S into S. ◇

11.1.2 Properties of Operations

Whenever an operation on a set is encountered, there are several properties that should immediately come to mind. To effectively make use of an operation, you should know which of these properties it has. By now, you should be familiar with most of these properties. We will list the most common ones here to refresh your memory and define them for the first time in a general setting.

First we list properties of a single binary operation.

Definition 11.1.5 Commutative Property. Let $*$ be a binary operation on a set S. We say that $*$ is **commutative** if and only if $a * b = b * a$ for all

11.1. OPERATIONS

$a, b \in S$. ◇

Definition 11.1.6 Associative Property. Let $*$ be a binary operation on a set S. We say that $*$ is **associative** if and only if $(a * b) * c = a * (b * c)$ for all $a, b, c \in S$. ◇

Definition 11.1.7 Identity Property. Let $*$ be a binary operation on a set S. We say that $*$ **has an identity** if and only if there exists an element, e, in S such that $a * e = e * a = a$ for all $a \in S$. ◇

The next property presumes that $*$ has the identity property.

Definition 11.1.8 Inverse Property. Let $*$ be a binary operation on a set S. We say that $*$ has the **inverse property** if and only if for each $a \in S$, there exists $b \in S$ such that $a * b = b * a = e$. We call b an inverse of a. ◇

Definition 11.1.9 Idempotent Property. Let $*$ be a binary operation on a set S. We say that $*$ is **idempotent** if and only if $a * a = a$ for all $a \in S$. ◇

Now we list properties that apply to two binary operations.

Definition 11.1.10 Left Distributive Property. Let $*$ and $\diamond$ be binary operations on a set S. We say that $\diamond$ is left distributive over $*$ if and only if $a \diamond (b * c) = (a \diamond b) * (a \diamond c)$ for all $a, b, c \in S$. ◇

Definition 11.1.11 Right Distributive Property. Let $*$ and $\diamond$ be binary operations on a set S. We say that $\diamond$ is right distributive over $*$ if and only if $(b * c) \diamond a = (b \diamond a) * (c \diamond a)$ for all $a, b, c \in S$. ◇

Definition 11.1.12 Distributive Property. Let $*$ and $\diamond$ be binary operations on a set S. We say that $\diamond$ is distributive over $*$ if and only if $\diamond$ is both left and right distributive over $*$. ◇

There is one significant property of unary operations.

Definition 11.1.13 Involution Property. Let $-$ be a unary operation on S. We say that $-$ has the involution property if $-(-a) = a$ for all $a \in S$. ◇

Finally, a property of sets, as they relate to operations.

Definition 11.1.14 Closure Property. Let T be a subset of S and let $*$ be a binary operation on S. We say that T is closed under $*$ if $a, b \in T$ implies that $a * b \in T$. ◇

In other words, T is closed under $*$ if by operating on elements of T with $*$, you can't get new elements that are outside of T.

Example 11.1.15 Some examples of closure and non-closure.

(a) The odd integers are closed under multiplication, but not under addition.

(b) Let p be a proposition over U and let A be the set of propositions over U that imply p. That is; $q \in A$ if $q \Rightarrow p$. Then A is closed under both conjunction and disjunction.

(c) The set of positive integers that are multiples of 5 is closed under both addition and multiplication.

□

It is important to realize that the properties listed above depend on both the set and the operation(s). Statements such as "Multiplication is commutative." or "The positive integers are closed." are meaningless on their own. Naturally, if we have established a context in which the missing set or operation is clearly implied, then they would have meaning.

11.1.3 Operation Tables

If the set on which a binary operation is defined is small, a table is often a good way of describing the operation. For example, we might want to define $\oplus$ on $\{0,1,2\}$ by $a \oplus b = \begin{cases} a+b & \text{if } a+b < 3 \\ a+b-3 & \text{if } a+b \geq 3 \end{cases}$ The table for $\oplus$ is

$\oplus$	0	1	2
0	0	1	2
1	1	2	0
2	2	0	1

The top row and left column of an operation table are the column and row headings, respectively. To determine $a \oplus b$, find the entry in the row labeled a and the column labeled b. The following operation table serves to define $*$ on $\{i, j, k\}$.

$*$	i	j	k
i	i	i	i
j	j	j	j
k	k	k	k

Note that $j*k = j$, yet $k*j = k$. Thus, $*$ is not commutative. Commutativity is easy to identify in a table: the table must be symmetric with respect to the diagonal going from the top left to lower right.

11.1.4 Exercises

1. Determine the properties that the following operations have on the positive integers.

   (a) addition

   (b) multiplication

   (c) M defined by $aMb = $ larger of a and b

   (d) m defined by $amb = $ smaller of a and b

   (e) @ defined by $a@b = a^b$

2. Determine the properties that the following operations have on given sets.

   (a) Intersection on the set of subsets of $\{1, 2, 3, 4\}$.

   (b) ω defined on the positive integers by $a\omega b = b$.

   (c) $*$ defined on the integers by $a * b = a + b - 2$

   (d) $\diamond$ defined by $a \diamond b = \frac{ab}{2}$.

   (e) Concatenation on the set of all strings of zeros and ones.

3. Let $*$ be an operation on a set S and $A, B \subseteq S$. Prove that if A and B are both closed under $*$, then $A \cap B$ is also closed under $*$, but $A \cup B$ need not be.

4. How can you pick out the identity of an operation from its table?

5. Define $a * b$ by $|a - b|$, the absolute value of $a - b$. Which properties does $*$ have on the set of natural numbers, $\mathbb{N}$?

6. Which pairs of operations in Exercise 1 are distributive over one another?

11.2 Algebraic Systems

An algebraic system is a mathematical system consisting of a set called the domain and one or more operations on the domain. If V is the domain and $*_1, *_2, \ldots, *_n$ are the operations, $[V; *_1, *_2, \ldots, *_n]$ denotes the mathematical system. If the context is clear, this notation is abbreviated to V.

11.2.1 Monoids at Two Levels

Consider the following two examples of algebraic systems.

(a) Let B^* be the set of all finite strings of 0's and 1's including the null (or empty) string, λ. An algebraic system is obtained by adding the operation of concatenation. The concatenation of two strings is simply the linking of the two strings together in the order indicated. The concatenation of strings a with b is denoted $a + b$. For example, $01101 + 101 = 01101101$ and $\lambda + 100 = 100$. Note that concatenation is an associative operation and that λ is the identity for concatenation.

A note on notation: There isn't a standard symbol for concatenation. We have chosen $+$ to be consistent with the notation used in Python and Sage for the concatenation.

(b) Let M be any nonempty set and let $*$ be any operation on M that is associative and has an identity in M. Any such system is called a **monoid**. We introduce monoids briefly here, but will discuss them further in Chapter 14, p. 129

Our second example might seem strange, but we include it to illustrate a point. The algebraic system $[B^*; +]$ is a special case of $[M; *]$. Most of us are much more comfortable with B^* than with M. No doubt, the reason is that the elements in B^* are more concrete. We know what they look like and exactly how they are combined. The description of M is so vague that we don't even know what the elements are, much less how they are combined. Why would anyone want to study M? The reason is related to this question: What theorems are of interest in an algebraic system? Answering this question is one of our main objectives in this chapter. Certain properties of algebraic systems are called algebraic properties, and any theorem that says something about the algebraic properties of a system would be of interest. The ability to identify what is algebraic and what isn't is one of the skills that you should learn from this chapter.

Now, back to the question of why we study M. Our answer is to illustrate the usefulness of M with a theorem about M.

Theorem 11.2.1 A Monoid Theorem. *If a, b are elements of M and $a * b = b * a$, then $(a * b) * (a * b) = (a * a) * (b * b)$.*
Proof.

$$\begin{aligned}
(a * b) * (a * b) &= a * (b * (a * b)) &&\text{Why?} \\
&= a * ((b * a) * b) &&\text{Why?} \\
&= a * ((a * b) * b) &&\text{Why?} \\
&= a * (a * (b * b)) &&\text{Why?} \\
&= (a * a) * (b * b) &&\text{Why?}
\end{aligned}$$

∎

The power of this theorem is that it can be applied to any algebraic system that M describes. Since B^* is one such system, we can apply Theorem 11.2.1, p. 25 to any two strings that commute. For example, 01 and 0101. Although a special case of this theorem could have been proven for B^*, it would not have been any easier to prove, and it would not have given us any insight into other special cases of M.

Example 11.2.2 More Concrete Monoids. Consider the set of 2×2 real matrices, $M_{2\times 2}(\mathbb{R})$, with the operation of matrix multiplication. In this context, Theorem 11.2.1, p. 25 can be interpreted as saying that if $AB = BA$, then $(AB)^2 = A^2 B^2$. One pair of matrices that this theorem applies to is $\begin{pmatrix} 2 & 1 \\ 1 & 2 \end{pmatrix}$ and $\begin{pmatrix} 3 & -4 \\ -4 & 3 \end{pmatrix}$.

For another pair of concrete monoids, we start with a universal set $U = \{1, 2, 3, 4, 5\}$ - although we could be a little less specific an imaging U to be any nonempty set. The power set of U with intersection, and the power set of U with union are both monoids. What the identities of these monoids? Are they really the same monoid? We will answer this last question in Section 11.7, p. 55. □

11.2.2 Levels of Abstraction

One of the fundamental tools in mathematics is abstraction. There are three levels of abstraction that we will identify for algebraic systems: concrete, axiomatic, and universal.

11.2.2.1 The Concrete Level

Almost all of the mathematics that you have done in the past was at the concrete level. As a rule, if you can give examples of a few typical elements of the domain and describe how the operations act on them, you are describing a concrete algebraic system. Two examples of concrete systems are B^* and $M_{2\times 2}(\mathbb{R})$. A few others are:

(a) The integers with addition. Of course, addition isn't the only standard operation that we could include. Technically, if we were to add multiplication, we would have a different system.

(b) The subsets of the natural numbers, with union, intersection, and complementation.

(c) The complex numbers with addition and multiplication.

11.2.2.2 The Axiomatic Level

The next level of abstraction is the axiomatic level. At this level, the elements of the domain are not specified, but certain axioms are stated about the number of operations and their properties. The system that we called M is an axiomatic system. Some combinations of axioms are so common that a name is given to any algebraic system to which they apply. Any system with the properties of M is called a monoid. The study of M would be called monoid theory. The assumptions that we made about M, associativity and the existence of an identity, are called the monoid axioms. One of your few brushes with the axiomatic level may have been in your elementary algebra course. Many algebra texts identify the properties of the real numbers with addition and multiplication as the field axioms. As we will see in Chapter 16, "Rings and Fields," the real numbers share these axioms with other concrete systems, all of which are called fields.

11.2.2.3 The Universal Level

The final level of abstraction is the universal level. There are certain concepts, called universal algebra concepts, that can be applied to the study of all algebraic systems. Although a purely universal approach to algebra would be much too abstract for our purposes, defining concepts at this level should make it easier to organize the various algebraic theories in your own mind. In this chapter, we will consider the concepts of isomorphism, subsystem, and direct product.

11.2.3 Groups

To illustrate the axiomatic level and the universal concepts, we will consider yet another kind of axiomatic system, the group. In Chapter 5 we noted that the simplest equation in matrix algebra that we are often called upon to solve is $AX = B$, where A and B are known square matrices and X is an unknown matrix. To solve this equation, we need the associative, identity, and inverse laws. We call the systems that have these properties groups.

Definition 11.2.3 Group. A group consists of a nonempty set G and a binary operation $*$ on G satisfying the properties

(a) $*$ is associative on G: $(a * b) * c = a * (b * c)$ for all $a, b, c \in G$.

(b) There exists an identity element, $e \in G$, such that $a * e = e * a = a$ for all $a \in G$.

(c) For all $a \in G$, there exists an inverse; that is, there exists $b \in G$ such that $a * b = b * a = e$.

$\diamond$

A group is usually denoted by its set's name, G, or occasionally by $[G; *]$ to emphasize the operation. At the concrete level, most sets have a standard operation associated with them that will form a group. As we will see below, the integers with addition is a group. Therefore, in group theory $\mathbb{Z}$ always stands for $[\mathbb{Z}; +]$.

Note 11.2.4 Generic Symbols. At the axiomatic and universal levels, there are often symbols that have a special meaning attached to them. In group theory, the letter e is used to denote the identity element of whatever group is being discussed. A little later, we will prove that the inverse of a group element, a, is unique and its inverse is usually denoted a^{-1} and is read "a inverse." When a concrete group is discussed, these symbols are dropped in favor of concrete symbols. These concrete symbols may or may not be similar to the generic symbols. For example, the identity element of the group of integers is 0, and the inverse of n is denoted by $-n$, the additive inverse of n.

The asterisk could also be considered a generic symbol since it is used to denote operations on the axiomatic level.

Example 11.2.5 Some concrete groups.

(a) The integers with addition is a group. We know that addition is associative. Zero is the identity for addition: $0 + n = n + 0 = n$ for all integers n. The additive inverse of any integer is obtained by negating it. Thus the inverse of n is $-n$.

(b) The integers with multiplication is not a group. Although multiplication is associative and 1 is the identity for multiplication, not all integers have a multiplicative inverse in $\mathbb{Z}$. For example, the multiplicative inverse of 10 is $\frac{1}{10}$, but $\frac{1}{10}$ is not an integer.

(c) The power set of any set U with the operation of symmetric difference, $\oplus$, is a group. If A and B are sets, then $A \oplus B = (A \cup B) - (A \cap B)$. We will leave it to the reader to prove that $\oplus$ is associative over $\mathcal{P}(U)$. The identity of the group is the empty set: $A \oplus \emptyset = A$. Every set is its own inverse since $A \oplus A = \emptyset$. Note that $\mathcal{P}(U)$ is not a group with union or intersection.

□

Definition 11.2.6 Abelian Group. A group is abelian if its operation is commutative. ◊

Abel. Most of the groups that we will discuss in this book will be abelian. The term abelian is used to honor the Norwegian mathematician N. Abel (1802-29), who helped develop group theory.

Figure 11.2.7 Norwegian Stamp honoring Abel

11.2.4 Exercises

1. Discuss the analogy between the terms generic and concrete for algebraic systems and the terms generic and trade for prescription drugs.

2. Discuss the connection between groups and monoids. Is every monoid a group? Is every group a monoid?

3. Which of the following are groups?

    (a) B^* with concatenation (see Subsection 11.2.1, p. 25).

    (b) $M_{2\times 3}(\mathbb{R})$ with matrix addition.

    (c) $M_{2\times 3}(\mathbb{R})$ with matrix multiplication.

    (d) The positive real numbers, $\mathbb{R}^+$, with multiplication.

    (e) The nonzero real numbers, $\mathbb{R}^*$, with multiplication.

    (f) $\{1, -1\}$ with multiplication.

    (g) The positive integers with the operation M defined by $aMb = $ the larger of a and b.

11.2. ALGEBRAIC SYSTEMS

4. Prove that, $\oplus$, defined by $A \oplus B = (A \cup B) - (A \cap B)$ is an associative operation on $\mathcal{P}(U)$.

5. The following problem supplies an example of a non-abelian group. A rook matrix is a matrix that has only 0's and 1's as entries such that each row has exactly one 1 and each column has exactly one 1. The term rook matrix is derived from the fact that each rook matrix represents the placement of n rooks on an $n \times n$ chessboard such that none of the rooks can attack one another. A rook in chess can move only vertically or horizontally, but not diagonally. Let R_n be the set of $n \times n$ rook matrices. There are six 3×3 rook matrices:

$$I = \begin{pmatrix} 1 & 0 & 0 \\ 0 & 1 & 0 \\ 0 & 0 & 1 \end{pmatrix} \quad R_1 = \begin{pmatrix} 0 & 1 & 0 \\ 0 & 0 & 1 \\ 1 & 0 & 0 \end{pmatrix} \quad R_2 = \begin{pmatrix} 0 & 0 & 1 \\ 1 & 0 & 0 \\ 0 & 1 & 0 \end{pmatrix}$$

$$F_1 = \begin{pmatrix} 1 & 0 & 0 \\ 0 & 0 & 1 \\ 0 & 1 & 0 \end{pmatrix} \quad F_2 = \begin{pmatrix} 0 & 0 & 1 \\ 0 & 1 & 0 \\ 1 & 0 & 0 \end{pmatrix} \quad F_3 = \begin{pmatrix} 0 & 1 & 0 \\ 1 & 0 & 0 \\ 0 & 0 & 1 \end{pmatrix}$$

   (a) List the 2×2 rook matrices. They form a group, R_2, under matrix multiplication. Write out the multiplication table. Is the group abelian?

   (b) Write out the multiplication table for R_3. This is another group. Is it abelian?

   (c) How many 4×4 rook matrices are there? How many $n \times n$ rook matrices are there?

6. For each of the following sets, identify the standard operation that results in a group. What is the identity of each group?

   (a) The set of all 2×2 matrices with real entries and nonzero determinants.

   (b) The set of 2×3 matrices with rational entries.

7. Let $V = \{e, a, b, c\}$. Let $*$ be defined (partially) by $x * x = e$ for all $x \in V$. Write a complete table for $*$ so that $[V; *]$ is a group.

8. Consider the following set of six algebraic expressions, each defining a function on the set of real numbers excluding the numbers 0 and 1.

$$\mathcal{H} = \left\{ x, 1-x, \frac{1}{1-x}, \frac{1}{x}, \frac{x-1}{x}, \frac{x}{x-1} \right\} = \{y_1, y_2, y_3, y_4, y_5, y_6\}$$

We can operate on any two of these expressions using function composition. For example,

$$(y_3 \circ y_4)(x) = y_3(y_4(x)) = y_3(\frac{1}{x}) = \frac{1}{1-\frac{1}{x}} = \frac{x}{x-1} = y_6(x)$$

Therefore, $y_3 \circ y_4 = y_6$. Complete the following operation table for function composition on $\mathcal{H}$.

∘	y_1	y_2	y_3	y_4	y_5	y_6
y_1	y_1	y_2	y_3	y_4	y_5	y_6
y_2	y_2	y_1	y_6	y_5	y_4	
y_3	y_3	y_4		y_6		
y_4		y_3	y_2			
y_5						
y_6						

Figure 11.2.8 Partially completed operation table for $\mathcal{H}$

Is $[\mathcal{H}, \circ]$ a monoid? Is it a group?

11.3 Some General Properties of Groups

In this section, we will present some of the most basic theorems of group theory. Keep in mind that each of these theorems tells us something about every group. We will illustrate this point with concrete examples at the close of the section.

11.3.1 First Theorems

Theorem 11.3.1 Identities are Unique. *The identity of a group is unique.*

One difficulty that students often encounter is how to get started in proving a theorem like this. The difficulty is certainly not in the theorem's complexity. It's too terse! Before actually starting the proof, we rephrase the theorem so that the implication it states is clear.

Theorem 11.3.2 Identities are Unique - Rephrased. *If $G = [G; *]$ is a group and e is an identity of G, then no other element of G is an identity of G.*
Proof. (Indirect): Suppose that $f \in G$, $f \neq e$, and f is an identity of G. We will show that $f = e$, which is a contradiction, completing the proof.

$$\begin{aligned} f &= f * e \quad \text{Since } e \text{ is an identity} \\ &= e \quad \text{Since } f \text{ is an identity} \end{aligned}$$

∎

Next we justify the phrase "... *the* inverse of an element of a group."

Theorem 11.3.3 Inverses are Unique. *The inverse of any element of a group is unique.*

The same problem is encountered here as in the previous theorem. We will leave it to the reader to rephrase this theorem. The proof is also left to the reader to write out in detail. Here is a hint: If b and c are both inverses of a, then you can prove that $b = c$. If you have difficulty with this proof, note that we have already proven it in a concrete setting in Chapter 5.

As mentioned above, the significance of Theorem 11.3.3, p. 30 is that we can refer to *the* inverse of an element without ambiguity. The notation for the inverse of a is usually a^{-1} (note the exception below).

Example 11.3.4 Some Inverses.

(a) In any group, e^{-1} is the inverse of the identity e, which always is e.

11.3. SOME GENERAL PROPERTIES OF GROUPS

(b) $\left(a^{-1}\right)^{-1}$ is the inverse of a^{-1}, which is always equal to a (see Theorem 11.3.5, p. 31 below).

(c) $(x * y * z)^{-1}$ is the inverse of $x * y * z$.

(d) In a concrete group with an operation that is based on addition, the inverse of a is usually written $-a$. For example, the inverse of $k - 3$ in the group $[\mathbb{Z}; +]$ is written $-(k - 3) = 3 - k$. In the group of 2×2 matrices over the real numbers under matrix addition, the inverse of $\begin{pmatrix} 4 & 1 \\ 1 & -3 \end{pmatrix}$ is written $-\begin{pmatrix} 4 & 1 \\ 1 & -3 \end{pmatrix}$, which equals $\begin{pmatrix} -4 & -1 \\ -1 & 3 \end{pmatrix}$.

□

Theorem 11.3.5 Inverse of Inverse Theorem. *If a is an element of group G, then $\left(a^{-1}\right)^{-1} = a$.*

Again, we rephrase the theorem to make it clear how to proceed.

Theorem 11.3.6 Inverse of Inverse Theorem (Rephrased). *If a has inverse b and b has inverse c, then $a = c$.*

Proof.

$$\begin{aligned} a &= a * e & & e \text{ is the identity of } G \\ &= a * (b * c) & & \text{because } c \text{ is the inverse of } b \\ &= (a * b) * c & & \text{why?} \\ &= e * c & & \text{why?} \\ &= c & & \text{by the identity property} \end{aligned}$$

∎

The next theorem gives us a formula for the inverse of $a * b$. This formula should be familiar. In Chapter 5 we saw that if A and B are invertible matrices, then $(AB)^{-1} = B^{-1}A^{-1}$.

Theorem 11.3.7 Inverse of a Product. *If a and b are elements of group G, then $(a * b)^{-1} = b^{-1} * a^{-1}$.*

Proof. Let $x = b^{-1} * a^{-1}$. We will prove that x inverts $a * b$. Since we know that the inverse is unique, we will have proved the theorem.

$$\begin{aligned} (a * b) * x &= (a * b) * \left(b^{-1} * a^{-1}\right) \\ &= a * \left(b * \left(b^{-1} * a^{-1}\right)\right) \\ &= a * \left(\left(b * b^{-1}\right) * a^{-1}\right) \\ &= a * \left(e * a^{-1}\right) \\ &= a * a^{-1} \\ &= e \end{aligned}$$

Similarly, $x * (a * b) = e$; therefore, $(a * b)^{-1} = x = b^{-1} * a^{-1}$ ∎

Theorem 11.3.8 Cancellation Laws. *If a, b, and c are elements of group G, then*

$$\begin{aligned} \text{left cancellation:} & \quad (a * b = a * c) \Rightarrow b = c \\ \text{right cancellation:} & \quad (b * a = c * a) \Rightarrow b = c \end{aligned}$$

Proof. We will prove the left cancellation law. The right law can be proved in exactly the same way. Starting with $a * b = a * c$, we can operate on both $a * b$

and $a*c$ on the left with a^{-1}:

$$a^{-1}*(a*b) = a^{-1}*(a*c)$$

Applying the associative property to both sides we get

$$(a^{-1}*a)*b = (a^{-1}*a)*c \Rightarrow e*b = e*c$$
$$\Rightarrow b = c$$

∎

Theorem 11.3.9 Linear Equations in a Group. *If G is a group and $a, b \in G$, the equation $a*x = b$ has a unique solution, $x = a^{-1}*b$. In addition, the equation $x*a = b$ has a unique solution, $x = b*a^{-1}$.*
Proof. We prove the theorem only for $a*x = b$, since the second statement is proven identically.

$$\begin{aligned} a*x = b &= e*b \\ &= (a*a^{-1})*b \\ &= a*(a^{-1}*b) \end{aligned}$$

By the cancellation law, we can conclude that $x = a^{-1}*b$.

If c and d are two solutions of the equation $a*x = b$, then $a*c = b = a*d$ and, by the cancellation law, $c = d$. This verifies that $a^{-1}*b$ is the only solution of $a*x = b$. ∎

Note 11.3.10 Our proof of Theorem 11.3.9, p. 32 was analogous to solving the concrete equation $4x = 9$ in the following way:

$$4x = 9 = \left(4 \cdot \frac{1}{4}\right)9 = 4\left(\frac{1}{4}9\right)$$

Therefore, by cancelling 4,

$$x = \frac{1}{4} \cdot 9 = \frac{9}{4}$$

11.3.2 Exponents

If a is an element of a group G, then we establish the notation that

$$a*a = a^2 \qquad a*a*a = a^3 \qquad \text{etc.}$$

In addition, we allow negative exponents and define, for example,

$$a^{-2} = \left(a^2\right)^{-1}$$

Although this should be clear, proving exponentiation properties requires a more precise recursive definition.

Definition 11.3.11 Exponentiation in Groups. For $n \geq 0$, define a^n recursively by $a^0 = e$ and if $n > 0, a^n = a^{n-1}*a$. Also, if $n > 1$, $a^{-n} = (a^n)^{-1}$.
◊

Example 11.3.12 Some concrete exponentiations.

(a) In the group of positive real numbers with multiplication,

$$5^3 = 5^2 \cdot 5 = \left(5^1 \cdot 5\right) \cdot 5 = \left(\left(5^0 \cdot 5\right) \cdot 5\right) \cdot 5 = ((1 \cdot 5) \cdot 5) \cdot 5 = 5 \cdot 5 \cdot 5 = 125$$

11.3. SOME GENERAL PROPERTIES OF GROUPS

and
$$5^{-3} = (125)^{-1} = \frac{1}{125}$$

(b) In a group with addition, we use a different form of notation, reflecting the fact that in addition repeated terms are multiples, not powers. For example, in $[\mathbb{Z}; +]$, $a + a$ is written as $2a$, $a + a + a$ is written as $3a$, etc. The inverse of a multiple of a such as $-(a + a + a + a + a) = -(5a)$ is written as $(-5)a$.

□

Although we define, for example, $a^5 = a^4 * a$, we need to be able to extract the single factor on the left. The following lemma justifies doing precisely that.

Lemma 11.3.13 *Let G be a group. If $b \in G$ and $n \geq 0$, then $b^{n+1} = b * b^n$, and hence $b * b^n = b^n * b$.*
Proof. (By induction): If $n = 0$,

$$\begin{aligned}
b^1 &= b^0 * b \text{ by the definition of exponentiation} \\
&= e * b \text{ by the basis for exponentiation} \\
&= b * e \text{ by the identity property} \\
&= b * b^0 \text{ by the basis for exponentiation}
\end{aligned}$$

Now assume the formula of the lemma is true for some $n \geq 0$.

$$\begin{aligned}
b^{(n+1)+1} &= b^{(n+1)} * b \text{ by the definition of exponentiation} \\
&= (b * b^n) * b \text{ by the induction hypothesis} \\
&= b * (b^n * b) \text{ associativity} \\
&= b * \left(b^{n+1}\right) \text{ definition of exponentiation}
\end{aligned}$$

■

Based on the definitions for exponentiation above, there are several properties that can be proven. They are all identical to the exponentiation properties from elementary algebra.

Theorem 11.3.14 Properties of Exponentiation. *If a is an element of a group G, and m and n are integers,*

(1) $a^{-n} = \left(a^{-1}\right)^n$ *and hence* $(a^n)^{-1} = \left(a^{-1}\right)^n$

(2) $a^{n+m} = a^n * a^m$

(3) $(a^n)^m = a^{nm}$

Proof. We will leave the proofs of these properties to the reader. All three parts can be done by induction. For example the proof of the second part would start by defining the proposition $p(m)$, $m \geq 0$, to be $a^{n+m} = a^n * a^m$ for all n. The basis is $p(0) : a^{n+0} = a^n * a^0$. ■

Our final theorem is the only one that contains a hypothesis about the group in question. The theorem only applies to finite groups.

Theorem 11.3.15 *If G is a finite group, $|G| = n$, and a is an element of G, then there exists a positive integer m such that $a^m = e$ and $m \leq n$.*
Proof. Consider the list $a, a^2, \ldots, a^{n+1}$. Since there are $n + 1$ elements of G in this list, there must be some duplication. Suppose that $a^p = a^q$, with $p < q$.

Let $m = q - p$. Then

$$a^m = a^{q-p}$$
$$= a^q * a^{-p}$$
$$= a^q * (a^p)^{-1}$$
$$= a^q * (a^q)^{-1}$$
$$= e$$

Furthermore, since $1 \leq p < q \leq n+1$, $m = q - p \leq n$. ∎

Consider the concrete group $[\mathbb{Z}; +]$. All of the theorems that we have stated in this section except for the last one say something about $\mathbb{Z}$. Among the facts that we conclude from the theorems about $\mathbb{Z}$ are:

- Since the inverse of 5 is -5, the inverse of -5 is 5.
- The inverse of $-6 + 71$ is $-(71) + -(-6) = -71 + 6$.
- The solution of $12 + x = 22$ is $x = -12 + 22$.
- $-4(6) + 2(6) = (-4 + 2)(6) = -2(6) = -(2)(6)$.
- $7(4(3)) = (7 \cdot 4)(3) = 28(3)$ (twenty-eight 3s).

11.3.3 Exercises

1. Let $[G; *]$ be a group and a be an element of G. Define $f : G \to G$ by $f(x) = a * x$.

    (a) Prove that f is a bijection.

    (b) On the basis of part a, describe a set of bijections on the set of integers.

2. Rephrase Theorem 11.3.3, p. 30 and write out a clear proof.

3. Prove by induction on n that if $a_1, a_2, \ldots, a_n$ are elements of a group G, $n \geq 2$, then $(a_1 * a_2 * \cdots * a_n)^{-1} = a_n^{-1} * \cdots * a_2^{-1} * a_1^{-1}$. Interpret this result in terms of $[\mathbb{Z}; +]$ and $[\mathbb{R}^*; \cdot]$.

4. True or false? If a, b, c are elements of a group G, and $a * b = c * a$, then $b = c$. Explain your answer.

5. Prove Theorem 11.3.14, p. 33.

6. Each of the following facts can be derived by identifying a certain group and then applying one of the theorems of this section to it. For each fact, list the group and the theorem that are used.

    (a) $\left(\frac{1}{3}\right) 5$ is the only solution of $3x = 5$.

    (b) $-(-(-18)) = -18$.

    (c) If A, B, C are 3×3 matrices over the real numbers, with $A + B = A + C$, then $B = C$.

    (d) There is only one subset K of the natural numbers for which $K \oplus A = A$ for every subset A of the natural numbers.

11.4 Greatest Common Divisors and the Integers Modulo n

In this section introduce the greatest common divisor operation, and introduce an important family of concrete groups, the integers modulo n.

11.4.1 Greatest Common Divisors

We start with a theorem about integer division that is intuitively clear. We leave the proof as an exercise.

Theorem 11.4.1 The Division Property for Integers. *If $m, n \in \mathbb{Z}$, $n > 0$, then there exist two unique integers, q (the quotient) and r (the remainder), such that $m = nq + r$ and $0 \leq r < n$.*

Note 11.4.2 The division property says that if m is divided by n, you will obtain a quotient and a remainder, where the remainder is less than n. This is a fact that most elementary school students learn when they are introduced to long division. In doing the division problem $1986 \div 97$, you obtain a quotient of 20 and a remainder of 46. This result could either be written $\frac{1986}{97} = 20 + \frac{46}{97}$ or $1986 = 97 \cdot 20 + 46$. The latter form is how the division property is normally expressed in higher mathematics.

> **List 11.4.3**
>
> We now remind the reader of some interchangeable terminology that is used when $r = 0$, i. e., $a = bq$. All of the following say the same thing, just from slightly different points of view.
>
> | | |
> |---:|:---|
> | **divides** | b divides a |
> | **multiple** | a is a multiple of b |
> | **factor** | b is a factor of a |
> | **divisor** | b is a divisor of a |
>
> We use the notation $b \mid a$ if b divides a.

For example $2 \mid 18$ and $9 \mid 18$, but $4 \nmid 18$.

Caution: Don't confuse the "divides" symbol with the "divided by" symbol. The former is vertical while the latter is slanted. Notice however that the statement $2 \mid 18$ is related to the fact that $18/2$ is a whole number.

Definition 11.4.4 Greatest Common Divisor. Given two integers, a and b, not both zero, the greatest common divisor of a and b is the positive integer $g = \gcd(a, b)$ such that $g \mid a$, $g \mid b$, and

$$c \mid a \text{ and } c \mid b \Rightarrow c \mid g$$

$\diamond$

A little simpler way to think of $\gcd(a, b)$ is as the largest positive integer that is a divisor of both a and b. However, our definition is easier to apply in proving properties of greatest common divisors.

For small numbers, a simple way to determine the greatest common divisor is to use factorization. For example if we want the greatest common divisor of 660 and 350, you can factor the two integers: $660 = 2^2 \cdot 3 \cdot 5 \cdot 11$ and $350 = 2 \cdot 5^2 \cdot 7$. Single factors of 2 and 5 are the only ones that appear in both factorizations, so the greatest common divisor is $2 \cdot 5 = 10$.

Some pairs of integers have no common divisors other than 1. Such pairs are called *relatively prime pairs*.

Definition 11.4.5 Relatively Prime. A pair of integers, a and b, are relatively prime if $\gcd(a,b) = 1$ ◊

For example, $128 = 2^7$ and $135 = 3^3 \cdot 5$ are relatively prime. Notice that neither 128 nor 135 are primes. In general, a and b need not be prime in order to be relatively prime. However, if you start with a prime, like 23, for example, it will be relatively prime to everything but its multiples. This theorem, which we prove later, generalizes this observation.

Theorem 11.4.6 *If p is a prime and a is any integer such that $p \nmid a$ then $\gcd(a,p) = 1$*

11.4.2 The Euclidean Algorithm

As early as Euclid's time it was known that factorization wasn't the best way to compute greatest common divisors.

The Euclidean Algorithm is based on the following properties of the greatest common divisor.

$$\gcd(a, 0) = a \text{ for } a \neq 0 \quad (11.4.1)$$

$$\gcd(a, b) = \gcd(b, r) \text{ if } b \neq 0 \text{ and } a = bq + r \quad (11.4.2)$$

To compute $\gcd(a, b)$, we divide b into a and get a remainder r such that $0 \leq r < |b|$. By the property above, $\gcd(a,b) = \gcd(b,r)$. We repeat the process until we get zero for a remainder. The last nonzero number that is the second entry in our pairs is the greatest common divisor. This is inevitable because the second number in each pair is smaller than the previous one. Table 11.4.7, p. 36 shows an example of how this calculation can be systematically performed.

Table 11.4.7 A Table to Compute $\gcd(99, 53)$

q	a	b
-	99	53
1	53	46
1	46	7
6	7	4
1	4	3
1	3	1
3	1	0

Here is a Sage computation to verify that $\gcd(99, 53) = 1$. At each line, the value of a is divided by the value of b. The quotient is placed on the next line along with the new value of a, which is the previous b; and the remainder, which is the new value of b. Recall that in Sage, a%b is the remainder when dividing b into a.

```
a=99
b=53
while b>0:
    print('computing gcd of '+str(a)+' and '+str(b))
    [a,b]=[b,a%b]
print('result is '+str(a))
```

```
computing gcd of 99 and 53
computing gcd of 53 and 46
computing gcd of 46 and 7
computing gcd of 7 and 4
computing gcd of 4 and 3
computing gcd of 3 and 1
result is 1
```

Investigation 11.4.1 If you were allowed to pick two integers less than 100, which would you pick in order to force Euclid to work hardest? Here's a hint: The size of the quotient at each step determines how quickly the numbers decrease.

Solution. If quotient in division is 1, then we get the slowest possible completion. If $a = b + r$, then working backwards, each remainder would be the sum of the two previous remainders. This described a sequence like the Fibonacci sequence and indeed, the greatest common divisor of two consecutive Fibonacci numbers will take the most steps to reach a final value of 1.

For fixed values of a and b, consider integers of the form $ax + by$ where x and y can be any two integers. For example if $a = 36$ and $b = 27$, some of these results are tabulated below with x values along the left column and the y values on top.

	-6	-5	-4	-3	-2	-1	0	1	2	3	4	5	6
-6	-378	-351	-324	-297	-270	-243	-216	-189	-162	-135	-108	-81	-54
-5	-342	-315	-288	-261	-234	-207	-180	-153	-126	-99	-72	-45	-18
-4	-306	-279	-252	-225	-198	-171	-144	-117	-90	-63	-36	-9	18
-3	-270	-243	-216	-189	-162	-135	-108	-81	-54	-27	0	27	54
-2	-234	-207	-180	-153	-126	-99	-72	-45	-18	9	36	63	90
-1	-198	-171	-144	-117	-90	-63	-36	-9	18	45	72	99	126
0	-162	-135	-108	-81	-54	-27	0	27	54	81	108	135	162
1	-126	-99	-72	-45	-18	9	36	63	90	117	144	171	198
2	-90	-63	-36	-9	18	45	72	99	126	153	180	207	234
3	-54	-27	0	27	54	81	108	135	162	189	216	243	270
4	-18	9	36	63	90	117	144	171	198	225	252	279	306
5	18	45	72	99	126	153	180	207	234	261	288	315	342
6	54	81	108	135	162	189	216	243	270	297	324	351	378

Figure 11.4.8 Linear combinations of 36 and 27

Do you notice any patterns? What is the smallest positive value that you see in this table? How is it connected to 36 and 27?

Theorem 11.4.9 Bézout's lemma. *If a and b are positive integers, the smallest positive value of $ax + by$ is the greatest common divisor of a and b, $\gcd(a, b)$.*

Proof. If $g = \gcd(a, b)$, since $g \mid a$ and $g \mid b$, we know that $g \mid (ax + by)$ for any integers x and y, so $ax + by$ can't be less than g. To show that g is exactly the least positive value, we show that g can be attained by extending the Euclidean Algorithm. Performing the extended algorithm involves building a table of numbers. The way in which it is built maintains an invariant, and by The Invariant Relation Theorem, p. 218, we can be sure that the desired values of x and y are produced. ∎

To illustrate the algorithm, Table 11.4.10, p. 38 displays how to compute $\gcd(152, 53)$. In the r column, you will find 152 and 53, and then the successive remainders from division. So each number in r after the first two is the remainder after dividing the number immediately above it into the next number up. To the left of each remainder is the quotient from the division. In this case the third row of the table tells us that $152 = 53 \cdot 2 + 46$. The last nonzero value in r is the greatest common divisor.

Table 11.4.10 The extended Euclidean algorithm to compute $\gcd(152, 53)$

q	r	s	t
--	152	1	0
--	53	0	1
2	46	1	-2
1	7	-1	3
6	4	7	-20
1	3	-8	23
1	1	15	-43
3	0	-53	152

The "s" and "t" columns are new. The values of s and t in each row are maintained so that $152s + 53t$ is equal to the number in the r column. Notice that

Table 11.4.11 Invariant in computing $\gcd(152, 53)$

$$152 = 152 \cdot 1 + 53 \cdot 0$$
$$53 = 152 \cdot 0 + 53 \cdot 1$$
$$46 = 152 \cdot 1 + 53 \cdot (-2)$$
$$\vdots$$
$$1 = 152 \cdot 15 + 53 \cdot (-43)$$
$$0 = 152 \cdot (-53) + 53 \cdot 152$$

The next-to-last equation is what we're looking for in the end! The main problem is to identify how to determine these values after the first two rows. The first two rows in these columns will always be the same. Let's look at the general case of computing $\gcd(a, b)$. If the s and t values in rows $i-1$ and $i-2$ are correct, we have

$$(A) \begin{cases} as_{i-2} + bt_{i-2} = r_{i-2} \\ as_{i-1} + bt_{i-1} = r_{i-1} \end{cases}$$

In addition, we know that

$$r_{i-2} = r_{i-1}q_i + r_i \Rightarrow r_i = r_{i-2} - r_{i-1}q_i$$

If you substitute the expressions for r_{i-1} and r_{i-2} from (A) into this last equation and then collect the a and b terms separately you get

$$r_i = a(s_{i-2} - q_i s_{i-1}) + b(t_{i-2} - q_i t_{i-1})$$

or

$$s_i = s_{i-2} - q_i s_{i-1} \text{ and } t_i = t_{i-2} - q_i t_{i-1}$$

Look closely at the equations for $r_i, s_i,$ and t_i. Their forms are all the same. With a little bit of practice you should be able to compute s and t values quickly.

11.4.3 Modular Arithmetic

We remind you of the relation on the integers that we call Congruence Modulo m, p. 12. If two integers, a and b, differ by a multiple of n, we say that they are congruent modulo n, denoted $a \equiv b \pmod{n}$. For example, $13 \equiv 38 \pmod 5$ because $13 - 38 = -25$, which is a multiple of 5.

Definition 11.4.12 Modular Addition. If n is a positive integer, we define addition modulo n ($+_n$) as follows. If $a, b \in \mathbb{Z}$,

$$a +_n b = \text{ the remainder after } a + b \text{ is divided by } n$$

◇

Definition 11.4.13 Modular Multiplication. If n is a positive integer, we define multiplication modulo n ($\times_n$) as follows. If $a, b \in \mathbb{Z}$,

$$a \times_n b = \text{ the remainder after } a \cdot b \text{ is divided by } n.$$

◇

Note 11.4.14

(a) The result of doing arithmetic modulo n is always an integer between 0 and $n - 1$, by the Division Property. This observation implies that $\{0, 1, \ldots, n - 1\}$ is closed under modulo n arithmetic.

(b) It is always true that $a +_n b \equiv (a + b) \pmod{n}$ and $a \times_n b \equiv (a \cdot b) \pmod{n}$. For example, $4 +_7 5 = 2 \equiv 9 \pmod 7$ and $4 \times_7 5 = 6 \equiv 20 \pmod 7$.

(c) We will use the notation $\mathbb{Z}_n$ to denote the set $\{0, 1, 2, \ldots, n - 1\}$. One interpretation of this set is that each element is a *representative* of its equivalence class with respect to congruence modula n. For example, if $n = 7$, the number 1 in $\mathbb{Z}_7$ really represents all numbers in $[1] = 1 + 7k : k \in \mathbb{Z}$. In doing modular arithmetic, we can temporarily replace elements of $\mathbb{Z}_n$ with other elements in their equivalence class modulo n.

Example 11.4.15 Some Examples.

(a) We are all somewhat familiar with $\mathbb{Z}_{12}$ since the hours of the day are counted using this group, except for the fact that 12 is used in place of 0. Military time uses the mod 24 system and does begin at 0. If someone started a four-hour trip at hour 21, the time at which she would arrive is $21 +_{24} 4 = 1$. If a satellite orbits the earth every four hours and starts its first orbit at hour 5, it would end its first orbit at time $5 +_{24} 4 = 9$. Its tenth orbit would end at $5 +_{24} 10 \times_{24} 4 = 21$ hours on the clock

(b) Virtually all computers represent unsigned integers in binary form with a fixed number of digits. A very small computer might reserve seven bits to store the value of an integer. There are only 2^7 different values that can be stored in seven bits. Since the smallest value is 0, represented as 0000000, the maximum value will be $2^7 - 1 = 127$, represented as 1111111. When a command is given to add two integer values, and the two values have a sum of 128 or more, overflow occurs. For example, if we try to add 56 and 95, the sum is an eight-digit binary integer 10010111. One common procedure is to retain the seven lowest-ordered digits. The result of adding 56 and 95 would be $0010111_{two} = 23 \equiv 56 + 95 \pmod{128}$. Integer arithmetic with this computer would actually be modulo 128 arithmetic.

□

11.4.4 Properties of Modular Arithmetic

Theorem 11.4.16 Additive Inverses in $\mathbb{Z}_n$. If $a \in \mathbb{Z}_n$, $a \neq 0$, then the additive inverse of a is $n - a$.

Proof. $a + (n-a) = n \equiv 0 (\bmod n)$, since $n = n \cdot 1 + 0$. Therefore, $a +_n (n-a) = 0$. ∎

Addition modulo n is always commutative and associative; 0 is the identity for $+_n$ and every element of $\mathbb{Z}_n$ has an additive inverse. These properties can be summarized by noting that for each $n \geq 1$, $[\mathbb{Z}_n; +_n]$ is a group.

Definition 11.4.17 The Additive Group of Integers Modulo n. The Additive Group of Integers Modulo n is the group with domain $\{0, 1, 2, \ldots, n-1\}$ and with the operation of mod n addition. It is denoted as $\mathbb{Z}_n$. ◊

Multiplication modulo n is always commutative and associative, and 1 is the identity for $\times_n$.

Notice that the algebraic properties of $+_n$ and $\times_n$ on $\mathbb{Z}_n$ are identical to the properties of addition and multiplication on $\mathbb{Z}$.

Notice that a group cannot be formed from the whole set $\{0, 1, 2, \ldots, n-1\}$ with mod n multiplication since zero never has a multiplicative inverse. Depending on the value of n there may be other restrictions. The following group will be explored in Exercise 9, p. 43.

Definition 11.4.18 The Multiplicative Group of Integers Modulo n. The Multiplicative Group of Integers Modulo n is the group with domain $\{k \in \mathbb{Z} | 1 \leq k \leq n-1 \text{ and } \gcd(n,k) = 1\}$ and with the operation of mod n multiplication. It is denoted as $\mathbb{U}_n$. ◊

Example 11.4.19 Some operation tables.

Here are examples of operation tables for modular groups. Notice that although 8 is greater than 5, the two groups $\mathbb{U}_5$ and $\mathbb{U}_8$ both have order 4. In the case of $\mathbb{U}_5$, since 5 is prime all of the nonzero elements of $\mathbb{Z}_5$ are included. Since 8 isn't prime we don't include integers that share a common factor with 8, the even integers in this case.

Table 11.4.20 Operation Table for the group $\mathbb{Z}_5$

$+_5$	0	1	2	3	4
0	0	1	2	3	4
1	1	2	3	4	0
2	2	3	4	0	1
3	3	4	0	1	2
4	4	0	1	2	3

Table 11.4.21 Operation table for the group $\mathbb{U}_5$

$\times_5$	1	2	3	4
1	1	2	3	4
2	2	4	1	3
3	3	1	4	2
4	4	3	2	1

Table 11.4.22 Operation table for the group $\mathbb{U}_8$

$\times_8$	1	3	5	7
1	1	3	5	7
3	3	1	7	5
5	5	7	1	3
7	7	5	3	1

□

Computing Modular Multiplicative Inverses. Unlike the nice neat formula for additive inverses mod n, multiplicative inverses can most easily computed by applying Bézout's lemma, p. 37. If a is an element of the group $\mathbb{U}_n$, then by definition $\gcd(n, a) = 1$, and so there exist integers s and t such that $1 = ns + at$. They can be computed with the Extended Euclidean Algorithm.

$$1 = ns + at \Rightarrow at = 1 + (-s)n \Rightarrow a \times_n t = 1$$

11.4. GREATEST COMMON DIVISORS AND THE INTEGERS MODULO N

Since t might not be in $\mathbb{U}_n$ you might need take the remainder after dividing it by n. Normally, that involves simply adding n to t.

For example, in $\mathbb{U}_{2048}$, if we want the muliplicative inverse of 1001, we run the Extended Euclidean Algorithm and find that

$$\gcd(2048, 1001) = 1 = 457 \cdot 2048 + (-935) \cdot 1001$$

Thus, the multiplicative inverse of 1001 is $2048 - 935 = 1113$. See the SageMath Note below to see how to run the Extended Euclidean Algorithm.

11.4.5 SageMath Note - Modular Arithmetic

Sage inherits the basic integer division functions from Python that compute a quotient and remainder in integer division. For example, here is how to divide 561 into 2017 and get the quotient and remainder.

```
a=2017
b=561
[q,r]=[a//b,a%b]
[q,r]
```

[3, 334]

In Sage, *gcd* is the greatest common divisor function. It can be used in two ways. For the gcd of 2343 and 4319 we can evaluate the expression $gcd(2343, 4319)$. If we are working with a fixed modulus m that has a value established in your Sage session, the expression $m.gcd(k)$ to compute the greatest common divisor of m and any integer value k. The extended Euclidean algorithm can also be called upon with *xgcd*:

```
a=2017
b=561
print(gcd(a,b))
print(xgcd(a,b))
```

1
(1, -173, 622)

Sage has some extremely powerful tool for working with groups. The integers modulo n are represented by the expression $Integers(n)$ and the addition and multiplications tables can be generated as follows.

```
R = Integers(6)
print(R.addition_table('elements'))
print(R.multiplication_table('elements'))
```

```
+  0 1 2 3 4 5
 +------------
0| 0 1 2 3 4 5
1| 1 2 3 4 5 0
2| 2 3 4 5 0 1
3| 3 4 5 0 1 2
4| 4 5 0 1 2 3
5| 5 0 1 2 3 4

*  0 1 2 3 4 5
 +------------
```

```
0| 0 0 0 0 0 0
1| 0 1 2 3 4 5
2| 0 2 4 0 2 4
3| 0 3 0 3 0 3
4| 0 4 2 0 4 2
5| 0 5 4 3 2 1
```

Once we have assigned R a value of $Integers(6)$, we can do calculations by wrapping $R()$ around the integers 0 through 5. Here is a list containing the mod 6 sum and product, respectively, of 5 and 4:

```
[R(5)+R(4), R(5)*R(4)]
```

[3, 2]

Generating the multiplication table for the family of groups $\mathbb{U}_n$ takes a bit more code. Here we restrict the allowed inputs to be integers from 2 to 64.

```
def U_table(n):
    if n.parent()!=2.parent() or n < 2 or n > 64:
        return "input_error/out_of_range"
    R=Integers(n)
    els=[]
    for k in filter(lambda k:gcd(n,k)==1,range(n)):
        els=els+[str(k)]
    return
        R.multiplication_table(elements=els,names="elements")

U_table(18)
```

```
 *   1  5  7 11 13 17
    +------------------
 1|  1  5  7 11 13 17
 5|  5  7 17  1 11 13
 7|  7 17 13  5  1 11
11| 11  1  5 13 17  7
13| 13 11  1 17  7  5
17| 17 13 11  7  5  1
```

11.4.6 Exercises

1. Determine the greatest common divisors of the following pairs of integers without using any computational assistance.

    (a) $2^3 \cdot 3^2 \cdot 5$ and $2^2 \cdot 3 \cdot 5^2 \cdot 7$

    (b) $7!$ and $3 \cdot 5 \cdot 7 \cdot 9 \cdot 11 \cdot 13$

    (c) 19^4 and 19^5

    (d) 12112 and 0

2. Find all possible values of the following, assuming that m is a positive integer.

    (a) $\gcd(m+1, m)$

    (b) $\gcd(m+2, m)$

    (c) $\gcd(m+4, m)$

3. Calculate:
   - (a) $7 +_8 3$
   - (b) $7 \times_8 3$
   - (c) $4 \times_8 4$
   - (d) $10 +_{12} 2$
   - (e) $6 \times_8 2 +_8 6 \times_8 5$
   - (f) $6 \times_8 (2 +_8 5)$
   - (g) $3 \times_5 3 \times_5 3 \times_5 3 \equiv 3^4 (\mod 5)$
   - (h) $2 \times_{11} 7$
   - (i) $2 \times_{14} 7$

4. List the additive inverses of the following elements:
   - (a) 4, 6, 9 in $\mathbb{Z}_{10}$
   - (b) 16, 25, 40 in $\mathbb{Z}_{50}$

5. In the group $\mathbb{Z}_{11}$, what are:
   - (a) $3(4)$?
   - (b) $36(4)$?
   - (c) How could you efficiently compute $m(4)$, $m \in \mathbb{Z}$?

6. Prove that $\{1, 2, 3, 4\}$ is a group under the operation $\times_5$.

7. A student is asked to solve the following equations under the requirement that all arithmetic should be done in $\mathbb{Z}_2$. List all solutions.
   - (a) $x^2 + 1 = 0$.
   - (b) $x^2 + x + 1 = 0$.

8. Determine the solutions of the same equations as in Exercise 5 in $\mathbb{Z}_5$.

9.
   - (a) Write out the operation table for $\times_8$ on $\{1, 3, 5, 7\}$, and convince your self that this is a group.
   - (b) Let $\mathbb{U}_n$ be the elements of $\mathbb{Z}_n$ that have inverses with respect to $\times_n$. Convince yourself that $\mathbb{U}_n$ is a group under $\times_n$.
   - (c) Prove that the elements of $\mathbb{U}_n$ are those elements $a \in \mathbb{Z}_n$ such that $\gcd(n, a) = 1$. You may use Theorem 11.4.9, p. 37 in this proof.

10. Prove the division property, Theorem 11.4.1, p. 35.

    Hint. Prove by induction on m that you can divide any positive integer into m. That is, let $p(m)$ be "For all n greater than zero, there exist unique integers q and r such that ... ." In the induction step, divide n into $m - n$.

11. Suppose $f : \mathbb{Z}_{17} \to \mathbb{Z}_{17}$ such $f(i) = a \times_{17} i +_{17} b$ where a and b are integer constants. Furthermore, assume that $f(1) = 11$ and $f(2) = 4$. Find a formula for $f(i)$ and also find a formula for the inverse of f.

12. Write out the operation table for mod 10 multiplication on $T = \{0, 2, 4, 6, 8\}$. Is $[T; \times_{10}]$ a monoid? Is it a group?

13. Given that $1 = 2021 \cdot (-169) + 450 \cdot 759$, explain why 450 is an element of the group $\mathbb{U}_{2021}$ and determine its inverse in that group.

14. Let $n = 2021$. Solve $450 \times_n x = 321$ for x in the group $\mathbb{U}_n$

15. Let p be an odd prime. Find all solutions to the equation $x^2 = x \times_n x = 1$ in the group $\mathbb{U}_p$.

16. It was observed above that in doing modular arithmetic, one can replace an element of $\mathbb{Z}_n$ with any other element of its equivalence class modulo n. For example, if one is computing $452 \times_{461} 7$, the alternative to multiplying

452 time 7 and then dividing by 461 to get the remainder in $\mathbb{Z}_{461}$, we can replace 452 with -9 and get a product of -63 which is conguent of 398. Use this "trick" to compute the following without the use of a calculator.

(a) $898 \times_{1001} 998$

(b) $77^{10} \mod 81$

(c) The solution to $196 \times_{197} x = 120$

11.5 Subsystems

11.5.1 Definition

The subsystem is a fundamental concept of algebra at the universal level.

Definition 11.5.1 Subsystem. If $[V; *_1, \ldots, *_n]$ is an algebraic system of a certain kind and W is a subset of V, then W is a subsystem of V if $[W; *_1, \ldots, *_n]$ is an algebraic system of the same kind as V. The usual notation for "W is a subsystem of V" is $W \leq V$. ◇

Since the definition of a subsystem is at the universal level, we can cite examples of the concept of subsystems at both the axiomatic and concrete level.

Example 11.5.2 Examples of Subsystems.

(a) (Axiomatic) If $[G; *]$ is a group, and H is a subset of G, then H is a subgroup of G if $[H; *]$ is a group.

(b) (Concrete) $U = \{-1, 1\}$ is a subgroup of $[\mathbb{R}^*; \cdot]$. Take the time now to write out the multiplication table of U and convince yourself that $[U; \cdot]$ is a group.

(c) (Concrete) The even integers, $2\mathbb{Z} = \{2k : k \text{ is an integer}\}$ is a subgroup of $[\mathbb{Z}; +]$. Convince yourself of this fact.

(d) (Concrete) The set of nonnegative integers is not a subgroup of $[\mathbb{Z}; +]$. All of the group axioms are true for this subset except one: no positive integer has a positive additive inverse. Therefore, the inverse property is not true. Note that every group axiom must be true for a subset to be a subgroup.

(e) (Axiomatic) If M is a monoid and P is a subset of M, then P is a submonoid of M if P is a monoid.

(f) (Concrete) If B^* is the set of strings of 0's and 1's of length zero or more with the operation of concatenation, then two examples of submonoids of B^* are: (i) the set of strings of even length, and (ii) the set of strings that contain no 0's. The set of strings of length less than 50 is not a submonoid because it isn't closed under concatenation. Why isn't the set of strings of length 50 or more a submonoid of B^*?

□

11.5.2 Subgroups

For the remainder of this section, we will concentrate on the properties of subgroups. The first order of business is to establish a systematic way of determining whether a subset of a group is a subgroup.

11.5. SUBSYSTEMS

Theorem 11.5.3 Subgroup Conditions. *To determine whether H, a subset of group $[G; *]$, is a subgroup, it is sufficient to prove:*

(a) *H is closed under $*$; that is, $a, b \in H \Rightarrow a * b \in H$;*

(b) *H contains the identity element for $*$; and*

(c) *H contains the inverse of each of its elements; that is, $a \in H \Rightarrow a^{-1} \in H$.*

Proof. Our proof consists of verifying that if the three properties above are true, then all the axioms of a group are true for $[H; *]$. By Condition (a), $*$ can be considered an operation on H. The associative, identity, and inverse properties are the axioms that are needed. The identity and inverse properties are true by conditions (b) and (c), respectively, leaving only the associative property. Since, $[G; *]$ is a group, $a * (b * c) = (a * b) * c$ for all $a, b, c \in G$. Certainly, if this equation is true for all choices of three elements from G, it will be true for all choices of three elements from H, since H is a subset of G. ∎

For every group with at least two elements, there are at least two subgroups: they are the whole group and $\{e\}$. Since these two are automatic, they are not considered very interesting and are called the improper subgroups of the group; $\{e\}$ is sometimes referred to as the trivial subgroup. All other subgroups, if there are any, are called proper subgroups.

We can apply Theorem 11.5.3, p. 45 at both the concrete and axiomatic levels.

Example 11.5.4 Applying Conditions for a Subgroup.

(a) (Concrete) We can verify that $2\mathbb{Z} \leq \mathbb{Z}$, as stated in Example 11.5.2, p. 44. Whenever you want to discuss a subset, you must find some convenient way of describing its elements. An element of $2\mathbb{Z}$ can be described as 2 times an integer; that is, $a \in 2\mathbb{Z}$ is equivalent to $(\exists k)_{\mathbb{Z}}(a = 2k)$. Now we can verify that the three conditions of Theorem 11.5.3, p. 45 are true for $2\mathbb{Z}$. First, if $a, b \in 2\mathbb{Z}$, then there exist $j, k \in \mathbb{Z}$ such that $a = 2j$ and $b = 2k$. A common error is to write something like $a = 2j$ and $b = 2j$. This would mean that $a = b$, which is not necessarily true. That is why two different variables are needed to describe a and b. Returning to our proof, we can add a and b: $a + b = 2j + 2k = 2(j + k)$. Since $j + k$ is an integer, $a + b$ is an element of $2\mathbb{Z}$. Second, the identity, 0, belongs to $2\mathbb{Z}$ ($0 = 2(0)$). Finally, if $a \in 2\mathbb{Z}$ and $a = 2k$, $-a = -(2k) = 2(-k)$, and $-k \in \mathbb{Z}$, therefore, $-a \in 2\mathbb{Z}$. By Theorem 11.5.3, p. 45, $2\mathbb{Z} \leq \mathbb{Z}$. How would this argument change if you were asked to prove that $3\mathbb{Z} \leq \mathbb{Z}$? or $n\mathbb{Z} \leq \mathbb{Z}, n \geq 2$?

(b) (Concrete) We can prove that $H = \{0, 3, 6, 9\}$ is a subgroup of $\mathbb{Z}_{12}$. First, for each ordered pair $(a, b) \in H \times H$, $a +_{12} b$ is in H. This can be checked without too much trouble since $|H \times H| = 16$. Thus we can conclude that H is closed under $+_{12}$. Second, $0 \in H$. Third, $-0 = 0$, $-3 = 9$, $-6 = 6$, and $-9 = 3$. Therefore, the inverse of each element of H is in H.

(c) (Axiomatic) If H and K are both subgroups of a group G, then $H \cap K$ is a subgroup of G. To justify this statement, we have no concrete information to work with, only the facts that $H \leq G$ and $K \leq G$. Our proof that $H \cap K \leq G$ reflects this and is an exercise in applying the definitions of intersection and subgroup, (i) If a and b are elements of $H \cap K$, then a and b both belong to H, and since $H \leq G$, $a * b$ must be an element of H. Similarly, $a * b \in K$; therefore, $a * b \in H \cap K$. (ii) The identity of G must belong to both H and K; hence it belongs to $H \cap K$. (iii) If $a \in H \cap K$, then $a \in H$, and since $H \leq G$, $a^{-1} \in H$. Similarly, $a^{-1} \in K$. Hence, by the theorem, $H \cap K \leq G$. Now that this fact has been established, we can apply it to any pair of subgroups of any group. For example, since $2\mathbb{Z}$ and $3\mathbb{Z}$ are both subgroups of $[\mathbb{Z};+]$, $2\mathbb{Z} \cap 3\mathbb{Z}$ is also a subgroup of $\mathbb{Z}$. Note that if $a \in 2\mathbb{Z} \cap 3\mathbb{Z}$, a must have a factor of 3; that is, there exists $k \in \mathbb{Z}$ such that $a = 3k$. In addition, a must be even, therefore k must be even. There exists $j \in \mathbb{Z}$ such that $k = 2j$, therefore $a = 3(2j) = 6j$. This shows that $2\mathbb{Z} \cap 3\mathbb{Z} \subseteq 6\mathbb{Z}$. The opposite containment can easily be established; therefore, $2\mathbb{Z} \cap 3\mathbb{Z} = 6\mathbb{Z}$.

□

Given a finite group, we can apply Theorem 11.3.15, p. 33 to obtain a simpler condition for a subset to be a subgroup.

Theorem 11.5.5 Condition for a Subgroup of Finite Group. *Given that $[G;*]$ is a finite group and H is a nonempty subset of G, if H is closed under $*$, then H is a subgroup of G.*

Proof. In this proof, we demonstrate that Conditions (b) and (c) of Theorem 11.5.3, p. 45 follow from the closure of H under $*$, which is condition (a) of the theorem. First, select any element of H; call it β. The powers of β: $\beta^1, \beta^2, \beta^3, \ldots$ are all in H by the closure property. By Theorem 11.3.15, p. 33, there exists m, $m \leq |G|$, such that $\beta^m = e$; hence $e \in H$. To prove that (c) is true, we let a be any element of H. If $a = e$, then a^{-1} is in H since $e^{-1} = e$. If $a \neq e$, $a^q = e$ for some q between 2 and $|G|$ and

$$e = a^q = a^{q-1} * a$$

Therefore, $a^{-1} = a^{q-1}$, which belongs to H since $q - 1 \geq 1$. ■

11.5.3 Sage Note - Applying the condition for a subgroup of a finite group

To determine whether $H_1 = \{0, 5, 10\}$ and $H_2 = \{0, 4, 8, 12\}$ are subgroups of $\mathbb{Z}_{15}$, we need only write out the addition tables (modulo 15) for these sets. This is easy to do with a bit of Sage code that we include below and then for any modulus and subset, we can generate the body of an addition table. The code is set up for H_1 but can be easily adjusted for H_2.

```
def addition_table(n,H):
    for a in H:
        line=[]
        for b in H:
            line+=[(a+b)%n]
        print(line)
addition_table(15,Set([0,5,10]))
```

```
[0, 10, 5]
[10, 5, 0]
[5, 0, 10]
```

Note that H_1 is a subgroup of $\mathbb{Z}_{15}$. Since the interior of the addition table for H_2 contains elements that are outside of H_2, H_2 is not a subgroup of $\mathbb{Z}_{15}$.

11.5.4 Cyclic Subgroups

One kind of subgroup that merits special mention due to its simplicity is the cyclic subgroup.

11.5. SUBSYSTEMS

Definition 11.5.6 Cyclic Subgroup. If G is a group and $a \in G$, the cyclic subgroup generated by a, $\langle a \rangle$, is the set of all powers of a:

$$\langle a \rangle = \{a^n : n \in \mathbb{Z}\}.$$

We refer to a as a generator of subgroup $\langle a \rangle$.

A subgroup H of a group G is cyclic if there exists $a \in H$ such that $H = \langle a \rangle$.
◇

Definition 11.5.7 Cyclic Group. A group G is cyclic if there exists $\beta \in G$ such that $\langle \beta \rangle = G$.
◇

Note 11.5.8 If the operation on G is additive, then $\langle a \rangle = \{(n)a : n \in \mathbb{Z}\}$.

Definition 11.5.9 Order of a Group Element. The order of an element a of group G is the number of elements in the cyclic subgroup of G generated by a. The order of a is denoted $ord(a)$.
◇

Example 11.5.10

(a) In $[\mathbb{R}^*; \cdot]$, $\langle 2 \rangle = \{2^n : n \in \mathbb{Z}\} = \{\ldots, \frac{1}{16}, \frac{1}{8}, \frac{1}{4}, \frac{1}{2}, 1, 2, 4, 8, 16, \ldots\}$.

(b) In $\mathbb{Z}_{15}$, $\langle 6 \rangle = \{0, 3, 6, 9, 12\}$. Here is why: If G is finite, you need list only the positive powers (or multiples) of a up to the first occurrence of the identity to obtain all of $\langle a \rangle$. In $\mathbb{Z}_{15}$, the multiples of 6 are 6, $(2)6 = 12$, $(3)6 = 3$, $(4)6 = 9$, and $(5)6 = 0$. Note that $\{0, 3, 6, 9, 12\}$ is also $\langle 3 \rangle$, $\langle 9 \rangle$, and $\langle 12 \rangle$. This shows that a cyclic subgroup can have different generators.

□

If you want to list the cyclic subgroups of a group, the following theorem can save you some time.

Theorem 11.5.11 If a is an element of group G, then $\langle a \rangle = \langle a^{-1} \rangle$.

This is an easy way of seeing, for example, that $\langle 9 \rangle$ in $\mathbb{Z}_{15}$ equals $\langle 6 \rangle$, since $-6 = 9$.

11.5.5 Exercises

1. Which of the following subsets of the real numbers is a subgroup of $[\mathbb{R}; +]$?

   (a) the rational numbers

   (b) the positive real numbers

   (c) $\{k/2 \mid k \text{ is an integer}\}$

   (d) $\{2^k \mid k \text{ is an integer}\}$

   (e) $\{x \mid -100 \leq x \leq 100\}$

2. Describe in simpler terms the following subgroups of $\mathbb{Z}$:

   (a) $5\mathbb{Z} \cap 4\mathbb{Z}$

   (b) $4\mathbb{Z} \cap 6\mathbb{Z}$ (be careful)

   (c) the only finite subgroup of $\mathbb{Z}$

3. Find at least two proper subgroups of R_3, the set of 3×3 rook matrices (see Exercise 11.2.4.5, p. 29).

4. Let H and K be subgroups of G with elements $a, x, y \in G$ located in the following Venn diagram. Where should you place the following elements in Figure 11.5.12, p. 48?

(a) e

(b) a^{-1}

(c) $x * y$

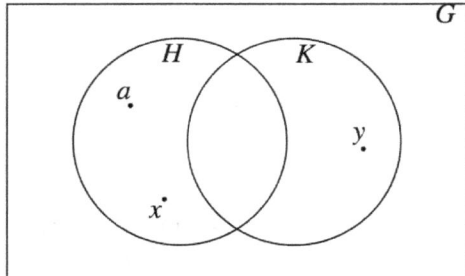

Figure 11.5.12 Figure for exercise 4

5.
   (a) List the cyclic subgroups of $\mathbb{Z}_6$ and draw an ordering diagram for the relation "is a subset of" on these subgroups.

   (b) Do the same for $\mathbb{Z}_{12}$.

   (c) Do the same for $\mathbb{Z}_8$.

   (d) On the basis of your results in parts a, b, and c, what would you expect if you did the same with $\mathbb{Z}_{24}$?

6. **Subgroups generated by subsets of a group.** The concept of a cyclic subgroup is a special case of the concept that we will discuss here. Let $[G; *]$ be a group and S a nonempty subset of G. Define the set $\langle S \rangle$ recursively by:

   - If $a \in S$, then $a \in \langle S \rangle$.
   - If $a, b \in \langle S \rangle$, then $a * b \in \langle S \rangle$, and
   - If $a \in \langle S \rangle$, then $a^{-1} \in \langle S \rangle$.

   (a) By its definition, $\langle S \rangle$ has all of the properties needed to be a subgroup of G. The only thing that isn't obvious is that the identity of G is in $\langle S \rangle$. Prove that the identity of G is in $\langle S \rangle$.

   (b) What is $\langle \{9, 15\} \rangle$ in $[\mathbb{Z}; +]$?

   (c) Prove that if $H \leq G$ and $S \subseteq H$, then $\langle S \rangle \leq H$. This proves that $\langle S \rangle$ is contained in every subgroup of G that contains S; that is, $\langle S \rangle = \bigcap_{S \subseteq H, H \leq G} H$.

   (d) Describe $\langle \{0.5, 3\} \rangle$ in $[\mathbb{R}^+; \cdot]$ and in $[\mathbb{R}; +]$.

   (e) If $j, k \in \mathbb{Z}$, $\langle \{j, k\} \rangle$ is a cyclic subgroup of $\mathbb{Z}$. In terms of j and k, what is a generator of $\langle \{j, k\} \rangle$?

7. Prove that if $H, K \leq G$, and $H \cup K = G$, then $H = G$ or $K = G$.

   Hint. Use an indirect argument.

8. Prove that the order of an element, a of a group is the least positive integer, k, such that a^k is the identity of the group.

11.6 Direct Products

11.6.1 Definition

Our second universal algebraic concept lets us look in the opposite direction from subsystems. Direct products allow us to create larger systems. In the following definition, we avoid complicating the notation by not specifying how many operations the systems have.

Definition 11.6.1 Direct Product. If $[V_i; *_i, \diamond_i, \ldots]$, $i = 1, 2, \ldots, n$ are algebraic systems of the same kind, then the direct product of these systems is $V = V_1 \times V_2 \times \cdots \times V_n$, with operations defined below. The elements of V are n-tuples of the form $(a_1, a_2, \ldots, a_n)$, where $a_k \in V_k$, $k = 1, \ldots, n$. The systems $V_1, V_2, \ldots, V_n$ are called the factors of V. There are as many operations on V as there are in the factors. Each of these operations is defined componentwise:
If $(a_1, a_2, \ldots, a_n), (b_1, b_2, \ldots, b_n) \in V$,

$$(a_1, a_2, \ldots, a_n) * (b_1, b_2, \ldots, b_n) = (a_1 *_1 b_1, a_2 *_2 b_2, \ldots, a_n *_n b_n)$$
$$(a_1, a_2, \ldots, a_n) \diamond (b_1, b_2, \ldots, b_n) = (a_1 \diamond_1 b_1, a_2 \diamond_2 b_2, \ldots, a_n \diamond_n b_n)$$
$$\text{etc.}$$

$\diamond$

Example 11.6.2 A Direct Product of Monoids. Consider the monoids $\mathbb{N}$ (the set of natural numbers with addition) and B^* (the set of finite strings of 0's and 1's with concatenation). The direct product of $\mathbb{N}$ with B^* is a monoid. We illustrate its operation, which we will denote by $*$, with examples:

$$(4, 001) * (3, 11) = (4 + 3, 001 + 11) = (7, 00111)$$
$$(0, 11010) * (3, 01) = (3, 1101001)$$
$$(0, \lambda) * (129, 00011) = (0 + 129, \lambda + 00011) = (129, 00011)$$
$$(2, 01) * (8, 10) = (10, 0110), \text{ and}$$
$$(8, 10) * (2, 01) = (10, 1001)$$

Note that our new monoid is not commutative. What is the identity for $*$? $\square$

The definiton of a Direct Product, p. 49 is quite general and may be confusing to some. Here is the definiton of the direct product of two groups. The definition extends easily to the direct product of three or more groups.

Definition 11.6.3 Direct Product of Two Groups. Let $[G_1; *_1]$ and $[G_2; *_2]$ be two groups. Their direct product is the system $[G_1 \times G_2; *]$ with domain equal to the Cartesian product of the domains of the two groups and with the coordinatewise operation $*$ defined by

$$(a_1, b_1) * (a_2, b_2) = (a_1 *_1 a_2, b_1 *_2 b_2)$$

for $(a_1, b_1), (a_2, b_2) \in G_1 \times G_2$. $\diamond$

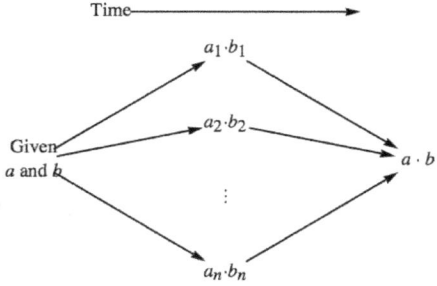

Figure 11.6.4 Concurrent calculation in a direct product

Note 11.6.5

(a) On notation. If two or more consecutive factors in a direct product are identical, it is common to combine them using exponential notation. For example, $\mathbb{Z} \times \mathbb{Z} \times \mathbb{R}$ can be written $\mathbb{Z}^2 \times \mathbb{R}$, and $\mathbb{R} \times \mathbb{R} \times \mathbb{R} \times \mathbb{R}$ can be written $\mathbb{R}^4$. This is purely a notational convenience; no exponentiation is really taking place.

(b) We call the operations in a direct product **componentwise operations**, and they are indeed operations on V. If two n-tuples, a and b, are selected from V, the first components of a and b, a_1 and b_1, are operated on with $*_1$ to obtain $a_1 *_1 b_1$, the first component of $a * b$. Note that since $*_1$ is an operation on V_1, $a_1 *_1 b_1$ is an element of V_1. Similarly, all other components of $a * b$, as they are defined, belong to their proper sets.

(c) One significant fact about componentwise operations is that the components of the result can all be computed at the same time (concurrently). The time required to compute in a direct product can be reduced to a length of time that is not much longer than the maximum amount of time needed to compute in the factors.

(d) A direct product of algebraic systems is not always an algebraic system of the same type as its factors. This is due to the fact that certain axioms that are true for the factors may not be true for the set of n-tuples. This situation does not occur with groups however. You will find that whenever a new type of algebraic system is introduced, call it type T, one of the first theorems that is usually proven, if possible, is that the direct product of two or more systems of type T is a system of type T.

11.6.2 Direct Products of Groups

We will explore properties of direct products of groups and examine some concrete examples

Theorem 11.6.6 The Direct Product of Groups is a Group. *The direct product of two or more groups is a group; that is, the algebraic properties of a system obtained by taking the direct product of two or more groups includes the group axioms.*

Proof. We will only present the proof of this theorem for the direct product of two groups. Some slight revisions can be made to produce a proof for any number of factors.

Stating that the direct product of two groups is a group is a short way of saying that if $[G_1; *_1]$ and $[G_2; *_2]$ are groups, then $[G_1 \times G_2; *]$ is also a group, where $*$ is the componentwise operation on $G_1 \times G_2$. Associativity of $*$: If

11.6. DIRECT PRODUCTS

$a, b, c \in G_1 \times G_2$,

$$\begin{aligned}
a * (b * c) &= (a_1, a_2) * ((b_1, b_2) * (c_1, c_2)) \\
&= (a_1, a_2) * (b_1 *_1 c_1, b_2 *_2 c_2) \\
&= (a_1 *_1 (b_1 *_1 c_1), a_2 *_2 (b_2 *_2 c_2)) \\
&= ((a_1 *_1 b_1) *_1 c_1, (a_2 *_2 b_2) *_2 c_2) \\
&= (a_1 *_1 b_1, a_2 *_2 b_2) * (c_1, c_2) \\
&= ((a_1, a_2) * (b_1, b_2)) * (c_1, c_2) \\
&= (a * b) * c
\end{aligned}$$

Notice how the associativity property hinges on the associativity in each factor. An identity for $*$: As you might expect, if e_1 and e_2 are identities for G_1 and G_2, respectively, then $e = (e_1, e_2)$ is the identity for $G_1 \times G_2$. If $a \in G_1 \times G_2$,

$$\begin{aligned}
a * e &= (a_1, a_2) * (e_1, e_2) \\
&= (a_1 *_1 e_1, a_2 *_2 e_2) \\
&= (a_1, a_2) = a
\end{aligned}$$

Similarly, $e * a = a$.

Inverses in $G_1 \times G_2$: The inverse of an element is determined componentwise $a^{-1} = (a_1, a_2)^{-1} = (a_1^{-1}, a_2^{-1})$. To verify, we compute $a * a^{-1}$:

$$\begin{aligned}
a * a^{-1} &= (a_1, a_2) * (a_1^{-1}, a_2^{-1}) \\
&= (a_1 *_1 a_1^{-1}, a_2 *_2 a_2^{-1}) \\
&= (e_1, e_2) = e
\end{aligned}$$

Similarly, $a^{-1} * a = e$. ∎

Example 11.6.7 Some New Groups.

(a) If $n \geq 2$, $\mathbb{Z}_2^n$, the direct product of n factors of $\mathbb{Z}_2$, is a group with 2^n elements. We will take a closer look at $\mathbb{Z}_2^3 = \mathbb{Z}_2 \times \mathbb{Z}_2 \times \mathbb{Z}_2$. The elements of this group are triples of zeros and ones. Since the operation on $\mathbb{Z}_2$ is $+_2$, we will use the symbol $+$ for the operation on $\mathbb{Z}_2^3$. Two of the eight triples in the group are $a = (1, 0, 1)$ and $b = (0, 0, 1)$. Their "sum" is $a + b = (1 +_2 0, 0 +_2 0, 1 +_2 1) = (1, 0, 0)$. One interesting fact about this group is that each element is its own inverse. For example $a + a = (1, 0, 1) + (1, 0, 1) = (0, 0, 0)$; therefore $-a = a$. We use the additive notation for the inverse of a because we are using a form of addition. Note that $\{(0, 0, 0), (1, 0, 1)\}$ is a subgroup of $\mathbb{Z}_2^3$. Write out the "addition" table for this set and apply Theorem 11.5.5, p. 46. The same can be said for any set consisting of $(0, 0, 0)$ and another element of $\mathbb{Z}_2^3$.

(b) The direct product of the positive real numbers with the integers modulo 4, $\mathbb{R}^+ \times \mathbb{Z}_4$ is an infinite group since one of its factors is infinite. The operations on the factors are multiplication and modular addition, so we will select the neutral symbol $\diamond$ for the operation on $\mathbb{R}^+ \times \mathbb{Z}_4$. If $a = (4, 3)$ and $b = (0.5, 2)$, then

$$\begin{aligned}
a \diamond b &= (4, 3) \diamond (0.5, 2) = (4 \cdot 0.5, 3 +_4 2) = (2, 1) \\
b^2 &= b \diamond b = (0.5, 2) \diamond (0.5, 2) = (0.25, 0) \\
a^{-1} &= (4^{-1}, -3) = (0.25, 1) \\
b^{-1} &= (0.5^{-1}, -2) = (2, 2)
\end{aligned}$$

It would be incorrect to say that $\mathbb{Z}_4$ is a subgroup of $\mathbb{R}^+ \times \mathbb{Z}_4$, but there is a subgroup of the direct product that closely resembles $\mathbb{Z}_4$. It is $\{(1,0),(1,1),(1,2),(1,3)\}$. Its table is

$\diamond$	$(1,0)$	$(1,1)$	$(1,2)$	$(1,3)$
$(1,0)$	$(1,0)$	$(1,1)$	$(1,2)$	$(1,3)$
$(1,1)$	$(1,1)$	$(1,2)$	$(1,3)$	$(1,0)$
$(1,2)$	$(1,2)$	$(1,3)$	$(1,0)$	$(1,1)$
$(1,3)$	$(1,3)$	$(1,0)$	$(1,1)$	$(1,2)$

Imagine erasing $(1,)$ throughout the table and writing $+_4$ in place of $\diamond$. What would you get? We will explore this phenomenon in detail in the next section.

The whole direct product could be visualized as four parallel half-lines labeled 0, 1, 2, and 3 as in Figure 11.6.8, p. 52. On the kth line, the point that lies x units to the right of the zero mark would be (x, k). The set $\{(2^n, (n)1) \mid n \in \mathbb{Z}\}$, which is depicted on the figure is a subgroup of $\mathbb{R}^+ \times \mathbb{Z}_4$. What cyclic subgroup is it?

The answer: $\langle(2,1)\rangle$ or $\langle(1/2,3)\rangle$. There are two different generators.

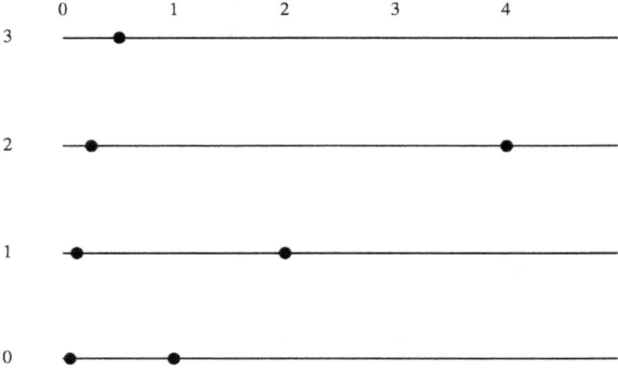

Figure 11.6.8 Visualization of the group $\mathbb{R}^+ \times \mathbb{Z}_4$

$\square$

A more conventional direct product is $\mathbb{R}^2$, the direct product of two factors of $[\mathbb{R}; +]$. The operation on $\mathbb{R}^2$ is componentwise addition; hence we will use $+$ as the operation symbol for this group. You should be familiar with this operation, since it is identical to addition of 2×1 matrices. The Cartesian coordinate system can be used to visualize $\mathbb{R}^2$ geometrically. We plot the pair (s, t) on the plane in the usual way: s units along the x axis and t units along the y axis. There is a variety of different subgroups of $\mathbb{R}^2$, a few of which are:

(a) $\{(x,0) \mid x \in \mathbb{R}\}$, all of the points on the x axis;

(b) $\{(x,y) \mid 2x - y = 0\}$, all of the points that are on the line 2x - y = 0;

(c) If $a, b \in \mathbb{R}$, $\{(x,y) \mid ax + by = 0\}$. The first two subgroups are special cases of this one, which represents any line that passes through the origin.

(d) $\{(x,y) \mid 2x - y = k, k \in \mathbb{Z}\}$, a union of a set of lines that are parallel to $2x - y = 0$.

(e) $\{(n, 3n) \mid n \in \mathbb{Z}\}$, which is the only countable subgroup that we have listed.

11.6. DIRECT PRODUCTS

We will leave it to the reader to verify that these sets are subgroups. We will only point out how the fourth example, call it H, is closed under "addition." If $a = (p,q)$ and $b = (s,t)$ and both belong to H, then $2p - q = j$ and $2s - t = k$, where both j and k are integers. $a + b = (p,q) + (s,t) = (p+s, q+t)$ We can determine whether $a+b$ belongs to H by deciding whether or not $2(p+s)-(q+t)$ is an integer:

$$\begin{aligned} 2(p+s) - (q+t) &= 2p + 2s - q - t \\ &= (2p - q) + (2s - t) \\ &= j + k \end{aligned}$$

Since j and k are integers, so is $j + k$. This completes a proof that H is closed under the operation of $\mathbb{R}^2$.

Several useful facts can be stated in regards to the direct product of two or more groups. We will combine them into one theorem, which we will present with no proof. Parts a and c were derived for $n = 2$ in the proof of Theorem 11.6.6, p. 50.

Theorem 11.6.9 Properties of Direct Products of Groups. *If $G = G_1 \times G_2 \times \cdots \times G_n$ is a direct product of n groups and $(a_1, a_2, \ldots, a_n) \in G$, then:*

(a) *The identity of G is $(e_1, e_2, \ldots, e_n)$, where e_k is the identity of G_k.*

(b) $(a_1, a_2, \ldots, a_n)^{-1} = \left(a_1^{-1}, a_2^{-1}, \ldots, a_n^{-1}\right)$.

(c) $(a_1, a_2, \ldots, a_n)^m = (a_1^m, a_2^m, \ldots, a_n^m)$ *for all $m \in \mathbb{Z}$.*

(d) *G is abelian if and only if each of the factors $G_1, G_2, \ldots, G_n$ is abelian.*

(e) *If $H_1, H_2, \ldots, H_n$ are subgroups of the corresponding factors, then $H_1 \times H_2 \times \cdots \times H_n$ is a subgroup of G.*

Not all subgroups of a direct product can be created using part e of Theorem 11.6.9, p. 53. For example, $\{(n,n) \mid n \in \mathbb{Z}\}$ is a subgroup of $\mathbb{Z}^2$, but is not a direct product of two subgroups of $\mathbb{Z}$.

Example 11.6.10 Linked Lists using a Direct Product - XOR Linked Lists. Using the identity $(x + y) + x = y$, in $\mathbb{Z}_2$, we can devise a scheme for representing a symmetrically linked list using only one link field. A symmetrically linked list is a list in which each node contains a pointer to its immediate successor and its immediate predecessor (see Figure 11.6.11, p. 54). If the pointers are n-digit binary addresses, then each pointer can be taken as an element of $\mathbb{Z}_2{}^n$. Lists of this type can be accomplished using cells with only one link. In place of a left and a right pointer, the only "link" is the value of the sum (left link) + (right link). All standard list operations (merge, insert, delete, traverse, and so on) are possible with this structure, provided that you know the value of the nil pointer and the address, f, of the first (i. e., leftmost) cell. Since first f. left is nil, we can recover f. right by adding the value of nil: f + nil = (nil + f.right) + nil = f.right, which is the address of the second item. Now if we temporarily retain the address, s, of the second cell, we can recover the address of the third item. The link field of the second item contains the sum s. left + s. right = first + third. Therefore

$$\begin{aligned} (\text{first} + \text{third}) + \text{first} &= s + s.\text{left} \\ &= (s.\text{left} + s.\text{right}) + s.\text{left} \\ &= s.\text{right} \\ &= \text{third} \end{aligned}$$

We no longer need the address of the first cell, only the second and third, to recover the fourth address, and so forth.

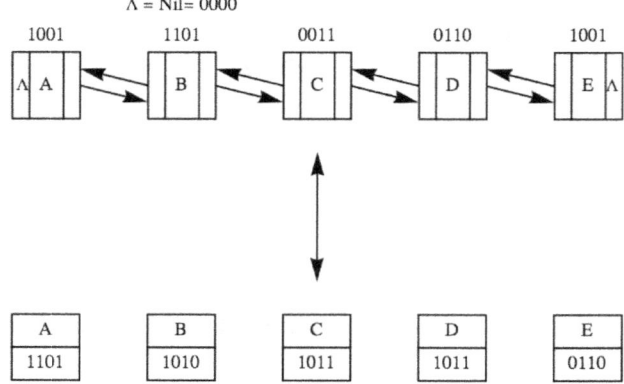

Figure 11.6.11 Symmetric Linked Lists

The following more formal algorithm uses names that reflects the timing of the visits.

Given a symmetric list, a traversal of the list is accomplished as follows, where *first* is the address of the first cell. We presume that each item has some information that is represented by item.info and a field called item.link that is the sum of the left and right links.

Table 11.6.12

    (1) yesterday =nil

    (2) today =first

    (3)   while today $\neq$ nil:

        (3.1)Write(today.info)

        (3.2)tomorrow = today.link + yesterday

        (3.3)yesterday = today

        (3.4)today = tomorrow.

At any point in this algorithm it would be quite easy to insert a cell between today and tomorrow. Can you describe how this would be accomplished?

This implementation of doubly linked lists is often referred to as an XOR linked list. For more information see the Wikipedia page en.wikipedia.org/wiki/XOR_linked_list. □

11.6.3 Exercises

1. Write out the group table of $\mathbb{Z}_2 \times \mathbb{Z}_3$ and find the two proper subgroups of this group.

2. List more examples of proper subgroups of $\mathbb{R}^2$ that are different from the ones listed in this section.

3. **Algebraic properties of the n-cube.**

    (a) The four elements of $\mathbb{Z}_2{}^2$ can be visualized geometrically as the four corners of the 2-cube. Algebraically describe the statements:

        (i) Corners a and b are adjacent.

        (ii) Corners a and b are diagonally opposite one another.

    (b) The eight elements of $\mathbb{Z}_2{}^3$ can be visualized as the eight corners of the 3-cube. One face contains $\mathbb{Z}_2 \times \mathbb{Z}_2 \times \{0\}$ and the opposite face contains the remaining four elements so that $(a, b, 1)$ is behind $(a, b, 0)$. As in part a, describe statements i and ii algebraically.

(c) If you could imagine a geometric figure similar to the square or cube in n dimensions, and its corners were labeled by elements of $\mathbb{Z}_2{}^n$ as in parts a and b, how would statements i and ii be expressed algebraically?

4.
   (a) Suppose that you were to be given a group $[G; *]$ and asked to solve the equation $x * x = e$. Without knowing the group, can you anticipate how many solutions there will be?

   (b) Answer the same question as part a for the equation $x * x = x$.

5. Which of the following sets are subgroups of $\mathbb{Z} \times \mathbb{Z}$? Give a reason for any negative answers.

   (a) $\{0\}$

   (b) $\{(2j, 2k) \mid j, k \in \mathbb{Z}\}$

   (c) $\{(2j + 1, 2k) \mid j, k \in \mathbb{Z}\}$

   (d) $\{(n, n^2) \mid n \in \mathbb{Z}\}$

   (e) $\{(j, k) \mid j + k \text{ is even}\}$

6. Determine the following values in the group $\mathbb{Z}_3 \times \mathbb{R}^*$:

   (a) $(2, 1) * (1, 2)$

   (b) the identity element

   (c) $(1, 1/2)^{-1}$

11.7 Isomorphisms

The following informal definition of isomorphic systems should be memorized. No matter how technical a discussion about isomorphic systems becomes, keep in mind that this is the essence of the concept.

Definition 11.7.1 Isomorphic Systems/Isomorphism - Informal Version. Two algebraic systems are isomorphic if there exists a translation rule between them so that any true statement in one system can be translated to a true statement in the other. ◊

Example 11.7.2 How to Do Greek Arithmetic. Imagine that you are a six-year-old child who has been reared in an English-speaking family, has moved to Greece, and has been enrolled in a Greek school. Suppose that your new teacher asks the class to do the following addition problem that has been written out in Greek.

$$\tau\rho\acute{\iota}\alpha \quad \sigma\upsilon\nu \quad \tau\acute{\epsilon}\sigma\sigma\epsilon\rho\alpha \quad \iota\sigma o\acute{\upsilon}\tau\alpha\iota \quad \underline{}$$

The natural thing for you to do is to take out your Greek-English/English-Greek dictionary and translate the Greek words to English, as outlined in Figure 11.7.3, p. 56 After you've solved the problem, you can consult the same dictionary to find the proper Greek word that the teacher wants. Although this is not the recommended method of learning a foreign language, it will surely yield the correct answer to the problem. Mathematically, we may say that the system of Greek integers with addition ($\sigma \upsilon \nu$) is isomorphic to English integers with addition (plus). The problem of translation between natural languages is more difficult than this though, because two complete natural languages are not isomorphic, or at least the isomorphism between them is not contained in a simple dictionary.

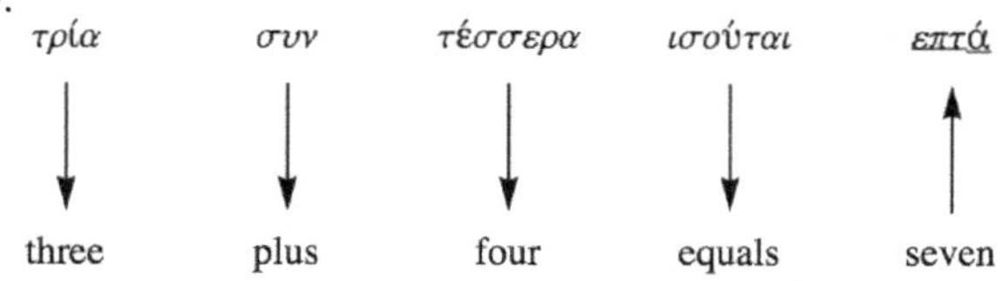

$$\begin{array}{ccccc} \tau\rho\iota\alpha & \sigma\upsilon\nu & \tau\epsilon\sigma\sigma\epsilon\rho\alpha & \iota\sigma o\upsilon\tau\alpha\iota & \epsilon\pi\tau\alpha \\ \downarrow & \downarrow & \downarrow & \downarrow & \uparrow \\ \text{three} & \text{plus} & \text{four} & \text{equals} & \text{seven} \end{array}$$

Figure 11.7.3 Solution of a Greek arithmetic problem

□

Example 11.7.4 Software Implementation of Sets. In this example, we will describe how set variables can be implemented on a computer. We will describe the two systems first and then describe the isomorphism between them.

System 1: The power set of $\{1, 2, 3, 4, 5\}$ with the operation union, $\cup$. For simplicity, we will only discuss union. However, the other operations are implemented in a similar way.

System 2: Strings of five bits of computer memory with an OR gate. Individual bit values are either zero or one, so the elements of this system can be visualized as sequences of five 0's and 1's. An OR gate, Figure 11.7.5, p. 56, is a small piece of computer hardware that accepts two bit values at any one time and outputs either a zero or one, depending on the inputs. The output of an OR gate is one, except when the two bit values that it accepts are both zero, in which case the output is zero. The operation on this system actually consists of sequentially inputting the values of two bit strings into the OR gate. The result will be a new string of five 0's and 1's. An alternate method of operating in this system is to use five OR gates and to input corresponding pairs of bits from the input strings into the gates concurrently.

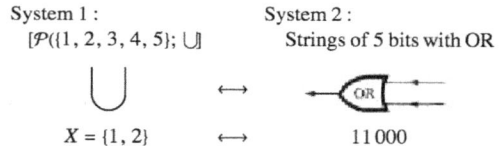

Figure 11.7.5 Translation between sets and strings of bits

The Isomorphism: Since each system has only one operation, it is clear that union and the OR gate translate into one another. The translation between sets and bit strings is easiest to describe by showing how to construct a set from a bit string. If $a_1 a_2 a_3 a_4 a_5$, is a bit string in System 2, the set that it translates to contains the number k if and only if a_k equals 1. For example, 10001 is translated to the set $\{1, 5\}$, while the set $\{1, 2\}$ is translated to 11000. Now imagine that your computer is like the child who knows English and must do a Greek problem. To execute a program that has code that includes the set expression $\{1, 2\} \cup \{1, 5\}$, it will follow the same procedure as the child to obtain the result, as shown in Figure 11.7.6, p. 56.

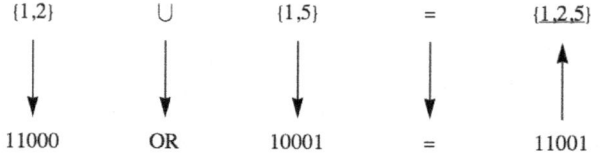

Figure 11.7.6 Translation of a problem in set theory

11.7.1 Group Isomorphisms

Example 11.7.7 Multiplying without doing multiplication. This isomorphism is between $[\mathbb{R}^+; \cdot]$ and $[\mathbb{R}; +]$. Until the 1970s, when the price of calculators dropped, multiplication and exponentiation were performed with an isomorphism between these systems. The isomorphism ($\mathbb{R}^+$ to $\mathbb{R}$) between the two groups is that $\cdot$ is translated into $+$ and any positive real number a is translated to the logarithm of a. To translate back from $\mathbb{R}$ to $\mathbb{R}^+$, you invert the logarithm function. If base ten logarithms are used, an element of $\mathbb{R}$, b, will be translated to 10^b. In pre-calculator days, the translation was done with a table of logarithms or with a slide rule. An example of how the isomorphism is used appears in Figure 11.7.8, p. 57.

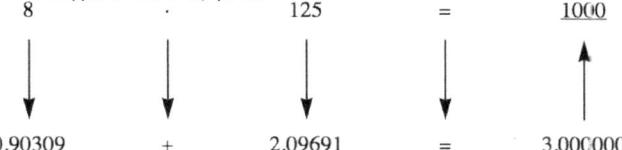

Figure 11.7.8 Multiplication using logarithms

The following definition of an isomorphism between two groups is a more formal one that appears in most abstract algebra texts. At first glance, it appears different, it is really a slight variation on the informal definition. It is the common definition because it is easy to apply; that is, given a function, this definition tells you what to do to determine whether that function is an isomorphism.

Definition 11.7.9 Group Isomorphism. If $[G_1; *_1]$ and $[G_2; *_2]$ are groups, $f : G_1 \to G_2$ is an isomorphism from G_1 into G_2 if:

(1) f is a bijection, and

(2) $f(a *_1 b) = f(a) *_2 f(b)$ for all $a, b \in G_1$

If such a function exists, then we say G_1 is isomorphic to G_2, denoted $G_1 \cong G_2$. ◊

We should note that "is isomorphic to" is an equivalence relation on the set of all groups. We leave it to the reader to verify the following.

- The identity function on a group G is an isomorphism.

- Bijections have inverses, the inverse of an isomorphism is an isomorphism.

- The composition of any two isomorphisms that can be composed is an isomorphism.

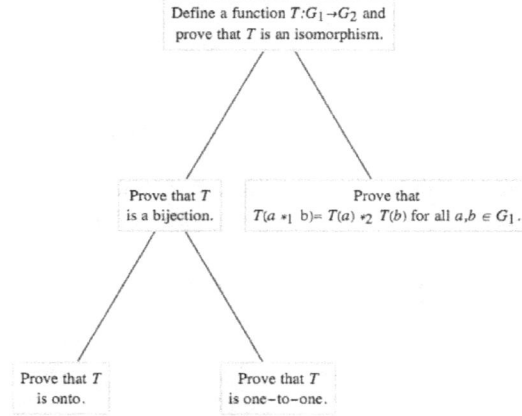

Figure 11.7.10 Steps in proving that G_1 and G_2 are isomorphic

Note 11.7.11

(a) There could be several different isomorphisms between the same pair of groups. Thus, if you are asked to demonstrate that two groups are isomorphic, your answer need not be unique.

(b) Any application of this definition requires a procedure outlined in Figure 11.7.10, p. 58. The first condition, that an isomorphism be a bijection, reflects the fact that every true statement in the first group should have exactly one corresponding true statement in the second group. This is exactly why we run into difficulty in translating between two natural languages. To see how Condition (b) of the formal definition is consistent with the informal definition, consider the function $L : \mathbb{R}^+ \to \mathbb{R}$ defined by $L(x) = \log_{10} x$. The translation diagram between $\mathbb{R}^+$ and $\mathbb{R}$ for the multiplication problem $a \cdot b$ appears in Figure 11.7.12, p. 58. We arrive at the same result by computing $L^{-1}(L(a) + L(b))$ as we do by computing $a \cdot b$. If we apply the function L to the two results, we get the same image:

$$L(a \cdot b) = L\left(L^{-1}(L(a) + L(b))\right) = L(a) + L(b) \qquad (11.7.1)$$

since $L\left(L^{-1}(x)\right) = x$. Note that (11.7.1) is exactly Condition b of the formal definition applied to the two groups $\mathbb{R}^+$ and $\mathbb{R}$.

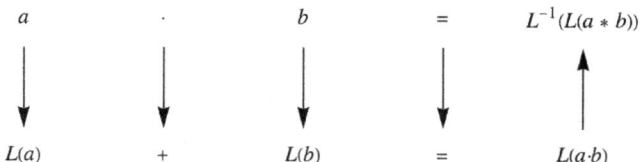

Figure 11.7.12 General Multiplication using logarithms

Example 11.7.13 Consider $G = \left\{ \begin{pmatrix} 1 & a \\ 0 & 1 \end{pmatrix} \middle| a \in \mathbb{R} \right\}$ with matrix multiplication. The group $[\mathbb{R}; +]$ is isomorphic to G. Our translation rule is the function $f : \mathbb{R} \to G$ defined by $f(a) = \begin{pmatrix} 1 & a \\ 0 & 1 \end{pmatrix}$. Since groups have only one operation, there is no need to state explicitly that addition is translated to matrix multiplication. That f is a bijection is clear from its definition.

11.7. ISOMORPHISMS

If a and b are any real numbers,

$$f(a)f(b) = \begin{pmatrix} 1 & a \\ 0 & 1 \end{pmatrix} \begin{pmatrix} 1 & b \\ 0 & 1 \end{pmatrix}$$
$$= \begin{pmatrix} 1 & a+b \\ 0 & 1 \end{pmatrix}$$
$$= f(a+b)$$

We can apply this translation rule to determine the inverse of a matrix in G. We know that $a + (-a) = 0$ is a true statement in $\mathbb{R}$. Using f to translate this statement, we get

$$f(a)f(-a) = f(0)$$

or

$$\begin{pmatrix} 1 & a \\ 0 & 1 \end{pmatrix} \begin{pmatrix} 1 & -a \\ 0 & 1 \end{pmatrix} = \begin{pmatrix} 1 & 0 \\ 0 & 1 \end{pmatrix}$$

therefore,

$$\begin{pmatrix} 1 & a \\ 0 & 1 \end{pmatrix}^{-1} = \begin{pmatrix} 1 & -a \\ 0 & 1 \end{pmatrix}$$

□

The next theorem summarizes some of the general facts about group isomorphisms that are used most often in applications. We leave the proof to the reader.

Theorem 11.7.14 Properties of Isomorphisms. *If $[G; *]$ and $[H; \diamond]$ are groups with identities e and e', respectively, and $T : G \to H$ is an isomorphism from G into H, then:*

(a) $T(e) = e'$

(b) $T(a)^{-1} = T(a^{-1})$ for all $a \in G$, and

(c) If K is a subgroup of G, then $T(K) = \{T(a) : a \in K\}$ is a subgroup of H and is isomorphic to K.

"Is isomorphic to" is an equivalence relation on the set of all groups. Therefore, the set of all groups is partitioned into equivalence classes, each equivalence class containing groups that are isomorphic to one another.

11.7.2 The order sequence of a finite group

This topic is somewhat obscure. It doesn't appear in most texts, but is a nice companion to degree sequences in graph theory. Recall that every undirected graph has a degree sequence, and graphs with different degree sequences 9.0.2, p. 17 are not isomorphic. This is a convenient way to identify non-isomorphic graphs. We see below that order sequences play exactly the same role in identifying whether two finite groups are isomorphic. Furthermore, identical order sequences of two finite groups give an excellent set of hints for constructing an isomorphism, if one such exists. My colleague, Jim Propp, has been using this idea for a while in his classes and I "discovered" it later. Neither of us can claim originality. Much of the following discussion is paraphrased from Jim's notes.

Definition 11.7.15 Order Sequence. The order sequence of a finite group is the sequence whose terms are the respective orders of all the elements of the group, arranged in increasing order. ◇

In $\mathbb{Z}_3$ the element 0 has order 1, the element 1 has order 3, and the element 2 has order 3, so the order sequence of this group is 1,3,3.

In $\mathbb{Z}_4$ the element 0 has order 1, the element 1 has order 4, the element 2 has order 2, and the element 3 has order 4, so the order sequence of this group is 1,2,4,4. (Note that we have arranged the numbers 1,4,2,4 in increasing order.)

Theorem 11.7.16 *If G_1 and G_2 are finite groups and f is an isomorphism between them, with $g \in G_1$ and $f(g) \in G_2$, the order of g in G_1 equals the order of $f(g)$ in G_2.*

Consequently:

Corollary 11.7.17 *If two groups are isomorphic, they have the same order sequence.*

The theorem is a handy tools for proving that two particular groups are not isomorphic. Consider the group $\mathbb{Z}_2 \times \mathbb{Z}_2$; the element $(0,0)$ has order 1 while the other elements $(0,1)$, $(1,0)$, and $(1,1)$ each have order 2, implying that the order sequence is 1,2,2,2. Since this is different from the sequence 1,2,4,4, the group $\mathbb{Z}_2 \times \mathbb{Z}_2$ is not isomorphic to the group $\mathbb{Z}_4$.

Order sequences are also useful in helping one find isomorphisms. Consider the group $\mathbb{U}_5$ (the set $\{1,2,3,4\}$ with mod-5 multiplication). Its order sequence is $1,2,4,4$, which suggests that it might be isomorphic to $\mathbb{Z}_4$. In fact, any isomorphism f from $\mathbb{Z}_4$ to $\mathbb{U}_5$ must map 0 (the only element of order 1 in $\mathbb{Z}_4$) to 1 (the only element of order 1 in $\mathbb{U}_4$) and must map 2 (the only element of order 2 in $\mathbb{Z}_4$) to 4 (the only element of order 2 in $\mathbb{U}_4$). There are only two bijections f from $\mathbb{Z}_4$ to $\mathbb{U}_4$ satisfying $f(0) = 1$ and $f(2) = 4$, so these are the only two candidate isomorphisms (and both candidates turn out to be true isomorphisms).

The following code will compute the order sequence for the group of integers mod n. The default value of n is 12 and you can change it in the last line of input.

```
def order_sequence_Z(n):
        G = Integers(n)
        os=[ ]
        for a in G:
                os=os+[a.order()]
        print(sorted(os))

order_sequence_Z(12)
```

[1, 2, 3, 3, 4, 4, 6, 6, 12, 12, 12, 12]

11.7.3 Conditions for groups to not be isomorphic

How do you decide that two groups are not isomorphic to one another? The negation of "G and H are isomorphic" is that no translation rule between G and H exists. If G and H have different cardinalities, then no bijection from G into H can exist. Hence they are not isomorphic. Given that $|G| = |H|$, it is usually impractical to list all bijections from G into H and show that none of them satisfy Condition b of the formal definition. The best way to prove that two groups are not isomorphic is to find a true statement about one group that is not true about the other group. We illustrate this method in the following checklist that you can apply to most pairs of non-isomorphic groups in this book.

11.7. ISOMORPHISMS

Assume that $[G; *]$ and $[H; \diamond]$ are groups. The following are reasons for G and H to be not isomorphic.

(a) G and H do not have the same cardinality. For example, $\mathbb{Z}_{12} \times \mathbb{Z}_5$ can't be isomorphic to $\mathbb{Z}_{50}$ and $[\mathbb{R}; +]$ can't be isomorphic to $[\mathbb{Q}^+; \cdot]$.

(b) G is abelian and H is not abelian since $a * b = b * a$ is always true in G, but $T(a) \diamond T(b) = T(b) \diamond T(a)$ would not always be true. We have seen two groups with six elements that apply here. They are $\mathbb{Z}_6$ and the group of 3×3 rook matrices (see Exercise 11.2.4.5, p. 29). The second group is non-abelian, therefore it can't be isomorphic to $\mathbb{Z}_6$.

(c) G has a certain kind of subgroup that H doesn't have. Part (c) of Theorem 11.7.14, p. 59 states that this cannot happen if G is isomorphic to H. $[\mathbb{R}^*; \cdot]$ and $[\mathbb{R}^+; \cdot]$ are not isomorphic since $\mathbb{R}^*$ has a subgroup with two elements, $\{-1, 1\}$, while the proper subgroups of $\mathbb{R}^+$ are all infinite (convince yourself of this fact!).

(d) The number of solutions of $x * x = e$ in G is not equal to the number of solutions of $y \diamond y = e'$ in H. $\mathbb{Z}_8$ is not isomorphic to $\mathbb{Z}_2^3$ since $x +_8 x = 0$ has two solutions, 0 and 4, while $y + y = (0,0,0)$ is true for all $y \in \mathbb{Z}_2^3$. If the operation in G is defined by a table, then the number of solutions of $x * x = e$ will be the number of occurrences of e in the main diagonal of the table. The equations $x^3 = e$, $x^4 = e, \ldots$ can also be used in the same way to identify pairs of non-isomorphic groups.

(e) One of the cyclic subgroups of G equals G (i. e., G is cyclic), while none of H's cyclic subgroups equals H (i. e., H is noncyclic). This is a special case of Condition c. $\mathbb{Z}$ and $\mathbb{Z} \times \mathbb{Z}$ are not isomorphic since $\mathbb{Z} = \langle 1 \rangle$ and $\mathbb{Z} \times \mathbb{Z}$ is not cyclic.

11.7.4 Exercises

1. State whether each pair of groups below is isomorphic. For each pair that is, give an isomorphism; for those that are not, give your reason.

    (a) $\mathbb{Z} \times \mathbb{R}$ and $\mathbb{R} \times \mathbb{Z}$

    (b) $\mathbb{Z}_2 \times \mathbb{Z}$ and $\mathbb{Z} \times \mathbb{Z}$

    (c) $\mathbb{R}$ and $\mathbb{Q} \times \mathbb{Q}$

    (d) $\mathcal{P}(\{1,2\})$ with symmetric difference and $\mathbb{Z}_2^2$

    (e) $\mathbb{Z}_2^2$ and $\mathbb{Z}_4$

    (f) $\mathbb{R}^4$ and $M_{2\times 2}(\mathbb{R})$ with matrix addition

    (g) $\mathbb{R}^2$ and $\mathbb{R} \times \mathbb{R}^+$

    (h) $\mathbb{Z}_2$ and the 2×2 rook matrices

    (i) $\mathbb{Z}_6$ and $\mathbb{Z}_2 \times \mathbb{Z}_3$

2. If you know two natural languages, show that they are not isomorphic.

3. Prove that the relation "is isomorphic to" on groups is transitive.

4.
    (a) Write out the operation table for $G = [\{1, -1, i, -i\}; \cdot]$ where i is the complex number for which $i^2 = -1$. Show that G is isomorphic to $[\mathbb{Z}_4; +_4]$.

(b) Solve $x^2 = -1$ in G by first translating the equation to $\mathbb{Z}_4$, solving the equation in $\mathbb{Z}_4$, and then translating back to G.

5. The two groups $[\mathbb{Z}_4; +_4]$ and $[U_5; \times_5]$ are isomorphic. One isomorphism $T : \mathbb{Z}_4 \to U_5$ is partially defined by $T(1) = 3$. Determine the values of $T(0)$, $T(2)$, and $T(3)$.

6. Prove Theorem 11.7.14, p. 59.

7. Prove that all infinite cyclic groups are isomorphic to $\mathbb{Z}$.

8.
   (a) Prove that $\mathbb{R}^*$ is isomorphic to $\mathbb{Z}_2 \times \mathbb{R}$.

   (b) Describe how multiplication of nonzero real numbers can be accomplished doing only additions and translations.

9. Prove that if G is any group and g is some fixed element of G, then the function ϕ_g defined by $\phi_g(x) = g * x * g^{-1}$ is an isomorphism from G into itself. An isomorphism of this type is called an inner automorphism.

10. Prove that "is isomorphic to" is an equivalence relation on the set of all groups by expanding on the observations made immediately after the definiton of an isomorphism.

11. It can be shown that there are five non-isomorphic groups of order eight. You should be able to describe at least three of them. Do so without use of tables. Be sure to explain why they are not isomorphic.

12. In Section 11.2, p. 25 we posed the question of whether the two monoids $[\mathcal{P}(U); \cap]$ and $[\mathcal{P}(U); \cup]$, both monoids on the power set of some nonempty universal set U, are different or really the same. At the time we didn't have the notion of isomorphism to draw upon. Now that we do, determine whether they are isomorphic monoids.

13. Prove that the number of 3's in an order sequence is even.

14. Prove that the number of 5's an order sequence is a multiple of four.

Chapter 12

More Matrix Algebra

augmented matrix

> There's a Gaussian technique whose intent
> Is to solve the constraints you present
> As a matrix equation—
> Once you've had the occasion
> To write down your constants (**augment**).

Steve Ngai, The Omnificent English Dictionary In Limerick Form

In Chapter 5 we studied matrix operations and the algebra of sets and logic. We also made note of the strong resemblance of matrix algebra to elementary algebra. The reader should briefly review this material. In this chapter we shall look at a powerful matrix tool in the applied sciences, namely a technique for solving systems of linear equations. We will then use this process for determining the inverse of $n \times n$ matrices, $n \geq 2$, when they exist. We proceed with a development of the diagonalization process, with a discussion of several of its applications. Finally, we discuss the solution of linear equations over the integers modulo 2.

12.1 Systems of Linear Equations

12.1.1 Solutions

The method of solving systems of equations by matrices that we will look at is based on procedures involving equations that we are familiar with from previous mathematics courses. The main idea is to reduce a given system of equations to another simpler system that has the same solutions.

Definition 12.1.1 Solution Set. Given a system of equations involving real variables $x_1, x_2, \ldots, x_n$, the solution set of the system is the set of n-tuples in $\mathbb{R}^n$, $(a_1, a_2, \ldots, a_n)$ such that the substitutions $x_1 = a_1, x_2 = a_2, \ldots, x_n = a_n$ make all the equations true. $\Diamond$

In terms of logic, a solution set is a truth set of a system of equations, which is a proposition over n-tuples of real numbers.

In general, if the variables are from a set S, then the solution set will be a subset of S^n. For example, in number theory mathematicians study Diophantine equations, where the variables can only take on integer values instead of real values.

Definition 12.1.2 Equivalent Systems of Equations. Two systems of linear equations are called equivalent if they have the same set of solutions. ◊

Example 12.1.3 Two equivalent systems. The previous definition tells us that if we know that the system

$$4x_1 + 2x_2 + x_3 = 1$$
$$2x_1 + x_2 + x_3 = 4$$
$$2x_1 + 2x_2 + x_3 = 3$$

is equivalent to the system

$$x_1 + 0x_2 + 0x_3 = -1$$
$$0x_1 + x_2 + 0x_3 = -1$$
$$0x_1 + 0x_2 + x_3 = 7$$

then both systems have the solution set $\{(-1, -1, 7)\}$. In other words, the simultaneous values $x_1 = -1$, $x_2 = -1$, and $x_3 = 7$ are the only values of the variables that make all three equations in either system true. □

12.1.2 Elementary Operations on Equations

Theorem 12.1.4 Elementary Operations on Equations. *If any sequence of the following operations is performed on a system of equations, the resulting system is equivalent to the original system:*

(a) *Interchange any two equations in the system.*

(b) *Multiply both sides of any equation by a nonzero constant.*

(c) *Multiply both sides of any equation by a nonzero constant and add the result to a second equation in the system, with the sum replacing the latter equation.*

Let us now use the above theorem to work out the details of Example 12.1.3, p. 64 and see how we can arrive at the simpler system.

The original system:

$$\begin{aligned} 4x_1 + 2x_2 + x_3 &= 1 \\ 2x_1 + x_2 + x_3 &= 4 \\ 2x_1 + 2x_2 + x_3 &= 3 \end{aligned} \quad (12.1.1)$$

Step 1. We will first change the coefficient of x_1 in the first equation to one and then use it as a pivot to obtain 0's for the coefficients of x_1 in Equations 2 and 3.

- Multiply Equation 1 by $\frac{1}{4}$ to obtain

$$\begin{aligned} x_1 + \tfrac{x_2}{2} + \tfrac{x_3}{4} &= \tfrac{1}{4} \\ 2x_1 + x_2 + x_3 &= 4 \\ 2x_1 + 2x_2 + x_3 &= 3 \end{aligned} \quad (12.1.2)$$

- Multiply Equation 1 by -2 and add the result to Equation 2 to obtain

$$\begin{aligned} x_1 + \tfrac{x_2}{2} + \tfrac{x_3}{4} &= \tfrac{1}{4} \\ 0x_1 + 0x_2 + \tfrac{x_3}{2} &= \tfrac{7}{2} \\ 2x_1 + 2x_2 + x_3 &= 3 \end{aligned} \quad (12.1.3)$$

12.1. SYSTEMS OF LINEAR EQUATIONS

- Multiply Equation 1 by -2 and add the result to Equation 3 to obtain

$$x_1 + \frac{x_2}{2} + \frac{x_3}{4} = \frac{1}{4}$$
$$0x_1 + 0x_2 + \frac{x_3}{2} = \frac{7}{2} \quad (12.1.4)$$
$$0x_1 + x_2 + \frac{x_3}{2} = \frac{5}{2}$$

We've explicitly written terms with zero coefficients such as $0x_1$ to make a point that all variables can be thought of as being involved in all equations. After this example is complete, we will discontinue this practice in favor of the normal practice of making these terms "disappear."

Step 2. We would now like to proceed in a fashion analogous to Step 1; namely, multiply the coefficient of x_2 in the second equation by a suitable number so that the result is 1. Then use it as a pivot to obtain 0's as coefficients for x_2 in the first and third equations. This is clearly impossible (Why?), so we will first interchange Equations 2 and 3 and proceed as outlined above.

- Exchange Equations 2 and 3 to obtain

$$x_1 + \frac{x_2}{2} + \frac{x_3}{4} = \frac{1}{4}$$
$$0x_1 + x_2 + \frac{x_3}{2} = \frac{5}{2} \quad (12.1.5)$$
$$0x_1 + 0x_2 + \frac{x_3}{2} = \frac{7}{2}$$

- Multiply Equation 2 by $\frac{1}{2}$ and subtract the result from Equation 1 to obtain

$$x_1 + 0x_2 + 0x_3 = -1$$
$$0x_1 + x_2 + \frac{x_3}{2} = \frac{5}{2} \quad (12.1.6)$$
$$0x_1 + 0x_2 + \frac{x_3}{2} = \frac{7}{2}$$

Step 3. Next, we will change the coefficient of x_3 in the third equation to one and then use it as a pivot to obtain 0's for the coefficients of x_3 in Equations 1 and 2. Notice that the coefficient of x_3 is already zero in Equation 1, so we have been saved some work!

- Multiply Equation 3 by 2 to obtain

$$x_1 + 0x_2 + 0x_3 = -1$$
$$0x_1 + x_2 + \frac{x_3}{2} = \frac{5}{2}$$
$$0x_1 + 0x_2 + x_3 = 7$$

- Multiply Equation 3 by $-1/2$ and add the result to Equation 2 to obtain

$$x_1 + 0x_2 + 0x_3 = -1$$
$$0x_1 + x_2 + 0x_3 = -1 \quad (12.1.7)$$
$$0x_1 + 0x_2 + x_3 = 7$$

From the system of equations at the end of Step 3, we see that the solution to the original system is $x_1 = -1$, $x_2 = -1$, and $x_3 = 7$.

12.1.3 Transition to Matrices

In the above sequence of steps, we note that the variables serve the sole purpose of keeping the coefficients in the appropriate location. This we can effect by using matrices. The matrix of the original system in our example is

$$\left(\begin{array}{ccc|c} 4 & 2 & 1 & 1 \\ 2 & 1 & 1 & 4 \\ 2 & 2 & 1 & 3 \end{array}\right)$$

where the matrix of the first three columns is called the coefficient matrix and the complete matrix is referred to as the augmented matrix. Since we are now using matrices to solve the system, we will translate Theorem 12.1.4, p. 64 into matrix language.

12.1.4 Elementary Row Operations

Theorem 12.1.5 Elementary Row Operations. *If any sequence of the following operations is performed on the augmented matrix of a system of equations, the resulting matrix is a system that is equivalent to the original system. The following operations on a matrix are called elementary row operations:*

(1) Exchange any two rows of the matrix.

(2) Multiply any row of the matrix by a nonzero constant.

(3) Multiply any row of the matrix by a nonzero constant and add the result to a second row, with the sum replacing that second row.

Definition 12.1.6 Row Equivalent Matrices. Two matrices, A and B, are said to be row-equivalent if one can be obtained from the other by any sequence of zero or more elementary row operations. ◊

If we use the notation R_i to stand for Row i of a matrix and $\longrightarrow$ to stand for row equivalence, then

$$A \xrightarrow{cR_i + R_j} B$$

means that the matrix B is obtained from the matrix A by multiplying the Row i of A by c and adding the result to Row j. The operation of multiplying row i by c is indicated by

$$A \xrightarrow{cR_i} B$$

while exchanging rows i and j is denoted by

$$A \xrightarrow{R_i \leftrightarrow R_j} B.$$

The matrix notation for the system given in our first example, with the

12.1. SYSTEMS OF LINEAR EQUATIONS

subsequent steps, is:

$$\begin{pmatrix} 4 & 2 & 1 & | & 1 \\ 2 & 1 & 1 & | & 4 \\ 2 & 2 & 1 & | & 3 \end{pmatrix} \xrightarrow{\frac{1}{4}R_1} \begin{pmatrix} 1 & \frac{1}{2} & \frac{1}{4} & | & \frac{1}{4} \\ 2 & 1 & 1 & | & 4 \\ 2 & 2 & 1 & | & 3 \end{pmatrix} \xrightarrow{-2R_1+R_2} \begin{pmatrix} 1 & \frac{1}{2} & \frac{1}{4} & | & \frac{1}{4} \\ 0 & 0 & \frac{1}{2} & | & \frac{7}{2} \\ 2 & 2 & 1 & | & 3 \end{pmatrix}$$

$$\xrightarrow{-2R_1+R_3} \begin{pmatrix} 1 & \frac{1}{2} & \frac{1}{4} & | & \frac{1}{4} \\ 0 & 0 & \frac{1}{2} & | & \frac{7}{2} \\ 0 & 1 & \frac{1}{2} & | & \frac{5}{2} \end{pmatrix} \xrightarrow{R_2 \leftrightarrow R_3} \begin{pmatrix} 1 & \frac{1}{2} & \frac{1}{4} & | & \frac{1}{4} \\ 0 & 1 & \frac{1}{2} & | & \frac{5}{2} \\ 0 & 0 & \frac{1}{2} & | & \frac{7}{2} \end{pmatrix}$$

$$\xrightarrow{-\frac{1}{2}R_2+R_1} \begin{pmatrix} 1 & 0 & 0 & | & -1 \\ 0 & 1 & \frac{1}{2} & | & \frac{5}{2} \\ 0 & 0 & \frac{1}{2} & | & \frac{7}{2} \end{pmatrix} \xrightarrow{2R_3} \begin{pmatrix} 1 & 0 & 0 & | & -1 \\ 0 & 1 & \frac{1}{2} & | & \frac{5}{2} \\ 0 & 0 & 1 & | & 7 \end{pmatrix}$$

$$\xrightarrow{-\frac{1}{2}R_3+R_2} \begin{pmatrix} 1 & 0 & 0 & | & -1 \\ 0 & 1 & 0 & | & -1 \\ 0 & 0 & 1 & | & 7 \end{pmatrix}$$

This again gives us the solution. This procedure is called the **Gauss-Jordan elimination method**.

It is important to remember when solving any system of equations via this or any similar approach that at any step in the procedure we can rewrite the matrix in "equation format" to help us to interpret the meaning of the augmented matrix.

In our first example we found a unique solution, only one triple, namely $(-1, -1, 7)$, which satisfies all three equations. For a system involving three unknowns, are there any other possible results? To answer this question, let's review some basic facts from analytic geometry.

The graph of a linear equation in three-dimensional space is a plane. So geometrically we can visualize the three linear equations as three planes in three-space. Certainly the three planes can intersect in a unique point, as in the first example, or two of the planes could be parallel. If two planes are parallel, there are no common points of intersection; that is, there are no triple of real numbers that will satisfy all three equations. Another possibility is that the three planes could intersect along a common axis or line. In this case, there would be an infinite number of real number triples in $\mathbb{R}^3$. Yet another possibility would be if the first two planes intersect in a line, but the line is parallel to, but not on, the third plane, giving us no solution. Finally if all three equations describe the same plane, the solution set would be that plane.

We can generalize these observations. In a system of n linear equations, n unknowns, there can be

(1) a unique solution,

(2) no solution, or

(3) an infinite number of solutions.

To illustrate these points, consider the following examples:

Example 12.1.7 A system with no solutions. Find all solutions to the system
$$\begin{aligned} x_1 + 3x_2 + x_3 &= 2 \\ x_1 + x_2 + 5x_3 &= 4 \\ 2x_1 + 2x_2 + 10x_3 &= 6 \end{aligned}$$

The reader can verify that the augmented matrix of this system, $\begin{pmatrix} 1 & 3 & 1 & | & 2 \\ 1 & 1 & 5 & | & 4 \\ 2 & 2 & 10 & | & 6 \end{pmatrix}$, reduces to $\begin{pmatrix} 1 & 3 & 1 & | & 2 \\ 1 & 1 & 5 & | & 4 \\ 0 & 0 & 0 & | & -2 \end{pmatrix}$.

We can attempt to row-reduce this matrix further if we wish. However, any further row-reduction will not substantially change the last row, which, in equation form, is $0x_1 + 0x_2 + 0x_3 = -2$, or simply $0 = -2$. It is clear that we cannot find real numbers x_1, x_2, and x_3 that will satisfy this equation. Hence we cannot find real numbers that will satisfy all three original equations simultaneously. When this occurs, we say that the system has no solution, or the solution set is empty. □

Example 12.1.8 A system with an infinite number of solutions. Next, let's attempt to find all of the solutions to:

$$\begin{aligned} x_1 + 6x_2 + 2x_3 &= 1 \\ 2x_1 + x_2 + 3x_3 &= 2 \\ 4x_1 + 2x_2 + 6x_3 &= 4 \end{aligned}$$

The augmented matrix for the system is

$$\begin{pmatrix} 1 & 6 & 2 & | & 1 \\ 2 & 1 & 3 & | & 2 \\ 4 & 2 & 6 & | & 4 \end{pmatrix} \quad (12.1.8)$$

which reduces to

$$\begin{pmatrix} 1 & 0 & \frac{16}{11} & | & 1 \\ 0 & 1 & \frac{1}{11} & | & 0 \\ 0 & 0 & 0 & | & 0 \end{pmatrix} \quad (12.1.9)$$

If we apply additional elementary row operations to this matrix, it will only become more complicated. In particular, we cannot get a one in the third row, third column. Since the matrix is in simplest form, we will express it in equation format to help us determine the solution set.

$$\begin{aligned} x_1 + \tfrac{16}{11}x_3 &= 1 \\ x_2 + \tfrac{1}{11}x_3 &= 0 \\ 0 &= 0 \end{aligned} \quad (12.1.10)$$

Any real numbers will satisfy the last equation. However, the first equation can be rewritten as $x_1 = 1 - \frac{16}{11}x_3$, which describes the coordinate x_1 in terms of x_3. Similarly, the second equation gives x_2 in terms of x_3. A convenient way of listing the solutions of this system is to use set notation. If we call the solution set of the system S, then

$$S = \left\{ \left(1 - \frac{16}{11}x_3, -\frac{1}{11}x_3, x_3\right) \mid x_3 \in \mathbb{R} \right\}.$$

What this means is that if we wanted to list all solutions, we would replace x_3 by all possible numbers. Clearly, there is an infinite number of solutions, two of which are $(1, 0, 0)$ and $(-15, -1, 11)$, when x_3 takes on the values 0 and 11, respectively.

A Word Of Caution: Frequently we may can get "different-looking" answers to the same problem when a system has an infinite number of solutions. Assume the solutions set in this example is reported to be $A =$

$\{(1 + 16x_2, x_2, -11x_3) \mid x_3 \in \mathbb{R}\}$. Certainly the result described by S looks different from that described by A. To see whether they indeed describe the same set, we wish to determine whether every solution produced in S can be generated in A. For example, the solution generated by S when $x_3 = 11$ is $(-15, -1, 11)$. The same triple can be produced by A by taking $x_2 = -1$. We must prove that every solution described in S is described in A and, conversely, that every solution described in A is described in S. (See Exercise 6 of this section.) □

To summarize the procedure in the Gauss-Jordan technique for solving systems of equations, we attempt to obtain 1's along the main diagonal of the coefficient matrix with 0's above and below the diagonal. We may find in attempting this that this objective cannot be completed, as in the last two examples we have seen. Depending on the way we interpret the results in equation form, we either recognize that no solution exists, or we identify "free variables" on which an infinite number of solutions are based. The final matrix forms that we have produced in our examples are referred to as **echelon forms**.

In practice, larger systems of linear equations are solved using computers. Generally, the Gauss-Jordan algorithm is the most useful; however, slight variations of this algorithm are also used. The different approaches share many of the same advantages and disadvantages. The two major concerns of all methods are:

(1) minimizing inaccuracies due to round-off errors, and

(2) minimizing computer time.

12.1.5 The Gauss-Jordan Algorithm

The accuracy of the Gauss-Jordan method can be improved by always choosing the element with the largest absolute value as the pivot element, as in the following algorithm.

Algorithm 12.1.9 The Gauss-Jordan Algorithm. *Given a matrix equation $Ax = b$, where A is $n \times m$, let C be the augmented matrix $[A|b]$. The process of row-reducing to echelon form involves performing the following algorithm where $C[i]$ is the ith row of C.*

(1) i = 1

(2) j = 1

(3) while i <= n and j <= m:

    (a) maxi=i

    (b) for $k = i+1$ to n:
        if abs(C[k,j])>abs(C[maxi,j]): *then* maxi=k

    (c) if C[maxi,j] != 0 *then:*

        (i) exchange rows i *and* maxi

        (ii) divide each entry in row i *by* C[i,j]

        (iii) for u = i+1 *to n:*
            subtract C[u,j]*C[i] *from* C[u]

        (iv) i = i+1

    (d) j=j+1

Note 12.1.10 At the end of this algorithm, with the final form of C you can revert back to the equation form of the system and a solution should be clear.

In general,

- If any row of C is all zeros, it can be ignored.

- If any row of C has all zero entries except for the entry in the $(m+1)^{\text{st}}$ position, the system has no solution. Otherwise, if a column has no pivot, the variable corresponding to it is a free variable. Variables corresponding to pivots are basic variables and can be expressed in terms of the free variables.

Example 12.1.11 If we apply The Gauss-Jordan Algorithm, p. 69 to the system
$$\begin{aligned} 5x_1 + x_2 + 2x_3 + x_4 &= 2 \\ 3x_1 + x_2 - 2x_3 &= 5 \\ x_1 + x_2 + 3x_3 - x_4 &= -1 \end{aligned}$$
the augmented matrix is
$$\begin{pmatrix} 5 & 1 & 2 & 1 & | & 2 \\ 3 & 1 & -2 & 0 & | & 5 \\ 1 & 1 & 3 & -1 & | & -1 \end{pmatrix}$$
is reduced to
$$\begin{pmatrix} 1 & 0 & 0 & \frac{1}{2} & | & \frac{1}{2} \\ 0 & 1 & 0 & -\frac{3}{2} & | & \frac{3}{2} \\ 0 & 0 & 1 & 0 & | & -1 \end{pmatrix}$$

Therefore, x_4 is a free variable in the solution and general solution of the system is
$$x = \begin{pmatrix} x_1 \\ x_2 \\ x_3 \\ x_4 \end{pmatrix} = \begin{pmatrix} \frac{1}{2} - \frac{1}{2}x_4 \\ \frac{3}{2} + \frac{3}{2}x_4 \\ -1 \\ x_4 \end{pmatrix}$$

This conclusion is easy to see if you revert back to the equations that the final value the reduced matrix represents. □

12.1.6 SageMath Note - Matrix Reduction

Given an augmented matrix, C, there is a matrix method called `echelon_form` that can be used to row reduce C. Here is the result for the system in Example 12.1.11, p. 70. In the assignment of a matrix value to C, notice that the first argument is QQ, which indicates that the entries should be rational numbers. As long as all the entries are rational, which is the case here since integers are rational, the row-reduced matrix will be all rational.

```
C = Matrix(QQ,[[5,1,2,1,2],[3,1,-2,0,5],[1,1,3,-1,-1]])
C.echelon_form()
```

```
[1    0    0    1/2    1/2]
[0    1    0   -3/2    3/2]
[0    0    1    0     -1]
```

If we don't specify the set from which entries are taken, it would assumed to be the integers and we do not get a fully row-reduced matrix. This is because the next step in working with the next output would involve multiplying row 2 by $\frac{1}{2}$ and row 3 by $\frac{1}{9}$, but these multipliers are not integers.

12.1. SYSTEMS OF LINEAR EQUATIONS

```
C2 = Matrix([[5,1,2,1,2],[3,1,-2,0,5],[1,1,3,-1,-1]])
C2.echelon_form()
```

```
[ 1  1  3 -1 -1]
[ 0  2  2 -3  1]
[ 0  0  9  0 -9]
```

If we specifying real entries, the result isn't as nice and clean as the rational output.

```
C3 = Matrix(RR,[[5,1,2,1,2],[3,1,-2,0,5],[1,1,3,-1,-1]])
C3.echelon_form()
```

```
[    1.000000     0.0000000    0.0000000     0.5000000
  0.500000000000000]
[    0.0000000     1.000000    0.0000000    -1.500000
  1.50000000000000]
[    0.0000000    0.0000000     1.000000    4.934324e-17
 -1.00000000000000]
```

The default number of decimal places may vary from what you see here, but it can be controlled. The single small number in row three column four isn't exactly zero because of round-off but we could just set it to zero.

12.1.7 Exercises

1. Solve the following systems by describing the solution sets completely:

   (a) $\begin{array}{l} 2x_1 + x_2 = 3 \\ x_1 - x_2 = 1 \end{array}$

   (b) $\begin{array}{l} 2x_1 + x_2 + 3x_3 = 5 \\ 4x_1 + x_2 + 2x_3 = -1 \\ 8x_1 + 2x_2 + 4x_3 = -2 \end{array}$

   (c) $\begin{array}{l} x_1 + x_2 + 2x_3 = 1 \\ x_1 + 2x_2 - x_3 = -1 \\ x_1 + 3x_2 + x_3 = 5 \end{array}$

   (d) $\begin{array}{l} x_1 - x_2 + 3x_3 = 7 \\ x_1 + 3x_2 + x_3 = 4 \end{array}$

2. Solve the following systems by describing the solution sets completely:

   (a) $\begin{array}{l} 2x_1 + 2x_2 + 4x_3 = 2 \\ 2x_1 + x_2 + 4x_3 = 0 \\ 3x_1 + 5x_2 + x_3 = 0 \end{array}$

   (b) $\begin{array}{l} 2x_1 + x_2 + 3x_3 = 2 \\ 4x_1 + x_2 + 2x_3 = -1 \\ 8x_1 + 2x_2 + 4x_3 = 4 \end{array}$

   (c) $\begin{array}{l} x_1 + x_2 + 2x_3 + x_4 = 3 \\ x_1 - x_2 + 3x_3 - x_4 = -2 \\ 3x_1 + 3x_2 + 6x_3 + 3x_4 = 9 \end{array}$

   (d) $\begin{array}{l} 6x_1 + 7x_2 + 2x_3 = 3 \\ 4x_1 + 2x_2 + x_3 = -2 \\ 6x_1 + x_2 + x_3 = 1 \end{array}$

   (e) $\begin{array}{l} x_1 + x_2 - x_3 + 2x_4 = 1 \\ x_1 + 2x_2 + 3x_3 + x_4 = 5 \\ x_1 + 3x_2 + 2x_3 - x_4 = -1 \end{array}$

3. Given the final augmented matrices below from the Gauss-Jordan Algorithm, identify the solutions sets. Identify the basic and free variables, and describe the solution set of the original system.

   (a) $\left(\begin{array}{cccc|c} 1 & 0 & -5 & 0 & 1.2 \\ 0 & 1 & 4 & 0 & 2.6 \\ 0 & 0 & 0 & 1 & 4.5 \end{array} \right)$

   (b) $\left(\begin{array}{ccc|c} 1 & 0 & 9 & 3 \\ 0 & 1 & 0 & 4 \\ 0 & 0 & 0 & 1 \end{array} \right)$

   (c) $\left(\begin{array}{ccc|c} 1 & 0 & 6 & 5 \\ 0 & 1 & -2 & 1 \\ 0 & 0 & 0 & 0 \end{array} \right)$

   (d) $\left(\begin{array}{ccc|c} 1 & 0 & 0 & -3 & 1 \\ 0 & 1 & 0 & 2 & 2 \\ 0 & 0 & 1 & -1 & 1 \end{array} \right)$

4.

    (a) Write out the details of Example 12.1.7, p. 67.

    (b) Write out the details of Example 12.1.8, p. 68.

    (c) Write out the details of Example 12.1.11, p. 70.

5. Solve the following systems using only mod 5 arithmetic. Your solutions should be n − tuples from $\mathbb{Z}_5$.

    (a) $\begin{array}{l} 2x_1 + x_2 = 3 \\ x_1 + 4x_2 = 1 \end{array}$ (compare your solution to the system in 5(a))

    (b) $\begin{array}{l} x_1 + x_2 + 2x_3 = 1 \\ x_1 + 2x_2 + 4x_3 = 4 \\ x_1 + 3x_2 + 3x_3 = 0 \end{array}$

6.

    (a) Use the solution set S of Example 12.1.8, p. 68 to list three different solutions to the given system. Then show that each of these solutions can be described by the set A in the same example.

    (b) Prove that $S = A$.

7. Given a system of n linear equations in n unknowns in matrix form $AX = b$, prove that if b is a matrix of all zeros, then the solution set of $AX = b$ is a subgroup of $\mathbb{R}^n$.

12.2 Matrix Inversion

12.2.1 Developing the Process

In Chapter 5 we defined the inverse of an $n \times n$ matrix. We noted that not all matrices have inverses, but when the inverse of a matrix exists, it is unique. This enables us to define the inverse of an $n \times n$ matrix A as the unique matrix B such that $AB = BA = I$, where I is the $n \times n$ identity matrix. In order to get some practical experience, we developed a formula that allowed us to determine the inverse of invertible 2×2 matrices. We will now use the Gauss-Jordan procedure for solving systems of linear equations to compute the inverses, when they exist, of $n \times n$ matrices, $n \geq 2$. The following procedure for a 3×3 matrix can be generalized for $n \times n$ matrices, $n \geq 2$.

Given the matrix $A = \begin{pmatrix} 1 & 1 & 2 \\ 2 & 1 & 4 \\ 3 & 5 & 1 \end{pmatrix}$, we want to find its inverse, the matrix $B = \begin{pmatrix} x_{11} & x_{12} & x_{13} \\ x_{21} & x_{22} & x_{23} \\ x_{31} & x_{32} & x_{33} \end{pmatrix}$, if it exists, such that $AB = I$ and $BA = I$. We will concentrate on finding a matrix that satisfies the first equation and then verify that B also satisfies the second equation.

The equation

$$\begin{pmatrix} 1 & 1 & 2 \\ 2 & 1 & 4 \\ 3 & 5 & 1 \end{pmatrix} \begin{pmatrix} x_{11} & x_{12} & x_{13} \\ x_{21} & x_{22} & x_{23} \\ x_{31} & x_{32} & x_{33} \end{pmatrix} = \begin{pmatrix} 1 & 0 & 0 \\ 0 & 1 & 0 \\ 0 & 0 & 1 \end{pmatrix}$$

12.2. MATRIX INVERSION

is equivalent to

$$\begin{pmatrix} x_{11}+x_{21}+2x_{31} & x_{12}+x_{22}+2x_{32} & x_{13}+x_{23}+2x_{33} \\ 2x_{11}+x_{21}+4x_{31} & 2x_{12}+x_{22}+4x_{32} & 2x_{13}+x_{23}+4x_{33} \\ 3x_{11}+5x_{21}+x_{31} & 3x_{12}+5x_{22}+x_{32} & 3x_{13}+5x_{23}+x_{33} \end{pmatrix} = \begin{pmatrix} 1 & 0 & 0 \\ 0 & 1 & 0 \\ 0 & 0 & 1 \end{pmatrix}$$

By definition of equality of matrices, this gives us three systems of equations to solve. The augmented matrix of one of the systems, the one equating the first columns of the two matrices is:

$$\begin{pmatrix} 1 & 1 & 2 & | & 1 \\ 2 & 1 & 4 & | & 0 \\ 3 & 5 & 1 & | & 0 \end{pmatrix} \qquad (12.2.1)$$

Using the Gauss-Jordan algorithm, we have:

$$\begin{pmatrix} 1 & 1 & 2 & | & 1 \\ 2 & 1 & 4 & | & 0 \\ 3 & 5 & 1 & | & 0 \end{pmatrix} \xrightarrow{-2R_1+R_2} \begin{pmatrix} 1 & 1 & 2 & | & 1 \\ 0 & -1 & 0 & | & -2 \\ 3 & 5 & 1 & | & 0 \end{pmatrix} \xrightarrow{-3R_1+R_3} \begin{pmatrix} 1 & 1 & 2 & | & 1 \\ 0 & -1 & 0 & | & -2 \\ 0 & 2 & -5 & | & -3 \end{pmatrix}$$

$$\xrightarrow{-1R_2} \begin{pmatrix} 1 & 1 & 2 & | & 1 \\ 0 & 1 & 0 & | & 2 \\ 0 & 2 & -5 & | & -3 \end{pmatrix}$$

$$\xrightarrow{-R_2+R_1 \text{ and} -2R_2+R_3} \begin{pmatrix} 1 & 0 & 2 & | & -1 \\ 0 & 1 & 0 & | & 2 \\ 0 & 0 & -5 & | & -7 \end{pmatrix}$$

$$\xrightarrow{-\frac{1}{5}R_3} \begin{pmatrix} 1 & 0 & 2 & | & -1 \\ 0 & 1 & 0 & | & 2 \\ 0 & 0 & 1 & | & 7/5 \end{pmatrix} \xrightarrow{-2R_3+R_1} \begin{pmatrix} 1 & 0 & 0 & | & -\frac{19}{5} \\ 0 & 1 & 0 & | & 2 \\ 0 & 0 & 1 & | & \frac{7}{5} \end{pmatrix}$$

So $x_{11} = -19/5, x_{21} = 2$ and $x_{31} = 7/5$, which gives us the first column of B.

The matrix form of the system to obtain x_{12}, x_{22}, and x_{32}, the second column of B, is:

$$\begin{pmatrix} 1 & 1 & 2 & | & 0 \\ 2 & 1 & 4 & | & 1 \\ 3 & 5 & 1 & | & 0 \end{pmatrix} \qquad (12.2.2)$$

which reduces to

$$\begin{pmatrix} 1 & 0 & 0 & | & \frac{9}{5} \\ 0 & 1 & 0 & | & -1 \\ 0 & 0 & 1 & | & -\frac{2}{5} \end{pmatrix} \qquad (12.2.3)$$

The critical thing to note here is that the coefficient matrix in (12.2.2) is the same as the matrix in (12.2.1), hence the sequence of row operations that we used in row reduction are the same in both cases.

To determine the third column of B, we reduce

$$\begin{pmatrix} 1 & 1 & 2 & | & 0 \\ 2 & 1 & 4 & | & 0 \\ 3 & 5 & 1 & | & 1 \end{pmatrix}$$

to obtain $x_{13} = 2/5, x_{23} = 0$ and $x_{33} = -1/5$. Here again it is important to note that the sequence of row operations used to solve this system is exactly the same as those we used in the first system. Why not save ourselves a considerable amount of time and effort and solve all three systems simultaneously? This we can do this by augmenting the coefficient matrix by the identity matrix I. We then have, by applying the same sequence of row operations as above,

$$\begin{pmatrix} 1 & 1 & 2 & | & 1 & 0 & 0 \\ 2 & 1 & 4 & | & 0 & 1 & 0 \\ 3 & 5 & 1 & | & 0 & 0 & 1 \end{pmatrix} \longrightarrow \begin{pmatrix} 1 & 0 & 0 & | & -\frac{19}{5} & \frac{9}{5} & \frac{2}{5} \\ 0 & 1 & 0 & | & 2 & -1 & 0 \\ 0 & 0 & 1 & | & \frac{7}{5} & -\frac{2}{5} & -\frac{1}{5} \end{pmatrix}$$

So that

$$B = \begin{pmatrix} -\frac{19}{5} & \frac{9}{5} & \frac{2}{5} \\ 2 & -1 & 0 \\ \frac{7}{5} & -\frac{2}{5} & -\frac{1}{5} \end{pmatrix}$$

The reader should verify that $BA = I$ so that $A^{-1} = B$.

12.2.2 The General Method for Computing Inverses

As the following theorem indicates, the verification that $BA = I$ is not necessary. The proof of the theorem is beyond the scope of this text. The interested reader can find it in most linear algebra texts.

Theorem 12.2.1 *Let A be an $n \times n$ matrix. If a matrix B can be found such that $AB = I$, then $BA = I$, so that $B = A^{-1}$. In fact, to find A^{-1}, we need only find a matrix B that satisfies one of the two conditions $AB = I$ or $BA = I$.*

It is clear from Chapter 5 and our discussions in this chapter that not all $n \times n$ matrices have inverses. How do we determine whether a matrix has an inverse using this method? The answer is quite simple: the technique we developed to compute inverses is a matrix approach to solving several systems of equations simultaneously.

Example 12.2.2 Recognition of a non-invertible matrix. The reader can verify that if $A = \begin{pmatrix} 1 & 2 & 1 \\ -1 & -2 & -1 \\ 0 & 5 & 8 \end{pmatrix}$ then the augmented matrix $\begin{pmatrix} 1 & 2 & 1 & | & 1 & 0 & 0 \\ -1 & -2 & -2 & | & 0 & 1 & 0 \\ 0 & 5 & 8 & | & 0 & 0 & 1 \end{pmatrix}$ reduces to

$$\begin{pmatrix} 1 & 2 & 1 & | & 1 & 0 & 0 \\ 0 & 0 & 0 & | & 1 & 1 & 0 \\ 0 & 5 & 8 & | & 0 & 0 & 1 \end{pmatrix} \tag{12.2.4}$$

Although this matrix can be row-reduced further, it is not necessary to do so since, in equation form, we have:

Table 12.2.3

$x_{11} + 2x_{21} + x_{31} = 1$ $x_{12} + 2x_{22} + x_{32} = 0$ $x_{13} + 2x_{23} + x_{33} = 0$
$0 = 1$ $0 = 1$ $0 = 0$
$5x_{21} + 8x_{31} = 0$ $5x_{22} + 8x_{32} = 0$ $5x_{23} + 8x_{33} = 1$

Clearly, there are no solutions to the first two systems, therefore A^{-1} does not exist. From this discussion it should be obvious to the reader that the zero row of the coefficient matrix together with the nonzero entry in the fourth column of that row in matrix (12.2.4) tells us that A^{-1} does not exist. □

12.2. MATRIX INVERSION

12.2.3 Exercises

1. In order to develop an understanding of the technique of this section, work out all the details of Example 12.2.2, p. 74.

2. Use the method of this section to find the inverses of the following matrices whenever possible. If an inverse does not exist, explain why.

    (a) $\begin{pmatrix} 1 & 2 \\ -1 & 3 \end{pmatrix}$

    (b) $\begin{pmatrix} 0 & 3 & 2 & 5 \\ 1 & -1 & 4 & 0 \\ 0 & 0 & 1 & 1 \\ 0 & 1 & 3 & -1 \end{pmatrix}$

    (c) $\begin{pmatrix} 2 & -1 & 0 \\ -1 & 2 & -1 \\ 0 & -1 & 2 \end{pmatrix}$

    (d) $\begin{pmatrix} 1 & 2 & -1 \\ -2 & -3 & 1 \\ 1 & 4 & -3 \end{pmatrix}$

    (e) $\begin{pmatrix} 6 & 7 & 2 \\ 4 & 2 & 1 \\ 6 & 1 & 1 \end{pmatrix}$

    (f) $\begin{pmatrix} 2 & 1 & 3 \\ 4 & 2 & 1 \\ 8 & 2 & 4 \end{pmatrix}$

3. Use the method of this section to find the inverses of the following matrices whenever possible. If an inverse does not exist, explain why.

    (a) $\begin{pmatrix} \frac{1}{3} & 2 \\ \frac{1}{5} & -1 \end{pmatrix}$

    (b) $\begin{pmatrix} 1 & 0 & 0 & 3 \\ 2 & -1 & 0 & 6 \\ 0 & 2 & 1 & 0 \\ 0 & -1 & 3 & 2 \end{pmatrix}$

    (c) $\begin{pmatrix} 1 & -1 & 0 \\ -1 & 2 & -1 \\ 0 & -1 & 1 \end{pmatrix}$

    (d) $\begin{pmatrix} 1 & 0 & 0 \\ 2 & 2 & -1 \\ 1 & -1 & 1 \end{pmatrix}$

    (e) $\begin{pmatrix} 2 & 3 & 4 \\ 3 & 4 & 5 \\ 4 & 5 & 6 \end{pmatrix}$

    (f) $\begin{pmatrix} 1 & \frac{1}{2} & \frac{1}{3} \\ \frac{1}{2} & \frac{1}{3} & \frac{1}{4} \\ \frac{1}{3} & \frac{1}{4} & \frac{1}{5} \end{pmatrix}$

4.
    (a) Find the inverses of the following matrices.

    (i) $\begin{pmatrix} 2 & 0 & 0 \\ 0 & 3 & 0 \\ 0 & 0 & 5 \end{pmatrix}$

    (ii) $\begin{pmatrix} -1 & 0 & 0 & 0 \\ 0 & \frac{5}{2} & 0 & 0 \\ 0 & 0 & \frac{1}{7} & 0 \\ 0 & 0 & 0 & \frac{3}{4} \end{pmatrix}$

    (b) If D is a diagonal matrix whose diagonal entries are nonzero, what is D^{-1}?

5. Express each system of equations in Exercise 12.1.7.1, p. 71 in the form $Ax = B$. When possible, solve each system by first finding the inverse of the matrix of coefficients.

12.3 An Introduction to Vector Spaces

12.3.1 Motivation for the study of vector spaces

When we encountered various types of matrices in Chapter 5, it became apparent that a particular kind of matrix, the diagonal matrix, was much easier to use in computations. For example, if $A = \begin{pmatrix} 2 & 1 \\ 2 & 3 \end{pmatrix}$, then A^5 can be found, but its computation is tedious. If $D = \begin{pmatrix} 1 & 0 \\ 0 & 4 \end{pmatrix}$ then

$$D^5 = \begin{pmatrix} 1 & 0 \\ 0 & 4 \end{pmatrix}^5 = \begin{pmatrix} 1^5 & 0 \\ 0 & 4^5 \end{pmatrix} = \begin{pmatrix} 1 & 0 \\ 0 & 1024 \end{pmatrix}$$

Even when presented with a non-diagonal matrix, we will see that it is sometimes possible to do a bit of work to be able to work with a diagonal matrix. This process is called **diagonalization**.

In a variety of applications it is beneficial to be able to diagonalize a matrix. In this section we will investigate what this means and consider a few applications. In order to understand when the diagonalization process can be performed, it is necessary to develop several of the underlying concepts of linear algebra.

12.3.2 Vector Spaces

By now, you realize that mathematicians tend to generalize. Once we have found a "good thing," something that is useful, we apply it to as many different concepts as possible. In doing so, we frequently find that the "different concepts" are not really different but only look different. Four sentences in four different languages might look dissimilar, but when they are translated into a common language, they might very well express the exact same idea.

Early in the development of mathematics, the concept of a vector led to a variety of applications in physics and engineering. We can certainly picture vectors, or "arrows," in the $xy-$ plane and even in the three-dimensional space. Does it make sense to talk about vectors in four-dimensional space, in ten-dimensional space, or in any other mathematical situation? If so, what is the essence of a vector? Is it its shape or the rules it follows? The shape in two- or three-space is just a picture, or geometric interpretation, of a vector. The essence is the rules, or properties, we wish vectors to follow so we can manipulate them algebraically. What follows is a definition of what is called a vector space. It is a list of all the essential properties of vectors, and it is the basic definition of the branch of mathematics called linear algebra.

Definition 12.3.1 Vector Space. Let V be any nonempty set of objects. Define on V an operation, called addition, for any two elements $\mathbf{x}, \mathbf{y} \in V$, and denote this operation by $\mathbf{x} + \mathbf{y}$. Let scalar multiplication be defined for a real number $a \in \mathbb{R}$ and any element $\mathbf{x} \in V$ and denote this operation by $a\mathbf{x}$. The set V together with operations of addition and scalar multiplication is called a vector space over $\mathbb{R}$ if the following hold for all $\mathbf{x}, \mathbf{y}, \mathbf{z} \in V$, and $a, b \in \mathbb{R}$:

- $\mathbf{x} + \mathbf{y} = \mathbf{y} + \mathbf{x}$

12.3. AN INTRODUCTION TO VECTOR SPACES

- $(\mathbf{x} + \mathbf{y}) + \mathbf{z} = \mathbf{x} + (\mathbf{y} + \mathbf{z})$

- There exists a vector $\mathbf{0} \in V$, such that $\mathbf{x} + \mathbf{0} = \mathbf{x}$ for all $x \in V$.

- For each vector $\mathbf{x} \in V$, there exists a unique vector $-\mathbf{x} \in V$, such that $-\mathbf{x} + \mathbf{x} = \mathbf{0}$.

These are the main properties associated with the operation of addition. They can be summarized by saying that $[V; +]$ is an abelian group.

The next four properties are associated with the operation of scalar multiplication and how it relates to vector addition.

- $a(\mathbf{x} + \mathbf{y}) = a\mathbf{x} + a\mathbf{y}$

- $(a + b)\mathbf{x} = a\mathbf{x} + b\mathbf{x}$

- $a(b\mathbf{x}) = (ab)\mathbf{x}$

- $1\mathbf{x} = \mathbf{x}$.

$\Diamond$

In a vector space it is common to call the elements of V vectors and those from $\mathbb{R}$ scalars. Vector spaces over the real numbers are also called real vector spaces.

Example 12.3.2 A Vector Space of Matrices. Let $V = M_{2\times 3}(\mathbb{R})$ and let the operations of addition and scalar multiplication be the usual operations of addition and scalar multiplication on matrices. Then V together with these operations is a real vector space. The reader is strongly encouraged to verify the definition for this example before proceeding further (see Exercise 3 of this section). Note we can call the elements of $M_{2\times 3}(\mathbb{R})$ vectors even though they are not arrows. □

Example 12.3.3 The Vector Space $\mathbb{R}^2$. Let $\mathbb{R}^2 = \{(a_1, a_2) \mid a_1, a_2 \in \mathbb{R}\}$. If we define addition and scalar multiplication the natural way, that is, as we would on 1×2 matrices, then $\mathbb{R}^2$ is a vector space over $\mathbb{R}$. See Exercise 12.3.3.4, p. 82 of this section.

In this example, we have the "bonus" that we can illustrate the algebraic concept geometrically. In mathematics, a "geometric bonus" does not always occur and is not necessary for the development or application of the concept. However, geometric illustrations are quite useful in helping us understand concepts and should be utilized whenever available.

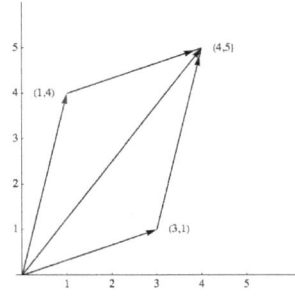

Figure 12.3.4 Sum of two vectors in $\mathbb{R}^2$

Let's consider some illustrations of the vector space $\mathbb{R}^2$. Let $\mathbf{x} = (1, 4)$ and $\mathbf{y} = (3, 1)$. We illustrate the vector (a_1, a_2) as a directed line segment, or "arrow," from the point $(0, 0)$ to the point (a_1, a_2). The vectors $\mathbf{x}$ and $\mathbf{y}$ are as shown in Figure 12.3.4, p. 77 together with $\mathbf{x} + \mathbf{y} = (1, 4) + (3, 1) = (4, 5)$. The vector $2\mathbf{x} = 2(1,4) = (2,8)$ is a vector in the same direction as $\mathbf{x}$, but with twice its length. □

Note 12.3.5

(1) The common convention is to use that boldface letters toward the end of the alphabet for vectors, while letters early in the alphabet are scalars.

(2) A common alternate notation for vectors is to place an arrow about a variable to indicate that it is a vector such as this: $\vec{x}$.

(3) The vector $(a_1, a_2, \ldots, a_n) \in \mathbb{R}^n$ is referred to as an n-tuple.

(4) For those familiar with vector calculus, we are expressing the vector $x = a_1\hat{\mathbf{i}} + a_2\hat{\mathbf{j}} + a_3\hat{\mathbf{k}} \in \mathbb{R}^3$ as (a_1, a_2, a_3). This allows us to discuss vectors in $\mathbb{R}^n$ in much simpler notation.

In many situations a vector space V is given and we would like to describe the whole vector space by the smallest number of essential reference vectors. An example of this is the description of $\mathbb{R}^2$, the xy-plane, via the x and y axes. Again our concepts must be algebraic in nature so we are not restricted solely to geometric considerations.

Definition 12.3.6 Linear Combination. A vector $\mathbf{y}$ in vector space V (over $\mathbb{R}$) is a linear combination of the vectors $\mathbf{x}_1, \mathbf{x}_2, \ldots, \mathbf{x}_n$ if there exist scalars $a_1, a_2, \ldots, a_n$ in $\mathbb{R}$ such that $\mathbf{y} = a_1\mathbf{x}_1 + a_2\mathbf{x}_2 + \ldots + a_n\mathbf{x}_n$ ◇

Example 12.3.7 A Basic Example. The vector $(2,3)$ in $\mathbb{R}^2$ is a linear combination of the vectors $(1,0)$ and $(0,1)$ since $(2,3) = 2(1,0) + 3(0,1)$. □

Example 12.3.8 A little less obvious example. Prove that the vector $(4,5)$ is a linear combination of the vectors $(3,1)$ and $(1,4)$.

By the definition we must show that there exist scalars a_1 and a_2 such that:

$$(4,5) = a_1(3,1) + a_2(1,4) \qquad \Rightarrow \qquad 3a_1 + a_2 = 4$$
$$= (3a_1 + a_2, a_1 + 4a_2) \qquad\qquad\qquad a_1 + 4a_2 = 5$$

This system has the solution $a_1 = 1$, $a_2 = 1$.

Hence, if we replace a_1 and a_2 both by 1, then the two vectors $(3, 1)$ and $(1, 4)$ produce, or generate, the vector $(4,5)$. Of course, if we replace a_1 and a_2 by different scalars, we can generate more vectors from $\mathbb{R}^2$. If, for example, $a_1 = 3$ and $a_2 = -2$, then

$$a_1(3,1) + a_2(1,4) = 3(3,1) + (-2)(1,4) = (9,3) + (-2,-8) = (7,-5)$$

□

Will the vectors $(3, 1)$ and $(1, 4)$ generate any vector we choose in $\mathbb{R}^2$? To see if this is so, we let (b_1, b_2) be an arbitrary vector in $\mathbb{R}^2$ and see if we can always find scalars a_1 and a_2 such that $a_1(3, 1) + a_2(1, 4) = (b_1, b_2)$. This is equivalent to solving the following system of equations:

$$3a_1 + a_2 = b_1$$
$$a_1 + 4a_2 = b_2$$

which always has solutions for a_1 and a_2, regardless of the values of the real numbers b_1 and b_2. Why? We formalize this situation in a definition:

Definition 12.3.9 Generation of a Vector Space. Let $\{\mathbf{x}_1, \mathbf{x}_2, \ldots, \mathbf{x}_n\}$ be a set of vectors in a vector space V over $\mathbb{R}$. This set is said to **generate**, or span, V if, for any given vector $\mathbf{y} \in V$, we can always find scalars $a_1, a_2, \ldots, a_n$

12.3. AN INTRODUCTION TO VECTOR SPACES

such that $\mathbf{y} = a_1\mathbf{x}_1 + a_2\mathbf{x}_2 + \ldots + a_n\mathbf{x}_n$. A set that generates a vector space is called a **generating set**. ◇

We now give a geometric interpretation of the previous examples.

We know that the standard coordinate system, x axis and y axis, were introduced in basic algebra in order to describe all points in the xy-plane algebraically. It is also quite clear that to describe any point in the plane we need exactly two axes.

We can set up a new coordinate system in the following way. Draw the vector $(3,1)$ and an axis from the origin through $(3, 1)$ and label it the x' axis. Also draw the vector $(1,4)$ and an axis from the origin through $(1,4)$ to be labeled the y' axis. Draw the coordinate grid for the axis, that is, lines parallel, and let the unit lengths of this "new" plane be the lengths of the respective vectors, $(3,1)$ and $(1,4)$, so that we obtain Figure 12.3.10, p. 79.

From Example 12.3.8, p. 78 and Figure 12.3.10, p. 79, we see that any vector on the plane can be described using the standard xy-axes or our new $x'y'$-axes. Hence the position which had the name $(3,1)$ in reference to the standard axes has the name $(1,0)$ with respect to the $x'y'$ axes, or, in the phraseology of linear algebra, the coordinates of the point $(1,4)$ with respect to the $x'y'$ axes are $(1,0)$.

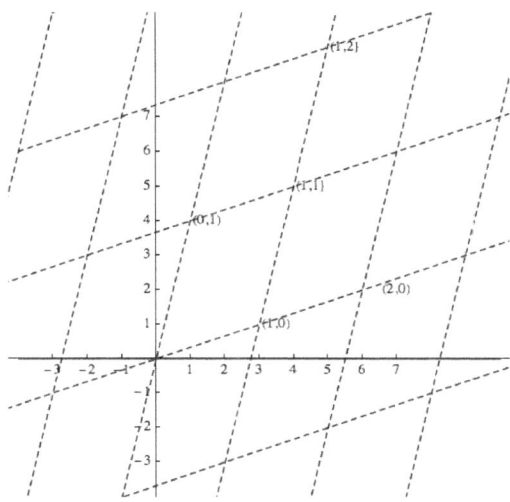

Figure 12.3.10 Two sets of axes for the plane

Example 12.3.11 One point, Two position descriptions. From Example 12.3.8, p. 78 we found that if we choose $a_1 = 1$ and $a_2 = 1$, then the two vectors $(3,1)$ and $(1,4)$ generate the vector $(4,5)$. Another geometric interpretation of this problem is that the coordinates of the position $(4,5)$ with respect to the $x'y'$ axes of Figure 12.3.10, p. 79 is $(1,1)$. In other words, a position in the plane has the name $(4,5)$ in reference to the xy-axes and the same position has the name $(1,1)$ in reference to the $x'y'$ axes.

From the above, it is clear that we can use different axes to describe points or vectors in the plane. No matter what choice we use, we want to be able to describe each position in a unique manner. This is not the case in Figure 12.3.12, p. 80. Any point in the plane could be described via the $x'y'$ axes, the $x'z'$ axes or the $y'z'$ axes. Therefore, in this case, a single point would have three different names, a very confusing situation.

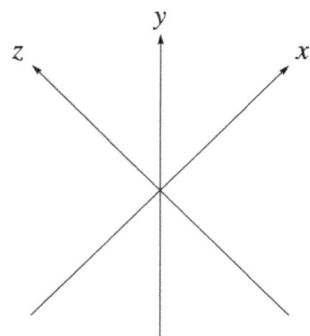

Figure 12.3.12 Three axes on a plane

□

We formalize the our observations in the previous examples in two definitions and a theorem.

Definition 12.3.13 Linear Independence/Linear Dependence. A set of vectors $\{\mathbf{x}_1, \mathbf{x}_2, \ldots, \mathbf{x}_n\}$ from a real vector space V is **linearly independent** if the only solution to the equation $a_1\mathbf{x}_1 + a_2\mathbf{x}_2 + \ldots + a_n\mathbf{x}_n = \mathbf{0}$ is $a_1 = a_2 = \ldots = a_n = 0$. Otherwise the set is called a **linearly dependent** set. ◊

Definition 12.3.14 Basis. A set of vectors $B = \{\mathbf{x}_1, \mathbf{x}_2, \ldots, \mathbf{x}_n\}$ is a basis for a vector space V if:

(1) B generates V, and

(2) B is linearly independent.

◊

Theorem 12.3.15 The fundamental property of a basis. *If $\{\mathbf{x}_1, \mathbf{x}_2, \ldots, \mathbf{x}_n\}$ is a basis for a vector space V over $\mathbb{R}$, then any vector $y \in V$ can be uniquely expressed as a linear combination of the $\mathbf{x}_i$'s.*

Proof. Assume that $\{\mathbf{x}_1, \mathbf{x}_2, \ldots, \mathbf{x}_n\}$ is a basis for V over $\mathbb{R}$. We must prove two facts:

(1) each vector $y \in V$ can be expressed as a linear combination of the $\mathbf{x}_i$'s, and

(2) each such expression is unique.

Part 1 is trivial since a basis, by its definition, must generate all of V.

The proof of part 2 is a bit more difficult. We follow the standard approach for any uniqueness facts. Let y be any vector in V and assume that there are two different ways of expressing y, namely

$$y = a_1\mathbf{x}_1 + a_2\mathbf{x}_2 + \ldots + a_n\mathbf{x}_n$$

and

$$y = b_1\mathbf{x}_1 + b_2\mathbf{x}_2 + \ldots + b_n\mathbf{x}_n$$

where at least one a_i is different from the corresponding b_i. Then equating these two linear combinations we get

$$a_1\mathbf{x}_1 + a_2\mathbf{x}_2 + \ldots + a_n\mathbf{x}_n = b_1\mathbf{x}_1 + b_2\mathbf{x}_2 + \ldots + b_n\mathbf{x}_n$$

so that

$$(a_1 - b_1)\mathbf{x}_1 + (a_2 - b_2)\mathbf{x}_2 + \ldots + (a_n - b_n)\mathbf{x}_n = \mathbf{0}$$

12.3. AN INTRODUCTION TO VECTOR SPACES

Now a crucial observation: since the $\mathbf{x}'_i s$ form a linearly independent set, the only solution to the previous equation is that each of the coefficients must equal zero, so $a_i - b_i = 0$ for $i = 1, 2, \ldots, n$. Hence $a_i = b_i$, for all i. This contradicts our assumption that at least one a_i is different from the corresponding b_i, so each vector $\mathbf{y} \in V$ can be expressed in one and only one way. ∎

This theorem, together with the previous examples, gives us a clear insight into the significance of linear independence, namely uniqueness in representing any vector.

Example 12.3.16 Another basis for $\mathbb{R}^2$. Prove that $\{(1,1), (-1,1)\}$ is a basis for $\mathbb{R}^2$ over $\mathbb{R}$ and explain what this means geometrically.

First we show that the vectors $(1,1)$ and $(-1,1)$ generate all of $\mathbb{R}^2$. We can do this by imitating Example 12.3.8, p. 78 and leave it to the reader (see Exercise 12.3.3.10, p. 82 of this section). Secondly, we must prove that the set is linearly independent.

Let a_1 and a_2 be scalars such that $a_1(1,1) + a_2(-1,1) = (0,0)$. We must prove that the only solution to the equation is that a_1 and a_2 must both equal zero. The above equation becomes $(a_1 - a_2, a_1 + a_2) = (0,0)$ which gives us the system

$$a_1 - a_2 = 0$$
$$a_1 + a_2 = 0$$

The augmented matrix of this system reduces in such way that the only solution is the trivial one of all zeros:

$$\begin{pmatrix} 1 & -1 & | & 0 \\ 1 & 1 & | & 0 \end{pmatrix} \longrightarrow \begin{pmatrix} 1 & 0 & | & 0 \\ 0 & 1 & | & 0 \end{pmatrix} \Rightarrow a_1 = a_2 = 0$$

Therefore, the set is linearly independent. □

To explain the results geometrically, note through Exercise 12, part a, that the coordinates of each vector $\mathbf{y} \in \mathbb{R}^2$ can be determined uniquely using the vectors $(1,1)$ and $(-1, 1)$. The concept of dimension is quite obvious for those vector spaces that have an immediate geometric interpretation. For example, the dimension of $\mathbb{R}^2$ is two and that of $\mathbb{R}^3$ is three. How can we define the concept of dimension algebraically so that the resulting definition correlates with that of $\mathbb{R}^2$ and $\mathbb{R}^3$? First we need a theorem, which we will state without proof.

Theorem 12.3.17 Basis Size is Constant. *If V is a vector space with a basis containing n elements, then all bases of V contain n elements.*

Definition 12.3.18 Dimension of a Vector Space. Let V be a vector space over $\mathbb{R}$ with basis $\{\mathbf{x}_1, \mathbf{x}_2, \ldots, \mathbf{x}_n\}$. Then the dimension of V is n. We use the notation $\dim V = n$ to indicate that V is n-dimensional. ◇

12.3.3 Exercises

1. If $a = 2$, $b = -3$, $A = \begin{pmatrix} 1 & 0 & -1 \\ 2 & 3 & 4 \end{pmatrix}$, $B = \begin{pmatrix} 2 & -2 & 3 \\ 4 & 5 & 8 \end{pmatrix}$, and $C = \begin{pmatrix} 1 & 0 & 0 \\ 3 & 2 & -2 \end{pmatrix}$ verify that all properties of the definition of a vector space are true for $M_{2 \times 3}(\mathbb{R})$ with these values.

2. Let $a = 3$, $b = 4$, $\mathbf{x} = (-1, 3)$, $\mathbf{y} = (2, 3)$, and $\mathbf{z} = (1, 0)$. Verify that all properties of the definition of a vector space are true for $\mathbb{R}^2$ for these values.

3.
   (a) Verify that $M_{2\times 3}(\mathbb{R})$ is a vector space over $\mathbb{R}$. What is its dimension?

   (b) Is $M_{m\times n}(\mathbb{R})$ a vector space over $\mathbb{R}$? If so, what is its dimension?

4.
   (a) Verify that $\mathbb{R}^2$ is a vector space over $\mathbb{R}$.

   (b) Is $\mathbb{R}^n$ a vector space over $\mathbb{R}$ for every positive integer n?

5. Let $P^3 = \{a_0 + a_1 x + a_2 x^2 + a_3 x^3 \mid a_0, a_1, a_2, a_3 \in \mathbb{R}\}$; that is, P^3 is the set of all polynomials in x having real coefficients with degree less than or equal to three. Verify that P^3 is a vector space over $\mathbb{R}$. What is its dimension?

6. For each of the following, express the vector **y** as a linear combination of the vectors x_1 and x_2.

   (a) $\mathbf{y} = (5,6)$, $\mathbf{x_1} = (1,0)$, and $\mathbf{x_2} = (0,1)$

   (b) $\mathbf{y} = (2,1)$, $\mathbf{x_1} = (2,1)$, and $\mathbf{x_2} = (1,1)$

   (c) $\mathbf{y} = (3,4)$, $\mathbf{x_1} = (1,1)$, and $\mathbf{x_2} = (-1,1)$

7. Express the vector $\begin{pmatrix} 1 & 2 \\ -3 & 3 \end{pmatrix} \in M_{2\times 2}(\mathbb{R})$, as a linear combination of
   $\begin{pmatrix} 1 & 1 \\ 1 & 1 \end{pmatrix}, \begin{pmatrix} -1 & 5 \\ 2 & 1 \end{pmatrix}, \begin{pmatrix} 0 & 1 \\ 1 & 1 \end{pmatrix}$ and $\begin{pmatrix} 0 & 0 \\ 0 & 1 \end{pmatrix}$

8. Express the vector $x^3 - 4x^2 + 3 \in P^3$ as a linear combination of the vectors $1, x, x^2$, and x^3.

9.
   (a) Show that the set $\{\mathbf{x_1}, \mathbf{x_2}\}$ generates $\mathbb{R}^2$ for each of the parts in Exercise 6 of this section.

   (b) Show that $\{\mathbf{x_1}, \mathbf{x_2}, \mathbf{x_3}\}$ generates $\mathbb{R}^2$ where $\mathbf{x_1} = (1,1)$, $\mathbf{x_2} = (3,4)$, and $\mathbf{x_3} = (-1,5)$.

   (c) Create a set of four or more vectors that generates $\mathbb{R}^2$.

   (d) What is the smallest number of vectors needed to generate $\mathbb{R}^2$? $\mathbb{R}^n$?

   (e) Show that the set

   $$\{A_1, A_2, A_3, A_4\} = \{\begin{pmatrix} 1 & 0 \\ 0 & 0 \end{pmatrix}, \begin{pmatrix} 0 & 1 \\ 0 & 0 \end{pmatrix}, \begin{pmatrix} 0 & 0 \\ 1 & 0 \end{pmatrix}, \begin{pmatrix} 0 & 0 \\ 0 & 1 \end{pmatrix}\}$$

   generates $M_{2\times 2}(\mathbb{R})$

   (f) Show that $\{1, x, x^2, x^3\}$ generates P^3.

10. Complete Example 12.3.16, p. 81 by showing that $\{(1,1), (-1,1)\}$ generates $\mathbb{R}^2$.

11.
   (a) Prove that $\{(4,1), (1,3)\}$ is a basis for $\mathbb{R}^2$ over $\mathbb{R}$.

   (b) Prove that $\{(1,0), (3,4)\}$ is a basis for $\mathbb{R}^2$ over $\mathbb{R}$.

   (c) Prove that $\{(1,0,-1), (2,1,1), (1,-3,-1)\}$ is a basis for $\mathbb{R}^3$ over $\mathbb{R}$.

   (d) Prove that the sets in Exercise 9, parts e and f, form bases of the respective vector spaces.

12.

    (a) Determine the coordinates of the points or vectors $(3,4)$, $(-1,1)$, and $(1,1)$ with respect to the basis $\{(1,1),(-1,1)\}$ of $\mathbb{R}^2$. Interpret your results geometrically.

    (b) Determine the coordinates of the points or vector $(3,5,6)$ with respect to the basis $\{(1,0,0),(0,1,0),(0,0,1)\}$. Explain why this basis is called the standard basis for $\mathbb{R}^3$.

13.

    (a) Let $\mathbf{y}_1 = (1,3,5,9)$, $\mathbf{y}_2 = (5,7,6,3)$, and $c = 2$. Find $\mathbf{y}_1 + \mathbf{y}_2$ and $c\mathbf{y}_1$.

    (b) Let $f_1(x) = 1 + 3x + 5x^2 + 9x^3$, $f_2(x) = 5 + 7x + 6x^2 + 3x^3$ and $c = 2$. Find $f_1(x) + f_2(x)$ and $cf_1(x)$.

    (c) Let $A = \begin{pmatrix} 1 & 3 \\ 5 & 9 \end{pmatrix}$, $B = \begin{pmatrix} 5 & 7 \\ 6 & 3 \end{pmatrix}$, and $c = 2$. Find $A + B$ and cA.

    (d) Are the vector spaces $\mathbb{R}^4$, P^3 and $M_{2\times 2}(\mathbb{R})$ isomorphic to each other? Discuss with reference to previous parts of this exercise.

12.4 The Diagonalization Process

12.4.1 Eigenvalues and Eigenvectors

We now have the background to understand the main ideas behind the diagonalization process.

Definition 12.4.1 Eigenvalue, Eigenvector. Let A be an $n \times n$ matrix over $\mathbb{R}$. λ is an eigenvalue of A if for some nonzero column vector $\mathbf{x} \in \mathbb{R}^n$ we have $A\mathbf{x} = \lambda\mathbf{x}$. $\mathbf{x}$ is called an **eigenvector** corresponding to the **eigenvalue** λ.
$\diamondsuit$

Example 12.4.2 Examples of eigenvalues and eigenvectors. Find the eigenvalues and corresponding eigenvectors of the matrix $A = \begin{pmatrix} 2 & 1 \\ 2 & 3 \end{pmatrix}$.

We want to find nonzero vectors $X = \begin{pmatrix} x_1 \\ x_2 \end{pmatrix}$ and real numbers λ such that

$$AX = \lambda X \Leftrightarrow \begin{pmatrix} 2 & 1 \\ 2 & 3 \end{pmatrix}\begin{pmatrix} x_1 \\ x_2 \end{pmatrix} = \lambda\begin{pmatrix} x_1 \\ x_2 \end{pmatrix}$$

$$\Leftrightarrow \begin{pmatrix} 2 & 1 \\ 2 & 3 \end{pmatrix}\begin{pmatrix} x_1 \\ x_2 \end{pmatrix} - \lambda\begin{pmatrix} x_1 \\ x_2 \end{pmatrix} = \begin{pmatrix} 0 \\ 0 \end{pmatrix}$$

$$\Leftrightarrow \begin{pmatrix} 2 & 1 \\ 2 & 3 \end{pmatrix}\begin{pmatrix} x_1 \\ x_2 \end{pmatrix} - \lambda\begin{pmatrix} 1 & 0 \\ 0 & 1 \end{pmatrix}\begin{pmatrix} x_1 \\ x_2 \end{pmatrix} = \begin{pmatrix} 0 \\ 0 \end{pmatrix} \quad (12.4.1)$$

$$\Leftrightarrow \left(\begin{pmatrix} 2 & 1 \\ 2 & 3 \end{pmatrix} - \lambda\begin{pmatrix} 1 & 0 \\ 0 & 1 \end{pmatrix}\right)\begin{pmatrix} x_1 \\ x_2 \end{pmatrix} = \begin{pmatrix} 0 \\ 0 \end{pmatrix}$$

$$\Leftrightarrow \begin{pmatrix} 2-\lambda & 1 \\ 2 & 3-\lambda \end{pmatrix}\begin{pmatrix} x_1 \\ x_2 \end{pmatrix} = \begin{pmatrix} 0 \\ 0 \end{pmatrix}$$

The last matrix equation will have nonzero solutions if and only if

$$\det\begin{pmatrix} 2-\lambda & 1 \\ 2 & 3-\lambda \end{pmatrix} = 0$$

or $(2-\lambda)(3-\lambda) - 2 = 0$, which simplifies to $\lambda^2 - 5\lambda + 4 = 0$. Therefore, the solutions to this quadratic equation, $\lambda_1 = 1$ and $\lambda_2 = 4$, are the eigenvalues of A. We now have to find eigenvectors associated with each eigenvalue.

Case 1. For $\lambda_1 = 1$, (12.4.1) becomes:

$$\begin{pmatrix} 2-1 & 1 \\ 2 & 3-1 \end{pmatrix} \begin{pmatrix} x_1 \\ x_2 \end{pmatrix} = \begin{pmatrix} 0 \\ 0 \end{pmatrix} \begin{pmatrix} 1 & 1 \\ 2 & 2 \end{pmatrix} \begin{pmatrix} x_1 \\ x_2 \end{pmatrix} = \begin{pmatrix} 0 \\ 0 \end{pmatrix}$$

which reduces to the single equation, $x_1 + x_2 = 0$. From this, $x_1 = -x_2$. This means the solution set of this equation is (in column notation)

$$E_1 = \left\{ \begin{pmatrix} -c \\ c \end{pmatrix} \middle| c \in \mathbb{R} \right\}$$

So any column vector of the form $\begin{pmatrix} -c \\ c \end{pmatrix}$ where c is any nonzero real number is an eigenvector associated with $\lambda_1 = 1$. The reader should verify that, for example,

$$\begin{pmatrix} 2 & 1 \\ 2 & 3 \end{pmatrix} \begin{pmatrix} 1 \\ -1 \end{pmatrix} = 1 \begin{pmatrix} 1 \\ -1 \end{pmatrix}$$

so that $\begin{pmatrix} 1 \\ -1 \end{pmatrix}$ is an eigenvector associated with eigenvalue 1.

Case 2. For $\lambda_2 = 4$ (12.4.1) becomes:

$$\begin{pmatrix} 2-4 & 1 \\ 2 & 3-4 \end{pmatrix} \begin{pmatrix} x_1 \\ x_2 \end{pmatrix} = \begin{pmatrix} 0 \\ 0 \end{pmatrix} \begin{pmatrix} -2 & 1 \\ 2 & -1 \end{pmatrix} \begin{pmatrix} x_1 \\ x_2 \end{pmatrix} = \begin{pmatrix} 0 \\ 0 \end{pmatrix}$$

which reduces to the single equation $-2x_1 + x_2 = 0$, so that $x_2 = 2x_1$. The solution set of the equation is

$$E_2 = \left\{ \begin{pmatrix} c \\ 2c \end{pmatrix} \middle| c \in \mathbb{R} \right\}$$

Therefore, all eigenvectors of A associated with the eigenvalue $\lambda_2 = 4$ are of the form $\begin{pmatrix} c \\ 2c \end{pmatrix}$, where c can be any nonzero number. $\square$

The following theorems summarize the most important aspects of the previous example.

Theorem 12.4.3 Characterization of Eigenvalues of a Square Matrix. *Let A be any $n \times n$ matrix over $\mathbb{R}$. Then $\lambda \in \mathbb{R}$ is an eigenvalue of A if and only if $\det(A - \lambda I) = 0$.*

The equation $\det(A - \lambda I) = 0$ is called the **characteristic equation**, and the left side of this equation is called the **characteristic polynomial** of A.

Theorem 12.4.4 Linear Independence of Eigenvectors. *Nonzero eigenvectors corresponding to distinct eigenvalues are linearly independent.*

The solution space of $(A - \lambda I)\mathbf{x} = \mathbf{0}$ is called the eigenspace of A corresponding to λ. This terminology is justified by Exercise 2 of this section.

12.4.2 Diagonalization

We now consider the main aim of this section. Given an $n \times n$ (square) matrix A, we would like to transform A into a diagonal matrix D, perform our tasks

12.4. THE DIAGONALIZATION PROCESS

with the simpler matrix D, and then describe the results in terms of the given matrix A.

Definition 12.4.5 Diagonalizable Matrix. An $n \times n$ matrix A is called diagonalizable if there exists an invertible $n \times n$ matrix P such that $P^{-1}AP$ is a diagonal matrix D. The matrix P is said to diagonalize the matrix A. ◊

Example 12.4.6 Diagonalization of a Matrix. We will now diagonalize the matrix A of Example 12.4.2, p. 83. We form the matrix P as follows: Let $P^{(1)}$ be the first column of P. Choose for $P^{(1)}$ any eigenvector from E_1. We may as well choose a simple vector in E_1 so $P^{(1)} = \begin{pmatrix} 1 \\ -1 \end{pmatrix}$ is our candidate.

Similarly, let $P^{(2)}$ be the second column of P, and choose for $P^{(2)}$ any eigenvector from E_2. The vector $P^{(2)} = \begin{pmatrix} 1 \\ 2 \end{pmatrix}$ is a reasonable choice, thus

$$P = \begin{pmatrix} 1 & 1 \\ -1 & 2 \end{pmatrix} \text{ and } P^{-1} = \frac{1}{3}\begin{pmatrix} 2 & -1 \\ 1 & 1 \end{pmatrix} = \begin{pmatrix} \frac{2}{3} & -\frac{1}{3} \\ \frac{1}{3} & \frac{1}{3} \end{pmatrix}$$

so that

$$P^{-1}AP = \frac{1}{3}\begin{pmatrix} 2 & -1 \\ 1 & 1 \end{pmatrix}\begin{pmatrix} 2 & 1 \\ 2 & 3 \end{pmatrix}\begin{pmatrix} 1 & 1 \\ -1 & 2 \end{pmatrix} = \begin{pmatrix} 1 & 0 \\ 0 & 4 \end{pmatrix}$$

Notice that the elements on the main diagonal of D are the eigenvalues of A, where D_{ii} is the eigenvalue corresponding to the eigenvector $P^{(i)}$. □

Note 12.4.7

(1) The first step in the diagonalization process is the determination of the eigenvalues. The ordering of the eigenvalues is purely arbitrary. If we designate $\lambda_1 = 4$ and $\lambda_2 = 1$, the columns of P would be interchanged and D would be $\begin{pmatrix} 4 & 0 \\ 0 & 1 \end{pmatrix}$ (see Exercise 3b of this section). Nonetheless, the final outcome of the application to which we are applying the diagonalization process would be the same.

(2) If A is an $n \times n$ matrix with distinct eigenvalues, then P is also an $n \times n$ matrix whose columns $P^{(1)}, P^{(2)}, \ldots, P^{(n)}$ are n linearly independent vectors.

Example 12.4.8 Diagonalization of a 3 by 3 matrix. Diagonalize the matrix

$$A = \begin{pmatrix} 1 & 12 & -18 \\ 0 & -11 & 18 \\ 0 & -6 & 10 \end{pmatrix}.$$

First, we find the eigenvalues of A.

$$\det(A - \lambda I) = \det\begin{pmatrix} 1-\lambda & 12 & -18 \\ 0 & -\lambda-11 & 18 \\ 0 & -6 & 10-\lambda \end{pmatrix}$$

$$= (1-\lambda)\det\begin{pmatrix} -\lambda-11 & 18 \\ -6 & 10-\lambda \end{pmatrix}$$

$$= (1-\lambda)((-\lambda-11)(10-\lambda)+108) = (1-\lambda)\left(\lambda^2+\lambda-2\right)$$

Hence, the equation $\det(A - \lambda I)$ becomes

$$(1-\lambda)\left(\lambda^2+\lambda-2\right) = -(\lambda-1)^2(\lambda+2)$$

Therefore, our eigenvalues for A are $\lambda_1 = -2$ and $\lambda_2 = 1$. We note that we do not have three distinct eigenvalues, but we proceed as in the previous example.

Case 1. For $\lambda_1 = -2$ the equation $(A - \lambda I)\mathbf{x} = \mathbf{0}$ becomes

$$\begin{pmatrix} 3 & 12 & -18 \\ 0 & -9 & 18 \\ 0 & -6 & 12 \end{pmatrix} \begin{pmatrix} x_1 \\ x_2 \\ x_3 \end{pmatrix} = \begin{pmatrix} 0 \\ 0 \\ 0 \end{pmatrix}$$

We can row reduce the matrix of coefficients to $\begin{pmatrix} 1 & 0 & 2 \\ 0 & 1 & -2 \\ 0 & 0 & 0 \end{pmatrix}$.

The matrix equation is then equivalent to the equations $x_1 = -2x_3$ and $x_2 = 2x_3$. Therefore, the solution set, or eigenspace, corresponding to $\lambda_1 = -2$ consists of vectors of the form

$$\begin{pmatrix} -2x_3 \\ 2x_3 \\ x_3 \end{pmatrix} = x_3 \begin{pmatrix} -2 \\ 2 \\ 1 \end{pmatrix}$$

Therefore $\begin{pmatrix} -2 \\ 2 \\ 1 \end{pmatrix}$ is an eigenvector corresponding to the eigenvalue $\lambda_1 = -2$, and can be used for our first column of P:

$$P = \begin{pmatrix} -2 & ? & ? \\ 2 & ? & ? \\ 1 & ? & ? \end{pmatrix}$$

Before we continue we make the observation: E_1 is a subspace of $\mathbb{R}^3$ with basis $\{P^{(1)}\}$ and $\dim E_1 = 1$.

Case 2. If $\lambda_2 = 1$, then the equation $(A - \lambda I)\mathbf{x} = \mathbf{0}$ becomes

$$\begin{pmatrix} 0 & 12 & -18 \\ 0 & -12 & 18 \\ 0 & -6 & 9 \end{pmatrix} \begin{pmatrix} x_1 \\ x_2 \\ x_3 \end{pmatrix} = \begin{pmatrix} 0 \\ 0 \\ 0 \end{pmatrix}$$

Without the aid of any computer technology, it should be clear that all three equations that correspond to this matrix equation are equivalent to $2x_2 - 3x_3 = 0$, or $x_2 = \frac{3}{2}x_3$. Notice that x_1 can take on any value, so any vector of the form

$$\begin{pmatrix} x_1 \\ \frac{3}{2}x_3 \\ x_3 \end{pmatrix} = x_1 \begin{pmatrix} 1 \\ 0 \\ 0 \end{pmatrix} + x_3 \begin{pmatrix} 0 \\ \frac{3}{2} \\ 1 \end{pmatrix}$$

will solve the matrix equation.

We note that the solution set contains two independent variables, x_1 and x_3. Further, note that we cannot express the eigenspace E_2 as a linear combination of a single vector as in Case 1. However, it can be written as

$$E_2 = \left\{ x_1 \begin{pmatrix} 1 \\ 0 \\ 0 \end{pmatrix} + x_3 \begin{pmatrix} 0 \\ \frac{3}{2} \\ 1 \end{pmatrix} \mid x_1, x_3 \in \mathbb{R} \right\}.$$

We can replace any vector in a basis is with a nonzero multiple of that vector. Simply for aesthetic reasons, we will multiply the second vector that generates E_2 by 2. Therefore, the eigenspace E_2 is a subspace of $\mathbb{R}^3$ with basis $\left\{ \begin{pmatrix} 1 \\ 0 \\ 0 \end{pmatrix}, \begin{pmatrix} 0 \\ 3 \\ 2 \end{pmatrix} \right\}$ and so dim $E_2 = 2$.

What this means with respect to the diagonalization process is that $\lambda_2 = 1$ gives us both Column 2 and Column 3 the diagonalizing matrix. The order is not important so we have

$$P = \begin{pmatrix} -2 & 1 & 0 \\ 2 & 0 & 3 \\ 1 & 0 & 2 \end{pmatrix}$$

The reader can verify (see Exercise 5 of this section) that $P^{-1} = \begin{pmatrix} 0 & 2 & -3 \\ 1 & 4 & -6 \\ 0 & -1 & 2 \end{pmatrix}$

and $P^{-1}AP = \begin{pmatrix} -2 & 0 & 0 \\ 0 & 1 & 0 \\ 0 & 0 & 1 \end{pmatrix}$ □

In doing Example 12.4.8, p. 85, the given 3×3 matrix A produced only two, not three, distinct eigenvalues, yet we were still able to diagonalize A. The reason we were able to do so was because we were able to find three linearly independent eigenvectors. Again, the main idea is to produce a matrix P that does the diagonalizing. If A is an $n \times n$ matrix, P will be an $n \times n$ matrix, and its n columns must be linearly independent eigenvectors. The main question in the study of diagonalizability is "When can it be done?" This is summarized in the following theorem.

Theorem 12.4.9 A condition for diagonalizability. *Let A be an $n \times n$ matrix. Then A is diagonalizable if and only if A has n linearly independent eigenvectors.*

Proof. Outline of a proof: ($\Longleftarrow$) Assume that A has linearly independent eigenvectors, $P^{(1)}, P^{(2)}, \ldots, P^{(n)}$, with corresponding eigenvalues $\lambda_1, \lambda_2, \ldots, \lambda_n$. We want to prove that A is diagonalizable. Column i of the $n \times n$ matrix AP is $AP^{(i)}$ (see Exercise 7 of this section). Then, since the $P^{(i)}$ is an eigenvector of A associated with the eigenvalue λ_i we have $AP^{(i)} = \lambda_i P^{(i)}$ for $i = 1, 2, \ldots, n$. But this means that $AP = PD$, where D is the diagonal matrix with diagonal entries $\lambda_1, \lambda_2, \ldots, \lambda_n$. If we multiply both sides of the equation by P^{-1} we get the desired $P^{-1}AP = D$.

($\Longrightarrow$) The proof in this direction involves a concept that is not covered in this text (rank of a matrix); so we refer the interested reader to virtually any linear algebra text for a proof. ■

We now give an example of a matrix that is not diagonalizable.

Example 12.4.10 A Matrix that is Not Diagonalizable. Let us attempt to diagonalize the matrix $A = \begin{pmatrix} 1 & 0 & 0 \\ 0 & 2 & 1 \\ 1 & -1 & 4 \end{pmatrix}$

First, we determine the eigenvalues.

$$\det(A - \lambda I) = \det \begin{pmatrix} 1-\lambda & 0 & 0 \\ 0 & 2-\lambda & 1 \\ 1 & -1 & 4-\lambda \end{pmatrix}$$

$$= (1-\lambda) \det \begin{pmatrix} 2-\lambda & 1 \\ -1 & 4-\lambda \end{pmatrix}$$

$$= (1-\lambda)((2-\lambda)(4-\lambda)+1)$$

$$= (1-\lambda)\left(\lambda^2 - 6\lambda + 9\right)$$

$$= (1-\lambda)(\lambda - 3)^2$$

Therefore there are two eigenvalues, $\lambda_1 = 1$ and $\lambda_2 = 3$. Since λ_1 is an eigenvalue of degree one, it will have an eigenspace of dimension 1. Since λ_2 is a double root of the characteristic equation, the dimension of its eigenspace must be 2 in order to be able to diagonalize.

Case 1. For $\lambda_1 = 1$, the equation $(A - \lambda I)\mathbf{x} = \mathbf{0}$ becomes

$$\begin{pmatrix} 0 & 0 & 0 \\ 0 & 1 & 1 \\ 1 & -1 & 3 \end{pmatrix} \begin{pmatrix} x_1 \\ x_2 \\ x_3 \end{pmatrix} = \begin{pmatrix} 0 \\ 0 \\ 0 \end{pmatrix}$$

Row reduction of this system reveals one free variable and eigenspace

$$\begin{pmatrix} x_1 \\ x_2 \\ x_3 \end{pmatrix} = \begin{pmatrix} -4x_3 \\ -x_3 \\ x_3 \end{pmatrix} = x_3 \begin{pmatrix} -4 \\ -1 \\ 1 \end{pmatrix}$$

Hence, $\left\{ \begin{pmatrix} -4 \\ -1 \\ 1 \end{pmatrix} \right\}$ is a basis for the eigenspace of $\lambda_1 = 1$.

Case 2. For $\lambda_2 = 3$, the equation $(A - \lambda I)\mathbf{x} = \mathbf{0}$ becomes

$$\begin{pmatrix} -2 & 0 & 0 \\ 0 & -1 & 1 \\ 1 & -1 & 1 \end{pmatrix} \begin{pmatrix} x_1 \\ x_2 \\ x_3 \end{pmatrix} = \begin{pmatrix} 0 \\ 0 \\ 0 \end{pmatrix}$$

Once again there is only one free variable in the row reduction and so the dimension of the eigenspace will be one:

$$\begin{pmatrix} x_1 \\ x_2 \\ x_3 \end{pmatrix} = \begin{pmatrix} 0 \\ x_3 \\ x_3 \end{pmatrix} = x_3 \begin{pmatrix} 0 \\ 1 \\ 1 \end{pmatrix}$$

Hence, $\left\{ \begin{pmatrix} 0 \\ 1 \\ 1 \end{pmatrix} \right\}$ is a basis for the eigenspace of $\lambda_2 = 3$. This means that $\lambda_2 = 3$ produces only one column for P. Since we began with only two eigenvalues, we had hoped that $\lambda_2 = 3$ would produce a vector space of dimension two, or, in matrix terms, two linearly independent columns for P. Since A does not have three linearly independent eigenvectors A cannot be diagonalized. □

12.4.3 SageMath Note - Diagonalization

We demonstrate how diagonalization can be done in Sage. We start by defining the matrix to be diagonalized, and also declare D and P to be variables.

```
var ('_D,_P')
A = Matrix (QQ, [[4, 1, 0], [1, 5, 1], [0, 1, 4]]);A
```

```
[4 1 0]
[1 5 1]
[0 1 4]
```

We have been working with "right eigenvectors" since the **x** in $A\mathbf{x} = \lambda\mathbf{x}$ is a column vector. It's not so common but still desirable in some situations to consider "left eigenvectors," so SageMath allows either one. The right_eigenmatrix method returns a pair of matrices. The diagonal matrix, D, with eigenvalues and the diagonalizing matrix, P, which is made up of columns that are eigenvectors corresponding to the eigenvectors of D.

```
(D,P)=A.right_eigenmatrix();(D,P)
```

```
([6 0 0]
 [0 4 0]
 [0 0 3],
 [ 1  1  1]
 [ 2  0 -1]
 [ 1 -1  1])
```

We should note here that P is not unique because even if an eigenspace has dimension one, any nonzero vector in that space will serve as an eigenvector. For that reason, the P generated by Sage isn't necessarily the same as the one computed by any other computer algebra system such as Mathematica. Here we verify the result for our Sage calculation. Recall that an asterisk is used for matrix multiplication in Sage.

```
P.inverse()*A*P
```

```
[6 0 0]
[0 4 0]
[0 0 3]
```

Here is a second matrix to diagonalize.

```
A2=Matrix(QQ,[[8,1,0],[1,5,1],[0,1,7]]);A2
```

```
[8 1 0]
[1 5 1]
[0 1 7]
```

Here we've already specified that the underlying system is the rational numbers. Since the eigenvalues are not rational, Sage will revert to approximate number by default. We'll just pull out the matrix of eigenvectors this time and display rounded entries.

```
P=A2.right_eigenmatrix()[1]
P.numerical_approx(digits=3)
print('-------------------')
D=(P.inverse()*A2*P);D.numerical_approx(digits=3)
```

```
[ 4.35 0.000 0.000]
[0.000  7.27 0.000]
[0.000 0.000  8.38]
```

Finally, we examine how Sage reacts to the matrix from Example 12.4.10, p. 87 that couldn't be diagonalized. Notice that the last column is a zero column, indicating the absence of one needed eigenvector.

```
A3=Matrix(QQ,[[1, 0, 0],[0,2,1],[1,-1,4]])
(D,P)=A3.right_eigenmatrix();(D,P)
```

```
([1 0 0]
 [0 3 0]
 [0 0 3],
 [   1    0    0]
 [ 1/4    1    0]
 [-1/4    1    0])
```

12.4.4 Exercises

1.

(a) List three different eigenvectors of $A = \begin{pmatrix} 2 & 1 \\ 2 & 3 \end{pmatrix}$, the matrix of Example 12.4.2, p. 83, associated with each of the two eigenvalues 1 and 4. Verify your results.

(b) Choose one of the three eigenvectors corresponding to 1 and one of the three eigenvectors corresponding to 4, and show that the two chosen vectors are linearly independent.

2.

(a) Verify that E_1 and E_2 in Example 12.4.2, p. 83 are vector spaces over $\mathbb{R}$. Since they are also subsets of $\mathbb{R}^2$, they are called subvector-spaces, or subspaces for short, of $\mathbb{R}^2$. Since these are subspaces consisting of eigenvectors, they are called eigenspaces.

(b) Use the definition of dimension in the previous section to find $\dim E_1$ and $\dim E_2$. Note that $\dim E_1 + \dim E_2 = \dim \mathbb{R}^2$. This is not a coincidence.

3.

(a) Verify that $P^{-1}AP$ is indeed equal to $\begin{pmatrix} 1 & 0 \\ 0 & 4 \end{pmatrix}$, as indicated in Example 12.4.6, p. 85.

(b) Choose $P^{(1)} = \begin{pmatrix} 1 \\ 2 \end{pmatrix}$ and $P^{(2)} = \begin{pmatrix} 1 \\ -1 \end{pmatrix}$ and verify that the new value of P satisfies $P^{-1}AP = \begin{pmatrix} 4 & 0 \\ 0 & 1 \end{pmatrix}$.

(c) Take two different (from the previous part) linearly independent eigenvectors of the matrix A of Example 12.4.6, p. 85 and verify that $P^{-1}AP$ is a diagonal matrix.

4.

(a) Let A be the matrix in Example 12.4.8, p. 85 and $P = \begin{pmatrix} 0 & 1 & 0 \\ 1 & 0 & 1 \\ 1 & 0 & 2 \end{pmatrix}$.

Without doing any actual matrix multiplications, determine the value of $P^{-1}AP$

(b) If you choose the columns of P in the reverse order, what is $P^{-1}AP$?

5. Diagonalize the following, if possible:

(a) $\begin{pmatrix} 1 & 2 \\ 3 & 2 \end{pmatrix}$ (c) $\begin{pmatrix} 3 & 0 \\ 0 & 4 \end{pmatrix}$ (e) $\begin{pmatrix} 6 & 0 & 0 \\ 0 & 7 & -4 \\ 9 & 1 & 3 \end{pmatrix}$

(b) $\begin{pmatrix} -2 & 1 \\ -7 & 6 \end{pmatrix}$ (d) $\begin{pmatrix} 1 & -1 & 4 \\ 3 & 2 & -1 \\ 2 & 1 & -1 \end{pmatrix}$ (f) $\begin{pmatrix} 1 & -1 & 0 \\ -1 & 2 & -1 \\ 0 & -1 & 1 \end{pmatrix}$

6. Diagonalize the following, if possible:

(a) $\begin{pmatrix} 0 & 1 \\ 1 & 1 \end{pmatrix}$ (c) $\begin{pmatrix} 2 & -1 \\ 1 & 0 \end{pmatrix}$ (e) $\begin{pmatrix} 1 & 1 & 0 \\ 1 & 0 & 1 \\ 0 & 1 & 1 \end{pmatrix}$

(b) $\begin{pmatrix} 2 & 1 \\ 4 & 2 \end{pmatrix}$ (d) $\begin{pmatrix} 1 & 3 & 6 \\ -3 & -5 & -6 \\ 3 & 3 & 6 \end{pmatrix}$ (f) $\begin{pmatrix} 2 & -1 & 0 \\ -1 & 2 & -1 \\ 0 & -1 & 2 \end{pmatrix}$

7. Let A and P be as in Example 12.4.8, p. 85. Show that the columns of the matrix AP can be found by computing $AP^{(1)}, AP^{(2)}, \ldots, AP^{(n)}$.

8. Prove that if P is an $n \times n$ matrix and D is a diagonal matrix with diagonal entries $d_1, d_2, \ldots, d_n$, then PD is the matrix obtained from P, by multiplying column i of P by d_i, $i = 1, 2, \ldots, n$.

12.5 Some Applications

A large and varied number of applications involve computations of powers of matrices. These applications can be found in science, the social sciences, economics, engineering, and, many other areas where mathematics is used. We will consider a few diverse examples the mathematics behind these applications here.

12.5.1 Diagonalization

We begin by developing a helpful technique for computing A^m, $m > 1$. If A can be diagonalized, then there is a matrix P such that $P^{-1}AP = D$, where D is a diagonal matrix and

$$A^m = PD^mP^{-1} \text{ for all } m \geq 1 \qquad (12.5.1)$$

The proof of this identity was an exercise in Section 5.4. The condition that D be a diagonal matrix is not necessary but when it is, the calculation on the right side is particularly easy to perform. Although the formal proof is done by induction, the reason why it is true is easily seen by writing out an example such as $m = 3$:

$$\begin{aligned}A^3 &= \left(PDP^{-1}\right)^3 \\ &= \left(PDP^{-1}\right)\left(PDP^{-1}\right)\left(PDP^{-1}\right) \\ &= PD\left(P^{-1}P\right)D\left(P^{-1}P\right)DP^{-1} \quad \text{by associativity of matrix multiplication} \\ &= PDIDIDP^{-1} \\ &= PDDDP^{-1} \\ &= PD^3P^{-1}\end{aligned}$$

Example 12.5.1 Application to Recursion: Matrix Computation of the Fibonnaci Sequence. Consider the computation of terms of the Fibonnaci Sequence. Recall that $F_0 = 1, F_1 = 1$ and $F_k = F_{k-1} + F_{k-2}$ for $k \geq 2$.

In order to formulate the calculation in matrix form, we introduced the "dummy equation" $F_{k-1} = F_{k-1}$ so that now we have two equations

$$F_k = F_{k-1} + F_{k-2}$$
$$F_{k-1} = F_{k-1}$$

If $A = \begin{pmatrix} 1 & 1 \\ 1 & 0 \end{pmatrix}$, these two equations can be expressed in matrix form as

$$\begin{aligned}\begin{pmatrix} F_k \\ F_{k-1} \end{pmatrix} &= \begin{pmatrix} 1 & 1 \\ 1 & 0 \end{pmatrix}\begin{pmatrix} F_{k-1} \\ F_{k-2} \end{pmatrix} \quad \text{if } k \geq 2 \\ &= A\begin{pmatrix} F_{k-1} \\ F_{k-2} \end{pmatrix} \\ &= A^2\begin{pmatrix} F_{k-2} \\ F_{k-3} \end{pmatrix} \quad \text{if } k \geq 3\end{aligned}$$

etc.

We can use induction to prove that if $k \geq 2$,

$$\begin{pmatrix} F_k \\ F_{k-1} \end{pmatrix} = A^{k-1}\begin{pmatrix} 1 \\ 1 \end{pmatrix}$$

Next, by diagonalizing A and using the fact that $A^m = PD^mP^{-1}$. we can show that

$$F_k = \frac{1}{\sqrt{5}}\left(\left(\frac{1+\sqrt{5}}{2}\right)^k - \left(\frac{1-\sqrt{5}}{2}\right)^k\right)$$

Some comments on this example:

(1) An equation of the form $F_k = aF_{k-1} + bF_{k-2}$, where a and b are given constants, is referred to as a linear homogeneous second-order difference equation. The conditions $F_0 = c_0$ and $F_1 = c_1$, where c_1 and c_2 are constants, are called initial conditions. Those of you who are familiar with differential equations may recognize that this language parallels what is used in differential equations. Difference (aka recurrence) equations move forward discretely; that is, in a finite number of positive steps. On the other hand, a differential equation moves continuously; that is, takes an infinite number of infinitesimal steps.

(2) A recurrence relationship of the form $S_k = aS_{k-1} + b$, where a and b are constants, is called a first-order difference equation. In order to write out

the sequence, we need to know one initial condition. Equations of this type can be solved similarly to the method outlined in the example by introducing the superfluous equation $1 = 0 \cdot F_{k-1} + 1$ to obtain in matrix equation:

$$\begin{pmatrix} F_k \\ 1 \end{pmatrix} = \begin{pmatrix} a & b \\ 0 & 1 \end{pmatrix} \begin{pmatrix} S_{k-1} \\ 1 \end{pmatrix} \Rightarrow \begin{pmatrix} F_k \\ 1 \end{pmatrix} = \begin{pmatrix} a & b \\ 0 & 1 \end{pmatrix}^k \begin{pmatrix} F_0 \\ 1 \end{pmatrix}$$

□

12.5.2 Path Counting

In the next example, we apply the following theorem, which can be proven by induction.

Theorem 12.5.2 Path Counting Theorem. *If A is the adjacency matrix of a graph with vertices $\{v_1, v_2, \ldots, v_n\}$, then the entry $(A^k)_{ij}$ is the number of paths of length k from node v_i to node v_j.*

Example 12.5.3 Counting Paths with Diagonalization. Consider the graph in Figure 12.5.4, p. 93.

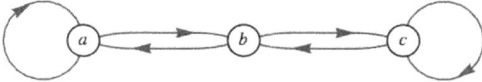

Figure 12.5.4 Counting Numbers of Paths

As we saw in Section 6.4, the adjacency matrix of this graph is $A = \begin{pmatrix} 1 & 1 & 0 \\ 1 & 0 & 1 \\ 0 & 1 & 1 \end{pmatrix}$.

Recall that A^k is the adjacency matrix of the relation r^k, where r is the relation $\{(a,a),(a,b),(b,a),(b,c),(c,b),(c,c)\}$ of the above graph. Also recall that in computing A^k, we used Boolean arithmetic. What happens if we use "regular" arithmetic? If we square A we get $A^2 = \begin{pmatrix} 2 & 1 & 1 \\ 1 & 2 & 1 \\ 1 & 1 & 2 \end{pmatrix}$

How can we interpret this? We note that $A_{33} = 2$ and that there are two paths of length two from c (the third node) to c. Also, $A_{13} = 1$, and there is one path of length 2 from a to c. The reader should verify these claims by examining the graph.

How do we compute A^k for possibly large values of k? From the discussion at the beginning of this section, we know that $A^k = P D^k P^{-1}$ if A is diagonalizable. We leave to the reader to show that $\lambda = 1, 2,$ and -1 are eigenvalues of A with eigenvectors

$$\begin{pmatrix} 1 \\ 0 \\ -1 \end{pmatrix}, \begin{pmatrix} 1 \\ 1 \\ 1 \end{pmatrix}, \text{ and } \begin{pmatrix} 1 \\ -2 \\ 1 \end{pmatrix}$$

Then

$$A^k = P \begin{pmatrix} 1 & 0 & 0 \\ 0 & 2^k & 0 \\ 0 & 0 & (-1)^k \end{pmatrix} P^{-1}$$

where $P = \begin{pmatrix} 1 & 1 & 1 \\ 0 & 1 & -2 \\ -1 & 1 & 1 \end{pmatrix}$ and $P^{-1} = \begin{pmatrix} \frac{1}{2} & 0 & -\frac{1}{2} \\ \frac{1}{3} & \frac{1}{3} & \frac{1}{3} \\ \frac{1}{6} & -\frac{1}{3} & \frac{1}{6} \end{pmatrix}$

See Exercise 12.5.5.5, p. 96 of this section for the completion of this example. □

12.5.3 Matrix Calculus

Example 12.5.5 Matrix Calculus - Exponentials. Those who have studied calculus recall that the Maclaurin series is a useful way of expressing many common functions. For example, $e^x = \sum_{k=0}^{\infty} \frac{x^k}{k!}$. Indeed, calculators and computers use these series for calculations. Given a polynomial $f(x)$, we defined the matrix-polynomial $f(A)$ for square matrices in Chapter 5. Hence, we are in a position to describe e^A for an $n \times n$ matrix A as a limit of polynomials, the partial sums of the series. Formally, we write

$$e^A = I + A + \frac{A^2}{2!} + \frac{A^3}{3!} + \cdots = \sum_{k=0}^{\infty} \frac{A^k}{k!}$$

Again we encounter the need to compute high powers of a matrix. Let A be an $n \times n$ diagonalizable matrix. Then there exists an invertible $n \times n$ matrix P such that $P^{-1}AP = D$, a diagonal matrix, so that

$$e^A = e^{PDP^{-1}}$$
$$= \sum_{k=0}^{\infty} \frac{\left(PDP^{-1}\right)^k}{k!}$$
$$= P\left(\sum_{k=0}^{\infty} \frac{D^k}{k!}\right) P^{-1}$$

The infinite sum in the middle of this final expression can be easily evaluated if D is diagonal. All entries of powers off the diagonal are zero and the i th entry of the diagonal is

$$\left(\sum_{k=0}^{\infty} \frac{D^k}{k!}\right)_{ii} = \sum_{k=0}^{\infty} \frac{D_{ii}^k}{k!} = e^{D_{ii}}$$

For example, if $A = \begin{pmatrix} 2 & 1 \\ 2 & 3 \end{pmatrix}$, the first matrix we diagonalized in Section 12.3, we found that $P = \begin{pmatrix} 1 & 1 \\ -1 & 2 \end{pmatrix}$ and $D = \begin{pmatrix} 1 & 0 \\ 0 & 4 \end{pmatrix}$.

Therefore,

$$e^A = \begin{pmatrix} 1 & 1 \\ -1 & 2 \end{pmatrix} \begin{pmatrix} e & 0 \\ 0 & e^4 \end{pmatrix} \begin{pmatrix} \frac{2}{3} & -\frac{1}{3} \\ \frac{1}{3} & \frac{1}{3} \end{pmatrix}$$
$$= \begin{pmatrix} \frac{2e}{3} + \frac{e^4}{3} & -\frac{e}{3} + \frac{e^4}{3} \\ -\frac{2e}{3} + \frac{2e^4}{3} & \frac{e}{3} + \frac{2e^4}{3} \end{pmatrix}$$
$$\approx \begin{pmatrix} 20.0116 & 17.2933 \\ 34.5866 & 37.3049 \end{pmatrix}$$

□

Remark 12.5.6 Many of the ideas of calculus can be developed using matrices. For example, if $A(t) = \begin{pmatrix} t^3 & 3t^2 + 8t \\ e^t & 2 \end{pmatrix}$ then $\frac{dA(t)}{dt} = \begin{pmatrix} 3t^2 & 6t + 8 \\ e^t & 0 \end{pmatrix}$

Many of the basic formulas in calculus are true in matrix calculus. For

example,
$$\frac{d(A(t)+B(t))}{dt} = \frac{dA(t)}{dt} + \frac{dB(t)}{dt}$$
and if A is a constant matrix,
$$\frac{de^{At}}{dt} = Ae^{At}$$

Matrix calculus can be used to solve systems of differential equations in a similar manner to the procedure used in ordinary differential equations.

12.5.4 SageMath Note - Matrix Exponential

Sage's matrix exponential method is exp.

```
A=Matrix(QQ,[[2,1],[2,3]])
A.exp()
```

```
[1/3*e^4 + 2/3*e   1/3*e^4 - 1/3*e]
[2/3*e^4 - 2/3*e   2/3*e^4 + 1/3*e]
```

12.5.5 Exercises

1.
    (a) Write out all the details of Example 12.5.1, p. 92 to show that the formula for F_k given in the text is correct.

    (b) Use induction to prove the assertion made in the example that
    $$\begin{pmatrix} F_k \\ F_{k-1} \end{pmatrix} = A^{k-1} \begin{pmatrix} 1 \\ 1 \end{pmatrix}$$

2.
    (a) Do Example 8.0.2, p. 15 using the method outlined in Example 12.5.1, p. 92. Note that the terminology characteristic equation, characteristic polynomial, and so on, introduced in Chapter 8, comes from the language of matrix algebra,

    (b) What is the significance of Algorithm 8.0.1, p. 15, part c, with respect to this section?

3. Solve $S(k) = 5S(k-1) + 4$, with $S(0) = 0$, using the method of this section.

4. How many paths are there of length 6 from vertex 1 to vertex 3 in Figure 12.5.7, p. 95? How many paths from vertex 2 to vertex 2 of length 6 are there?

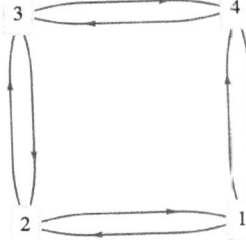

Figure 12.5.7 Graph for exercise 4

Hint. The characteristic polynomial of the adjacency matrix is $\lambda^4 - 4\lambda^2$.

5. Regarding Example 12.5.3, p. 93,

    (a) Use matrices to determine the number of paths of length 1 that exist from vertex a to each of the vertices in the given graph. Verify using the graph. Do the same for vertices b and c.

    (b) Verify all the details provided in the example.

    (c) Use matrices to determine the number of paths of length 4 there between each pair of nodes in the graph. Verify your results using the graph.

6. Let $A = \begin{pmatrix} 2 & -1 \\ -1 & 2 \end{pmatrix}$

    (a) Find e^A

    (b) Recall that $\sin x = \sum_{k=0}^{\infty} \frac{(-1)^k x^k}{(2k+1)!}$ and compute $\sin A$.

    (c) Formulate a reasonable definition of the natural logarithm of a matrix and compute $\ln A$.

7. We noted in Chapter 5 that since matrix algebra is not commutative under multiplication, certain difficulties arise. Let $A = \begin{pmatrix} 1 & 1 \\ 0 & 0 \end{pmatrix}$ and $B = \begin{pmatrix} 0 & 0 \\ 0 & 2 \end{pmatrix}$.

    (a) Compute e^A, e^B, and e^{A+B}. Compare $e^A e^B$, $e^B e^A$ and e^{A+B}.

    (b) Show that if $\mathbf{0}$ is the 2×2 zero matrix, then $e^{\mathbf{0}} = I$.

    (c) Prove that if A and B are two matrices that do commute, then $e^{A+B} = e^A e^B$, thereby proving that e^A and e^B commute.

    (d) Prove that for any matrix A, $\left(e^A\right)^{-1} = e^{-A}$.

8. Another observation for adjacency matrices: For the matrix in Example 12.5.3, p. 93, note that the sum of the elements in the row corresponding to the node a (that is, the first row) gives the outdegree of a. Similarly, the sum of the elements in any given column gives the indegree of the node corresponding to that column.

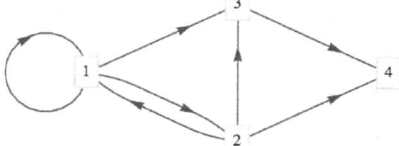

Figure 12.5.8 Graph for exercise 8

    (a) Using the matrix A of Example 12.5.3, p. 93, find the outdegree and the indegree of each node. Verify by the graph.

    (b) Repeat part (a) for the directed graphs in Figure 12.5.8, p. 96.

12.6 Linear Equations over the Integers Mod 2

12.6.1 Row reduction mod 2

The methods we have studied for solving systems of equations up to this point can be applied to systems in which all arithmetic is done over other algebraic systems, including the integers modulo 2. The mod 2 case will become particularly useful in our later study of coding theory.

When solving systems of equations with mod 2 arithmetic, the elementary row operations are still fundamental. However, since there is only one nonzero element, 1, you never need to multiply a row by a nonzero constant. One other big difference is that the number of possible solutions is always finite. If you have m linear equations in n unknowns, each unknown can only take on one of two values, 0 or 1. Therefore there are only 2^n possible n-tuples to from which to draw a solution set. Assuming $m \leq n$, you typically (but not always) will have m basic variables after row-reduction and $n - m$ free variable. If this is the case, and any solution exists, there will be 2^{n-m} different solutions.

Let's look at an example, which is coverted to matrix form immediately.

$$\begin{aligned} x_1 + x_2 + x_4 &= 1 \\ x_1 + x_3 + x_5 &= 0 \\ x_2 + x_3 + x_6 &= 0 \end{aligned}$$

The augmented matrix of the system is

$$\begin{pmatrix} 1 & 1 & 0 & 1 & 0 & 0 & | & 1 \\ 1 & 0 & 1 & 0 & 1 & 0 & | & 0 \\ 0 & 1 & 1 & 0 & 0 & 1 & | & 0 \end{pmatrix}$$

The steps in row-reducing this matrix follow. Entries on which we "pivot" are displayed in bold face to more easily identify the basic variables.

$$\begin{pmatrix} 1 & 1 & 0 & 1 & 0 & 0 & | & 1 \\ 1 & 0 & 1 & 0 & 1 & 0 & | & 0 \\ 0 & 1 & 1 & 0 & 0 & 1 & | & 0 \end{pmatrix} \xrightarrow{\text{add } R_1 \text{ to } R_2} \begin{pmatrix} \mathbf{1} & 1 & 0 & 1 & 0 & 0 & | & 1 \\ 0 & 1 & 1 & 1 & 1 & 0 & | & 1 \\ 0 & 1 & 1 & 0 & 0 & 1 & | & 0 \end{pmatrix}$$

$$\xrightarrow{\text{add } R_2 \text{ to } R_1} \begin{pmatrix} \mathbf{1} & 0 & 1 & 0 & 1 & 0 & | & 0 \\ 0 & \mathbf{1} & 1 & 1 & 1 & 0 & | & 1 \\ 0 & 1 & 1 & 0 & 0 & 1 & | & 0 \end{pmatrix}$$

$$\xrightarrow{\text{add } R_2 \text{ to } R_3} \begin{pmatrix} \mathbf{1} & 0 & 1 & 0 & 1 & 0 & | & 0 \\ 0 & \mathbf{1} & 1 & 1 & 1 & 0 & | & 1 \\ 0 & 0 & 0 & 1 & 1 & 1 & | & 1 \end{pmatrix}$$

Notice that at this point, we cannot pivot on the third row, third column since that entry is zero. Therefore we move over to the next column, making the x_4 basic.

$$\xrightarrow{\text{add } R_3 \text{ to } R_2} \begin{pmatrix} 1 & 0 & 1 & 0 & 1 & 0 & | & 0 \\ 0 & \mathbf{1} & 1 & 0 & 0 & 1 & | & 0 \\ 0 & 0 & 0 & \mathbf{1} & 1 & 1 & | & 1 \end{pmatrix}$$

This completes the row reduction and we can now identify the solution set. Keep in mind that since addition is subtraction, terms can be moved to either side of an equals sign without any change in sign. The basic variables are x_1,

x_2, and x_4, while the other three variables are free. The general solution of the system is

$$x_1 = x_3 + x_5$$
$$x_2 = x_3 + x_6$$
$$x_3 = x_3$$
$$x_4 = 1 + x_5 + x_6$$
$$x_5 = x_5$$
$$x_6 = x_6$$

With three free variables, there are $2^3 = 8$ solutions to this system. For example, one of them is obtained by setting $x_3 = 1$, $x_5 = 1$, and $x_6 = 0$, which produces $(x_1, x_2, x_3, x_4, x_5, x_6) = (0, 1, 1, 0, 1, 0)$.

We can check our row reduction with SageMath:

```
H=Matrix(Integers(2),[[1,1,0,1,0,0,1],
   [1,0,1,0,1,0,0],[0,1,1,0,0,1,0]])
H.echelon_form()
```

```
[1 0 1 0 1 0 0]
[0 1 1 0 0 1 0]
[0 0 0 1 1 1 1]
```

12.6.2 Exercises

In all of the exercises that follow, the systems of equations are over $\mathbb{Z}_2$, and so mod 2 arithmetic should be used in solving them.

1. Solve the following systems, describing the solution sets completely:

    (a) $\begin{aligned} x_1 + x_2 &= 0 \\ x_1 \phantom{{}+x_2} + x_3 &= 0 \end{aligned}$

    (b) $\begin{aligned} x_1 + x_2 \phantom{{}+x_3} &= 0 \\ x_2 + x_3 &= 0 \\ x_3 + x_4 &= 1 \\ x_1 + x_2 + x_3 &= 1 \end{aligned}$

2. This exercise provides an example in which the number of basic variables is less than the number of equations. The only difference between the two systems below is the right hand sides. You can start with an augmented matrix having two right side columns and do row reduction for both systems at the same time.

    (a) $\begin{aligned} x_1 + x_2 \phantom{{}+x_3} + x_4 &= 1 \\ x_1 \phantom{{}+x_2} + x_3 + x_4 &= 0 \\ x_2 + x_3 \phantom{{}+x_4} &= 1 \end{aligned}$

    (b) $\begin{aligned} x_1 + x_2 \phantom{{}+x_3} + x_4 &= 1 \\ x_1 \phantom{{}+x_2} + x_3 + x_4 &= 0 \\ x_2 + x_3 \phantom{{}+x_4} &= 0 \end{aligned}$

3. This exercise motivates the concept of a coset in Chapter 15.

    (a) Solve the following system and prove that the solution set is a linear combination of vectors in $\mathbb{Z}_2^5$ and also a subgroup of the group $\mathbb{Z}_2^5$

under coordinatewise mod 2 addition.

$$
\begin{aligned}
x_1 + x_2 + x_5 &= 0 \\
x_1 + x_3 + x_5 &= 0 \\
x_1 + x_3 + x_4 &= 0 \\
x_2 + x_3 + x_4 &= 0
\end{aligned}
$$

(b) Describe the solution set to the following system as it relates to the solution set to the system in the previous part of this exercise.

$$
\begin{aligned}
x_1 + x_2 + x_5 &= 1 \\
x_1 + x_3 + x_5 &= 0 \\
x_1 + x_3 + x_4 &= 1 \\
x_2 + x_3 + x_4 &= 0
\end{aligned}
$$

Chapter 13

Boolean Algebra

Figure 13.0.1 George Boole, 1815 - 1864

George Boole

George Boole wasn't idle a lot.
He churned out ideas on the spot,
Making marvellous use of
Inclusive/exclusive
Expressions like AND, OR, and NOT

Andrew Robinson, The Omnificent English Dictionary In Limerick Form

In this chapter we will develop a type of algebraic system, Boolean algebras, that is particularly important to computer scientists, as it is the mathematical foundation of computer design, or switching theory. The similarities of Boolean algebras and the algebra of sets and logic will be discussed, and we will discover properties of finite Boolean algebras.

In order to achieve these goals, we will recall the basic ideas of posets introduced in Chapter 6 and develop the concept of a lattice. The reader should view the development of the topics of this chapter as another example of an algebraic system. Hence, we expect to define first the elements in the system, next the operations on the elements, and then the common properties of the operations in the system.

13.1 Posets Revisited

We recall the definition a partially ordering:

Definition 13.1.1 Partial Ordering. Let $\preceq$ be a relation on a set L. We say that $\preceq$ is a partial ordering on L if it is reflexive, antisymmetric and transitive. That is:

(1) $\preceq$ is reflexive: $a \preceq a \quad \forall a \in L$

(2) $\preceq$ is antisymmetric: $a \preceq b$ and $a \neq b \Rightarrow b \not\preceq a \quad \forall a, b \in L$

(3) $\preceq$ is transitive: $a \preceq b$ and $b \preceq c \Rightarrow a \preceq c \quad \forall a, b, c \in L$

The set together with the relation $(L, \preceq)$ is called a poset. $\diamond$

Example 13.1.2 Some posets. We recall a few examples of posets:

(a) $(\mathbb{R}, \leq)$ is a poset. Notice that our generic symbol for the partial ordering, $\preceq$, is selected to remind us that a partial ordering is similar to "less than or equal to."

(b) Let $A = \{a, b\}$. Then $(\mathcal{P}(A), \subseteq)$ is a poset.

(c) Let $L = \{1, 2, 3, 6\}$. Then $(L, |)$ is a poset.

$\square$

The posets we will concentrate on in this chapter will be those which have upper and lower bounds in relation to any pair of elements. Next, we define this concept precisely.

Definition 13.1.3 Lower Bound, Upper Bound. Let $(L, \preceq)$ be a poset, and $a, b \in L$. Then $c \in L$ is a lower bound of a and b if $c \preceq a$ and $c \preceq b$. Also, $d \in L$ is an upper bound of a and b if $a \preceq d$ and $b \preceq d$. $\diamond$

In most of the posets that will interest us, every pair of elements have both upper and lower bounds, though there are posets for which this is not true.

Definition 13.1.4 Greatest Lower Bound. Let $(L, \preceq)$ be a poset. If $a, b \in L$, then $\ell \in L$ is a greatest lower bound of a and b if and only if

- $\ell \preceq a$
- $\ell \preceq b$
- If $\ell' \in L$ such that $\ell' \preceq a$ and $\ell' \preceq b$, then $\ell' \preceq \ell$.

$\diamond$

The last condition in the definition of Greatest Lower Bound says that if ℓ' is also a lower bound, then ℓ is "greater" in relation to $\preceq$ than ℓ'. The definition of a least upper bound is a mirror image of a greatest lower bound:

Definition 13.1.5 Least Upper Bound. Let $(L, \preceq)$ be a poset. If $a, b \in L$, then $u \in L$ is a least upper bound of a and b if and only if

- $a \preceq u$
- $b \preceq u$
- If $u' \in L$ such that if $a \preceq u'$ and $b \preceq u'$, then $u \preceq u'$.

$\diamond$

Notice that the two definitions above refer to "...a greatest lower bound" and "a least upper bound." Any time you define an object like these you need to have an open mind as to whether more than one such object can exist. In fact, we now can prove that there can't be two greatest lower bounds or two least upper bounds.

Theorem 13.1.6 Uniqueness of Least Upper and Greatest Lower Bounds. *Let $(L, \preceq)$ be a poset, and $a, b \in L$. If a greatest lower bound of a and b exists, then it is unique. The same is true of a least upper bound, if it exists.*

Proof. Let ℓ and ℓ' be greatest lower bounds of a and b. We will prove that $\ell = \ell'$.

(1) ℓ a greatest lower bound of a and b $\Rightarrow$ ℓ is a lower bound of a and b.

(2) ℓ' a greatest lower bound of a and b and ℓ a lower bound of a and b $\Rightarrow \ell \preceq \ell'$, by the definition of greatest lower bound.

(3) ℓ' a greatest lower bound of a and b $\Rightarrow \ell'$ is a lower bound of a and b.

(4) ℓ a greatest lower bound of a and b and ℓ' a lower bound of a and b. $\Rightarrow \ell' \preceq \ell$ by the definition of greatest lower bound.

(5) $\ell \preceq \ell'$ and $\ell' \preceq \ell \Rightarrow \ell = \ell'$ by the antisymmetry property of a partial ordering.

The proof of the second statement in the theorem is almost identical to the first and is left to the reader. ∎

Definition 13.1.7 Greatest Element, Least Element. Let $(L, \preceq)$ be a poset. $M \in L$ is called the greatest (maximum) element of L if, for all $a \in L$, $a \preceq M$. In addition, $m \in L$ is called the least (minimum) element of L if for all $a \in L$, $m \preceq a$. The greatest and least elements, when they exist, are frequently denoted by **1** and **0** respectively. $\diamond$

Example 13.1.8 Bounds on the divisors of 105. Consider the partial ordering "divides" on $L = \{1, 3, 5, 7, 15, 21, 35, 105\}$. Then $(L, |)$ is a poset. To determine the least upper bound of 3 and 7, we look for all $u \in L$, such that

13.1. POSETS REVISITED

$3|u$ and $7|u$. Certainly, both $u = 21$ and $u = 105$ satisfy these conditions and no other element of L does. Next, since $21|105$, 21 is the least upper bound of 3 and 7. Similarly, the least upper bound of 3 and 5 is 15. The greatest element of L is 105 since $a|105$ for all $a \in L$. To find the greatest lower bound of 15 and 35, we first consider all elements g of L such that $g \mid 15$. They are 1, 3, 5, and 15. The elements for which $g \mid 35$ are 1, 5, 7, and 35. From these two lists, we see that $\ell = 5$ and $\ell = 1$ satisfy the required conditions. But since $1|5$, the greatest lower bound is 5. The least element of L is 1 since $1|a$ for all $a \in L$. □

Definition 13.1.9 The Set of Divisors of an Integer. For any positive integer n, the divisors of n is the set of integers that divide evenly into n. We denote this set D_n. ◇

For example, the set L of Example 13.1.8, p. 104 is D_{105}.

Example 13.1.10 The power set of a three element set. Consider the poset $(\mathcal{P}(A), \subseteq)$, where $A = \{1, 2, 3\}$. The greatest lower bound of $\{1, 2\}$ and $\{1, 3\}$ is $\ell = \{1\}$. For any other element ℓ' which is a subset of $\{1, 2\}$ and $\{1, 3\}$ (there is only one; what is it?), $\ell' \subseteq \ell$. The least element of $\mathcal{P}(A)$ is $\emptyset$ and the greatest element is $A = \{1, 2, 3\}$. The Hasse diagram of this poset is shown in Figure 13.1.11, p. 105.

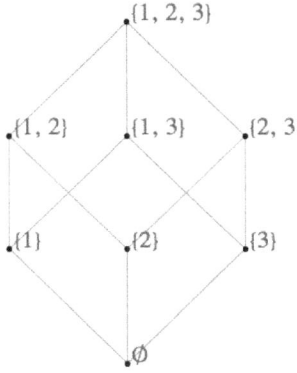

Figure 13.1.11 Power Set of $\{1, 2, 3\}$

□

The previous examples and definitions indicate that the least upper bound and greatest lower bound are defined in terms of the partial ordering of the given poset. It is not yet clear whether all posets have the property such that every pair of elements always has both a least upper bound and greatest lower bound. Indeed, this is not the case (see Exercise 13.1.3, p. 106).

Exercises

1. Consider the poset $(D_{30}, |)$, where $D_{30} = \{1, 2, 3, 5, 6, 10, 15, 30\}$.

    (a) Find all lower bounds of 10 and 15.

    (b) Find the greatest lower bound of 10 and 15.

    (c) Find all upper bounds of 10 and 15.

    (d) Determine the least upper bound of 10 and 15.

    (e) Draw the Hasse diagram for D_{30} with respect to $|$. Compare this Hasse diagram with that of Example 13.1.10, p. 105. Note that the two diagrams are structurally the same.

2. List the elements of the sets D_8, D_{50}, and D_{1001}. For each set, draw the Hasse diagram for "divides."

3. Figure 13.1.12, p. 106 contains Hasse diagrams of posets.

   (a) Determine the least upper bound and greatest lower bound of all pairs of elements when they exist. Indicate those pairs that do not have a least upper bound (or a greatest lower bound).

   (b) Find the least and greatest elements when they exist.

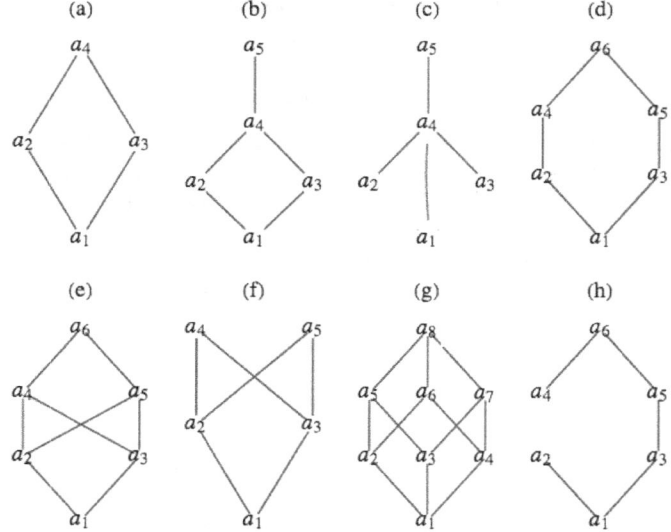

Figure 13.1.12 Figure for Exercise 3

4. For the poset $(\mathbb{N}, \leq)$, what are the greatest lower bound and least upper bound of two elements a and b? Are there least and/or greatest elements?

5.
   (a) Prove the second part of Theorem 13.1.6, p. 104, the least upper bound of two elements in a poset is unique, if one exists.

   (b) Prove that if a poset L has a least element, then that element is unique.

6. We naturally order the numbers in $A_m = \{1, 2, ..., m\}$ with "less than or equal to," which is a partial ordering. We define an ordering, $\preceq$ on the elements of $A_m \times A_n$ by

   $$(a, b) \preceq (a', b') \Leftrightarrow a \leq a' \text{ and } b \leq b'$$

   (a) Prove that $\preceq$ is a partial ordering on $A_m \times A_n$.

   (b) Draw the ordering diagrams for $\preceq$ on $A_2 \times A_2$, $A_2 \times A_3$, and $A_3 \times A_3$.

   (c) In general, how does one determine the least upper bound and greatest lower bound of two elements of $A_m \times A_n$, (a, b) and (a', b')?

   (d) Are there least and/or greatest elements in $A_m \times A_n$?

7. Let $\mathcal{P}_0$ be the set of all subsets T of $S = \{1, 2, \ldots, 9\}$ such that the sum of the elements in T is even. (Note that the empty set $\emptyset$ will be included as an element of $\mathcal{P}_0$.) For instance, $\{2, 3, 6, 7\}$ is in $\mathcal{P}_0$ because $2 + 3 + 6 + 7$ is even, but $\{1, 3, 5, 6\}$ is not in $\mathcal{P}_0$ because $1 + 3 + 5 + 6$ is odd. Consider

the poset $(\mathcal{P}_0, \subseteq)$. Let $A = \{1, 2, 3, 6\}$ and $B = \{2, 3, 6, 7\}$ be elements of $\mathcal{P}_0$.

(a) Explain why $A \cap B$ is not element of the poset.

(b) Use the definitions of the italicized terms and the given partial ordering to complete the following statements:

  (i) $R \in \mathcal{P}_0$ is an *upper bound* of A and B if _____

  (ii) $R \in \mathcal{P}_0$ is the *least element* of $\mathcal{P}_0$ if _____

(c) Find three different upper bounds of A and B.

(d) Find the least upper bound of A and B. If it doesn't exist, explain why not.

13.2 Lattices

In this section, we restrict our discussion to lattices, those posets for which every pair of elements has both a greatest lower bound and least upper bound. We first introduce some notation.

Definition 13.2.1 Join, Meet. Let $(L, \preceq)$ be a poset, and $a, b \in L$. We define:

- $a \vee b$, read "a join b", as the least upper bound of a and b, if it exists. and

- $a \wedge b$, read "a meet b", as the greatest lower bound of a and b, if it exists.

$\Diamond$

Since the join and meet produce a unique result in all cases where they exist, by Theorem 13.1.6, p. 104, we can consider them as binary operations on a set if they always exist. Thus the following definition:

Definition 13.2.2 Lattice. A lattice is a poset $(L, \preceq)$ for which every pair of elements has a greatest lower bound and least upper bound. Since a lattice L is an algebraic system with binary operations $\vee$ and $\wedge$, it is denoted by $[L; \vee, \wedge]$. If we want to make it clear what partial ordering the lattice is based on, we say it is a lattice under $\preceq$. $\Diamond$

Example 13.2.3 The power set of a three element set. Consider the poset $(\mathcal{P}(A), \subseteq)$ we examined in Example 13.1.10, p. 105. It isn't too surprising that every pair of sets had a greatest lower bound and least upper bound. Thus, we have a lattice in this case; and $A \vee B = A \cup B$ and $A \wedge B = A \cap B$. The reader is encouraged to write out the operation tables $[\mathcal{P}(A); \cup, \cap]$. □

Our first concrete lattice can be generalized to the case of any set A, producing the lattice $[\mathcal{P}(A); \vee, \wedge]$, where the join operation is the set operation of union and the meet operation is the operation intersection; that is, $\vee = \cup$ and $\wedge = \cap$.

It can be shown (see the exercises) that the commutative laws, associative laws, idempotent laws, and absorption laws are all true for any lattice. A concrete example of this is clearly $[\mathcal{P}(A); \cup, \cap]$, since these laws hold in the algebra of sets. This lattice also has distributive property in that join is distributive over meet and meet is distributive over join. However, this is not always the case for lattices in general.

Definition 13.2.4 Distributive Lattice. Let $\mathcal{L} = [L; \vee, \wedge]$ be a lattice under $\preceq$. $\mathcal{L}$ is called a distributive lattice if and only if the distributive laws hold; that is, for all $a, b, c \in L$ we have

$$a \vee (b \wedge c) = (a \vee b) \wedge (a \vee c)$$
$$\text{and}$$
$$a \wedge (b \vee c) = (a \wedge b) \vee (a \wedge c)$$

◇

Example 13.2.5 A Nondistributive Lattice. We now give an example of a lattice where the distributive laws do not hold. Let $L = \{\mathbf{0}, a, b, c, \mathbf{1}\}$. We define the partial ordering $\preceq$ on L by the set

$$\{(\mathbf{0},\mathbf{0}),(\mathbf{0},a),(\mathbf{0},b),(\mathbf{0},c),(\mathbf{0},\mathbf{1}),(a,a),(a,\mathbf{1}),(b,b),(b,\mathbf{1}),(c,c),(c,\mathbf{1}),(\mathbf{1},\mathbf{1})\}$$

The operation tables for $\vee$ and $\wedge$ on L are:

$\vee$	$\mathbf{0}$	a	b	c	$\mathbf{1}$		$\wedge$	$\mathbf{0}$	a	b	c	$\mathbf{1}$
$\mathbf{0}$	$\mathbf{0}$	a	b	c	$\mathbf{1}$		$\mathbf{0}$	$\mathbf{0}$	$\mathbf{0}$	$\mathbf{0}$	$\mathbf{0}$	$\mathbf{0}$
a	a	a	$\mathbf{1}$	$\mathbf{1}$	$\mathbf{1}$		a	$\mathbf{0}$	a	$\mathbf{0}$	$\mathbf{0}$	a
b	b	$\mathbf{1}$	b	$\mathbf{1}$	$\mathbf{1}$		b	$\mathbf{0}$	$\mathbf{0}$	b	$\mathbf{0}$	b
c	c	$\mathbf{1}$	$\mathbf{1}$	c	$\mathbf{1}$		c	$\mathbf{0}$	$\mathbf{0}$	$\mathbf{0}$	c	c
$\mathbf{1}$	$\mathbf{1}$	$\mathbf{1}$	$\mathbf{1}$	$\mathbf{1}$	$\mathbf{1}$		$\mathbf{1}$	$\mathbf{0}$	a	b	c	$\mathbf{1}$

Since every pair of elements in L has both a join and a meet, $[L; \vee, \wedge]$ is a lattice (under divides). Is this lattice distributive? We note that: $a \vee (c \wedge b) = a \vee \mathbf{0} = a$ and $(a \vee c) \wedge (a \vee b) = \mathbf{1} \wedge \mathbf{1} = \mathbf{1}$. Therefore, $a \vee (b \wedge c) \neq (a \vee b) \wedge (a \vee c)$ for some values of $a, b, c \in L$. Thus, this lattice is not distributive. □

Our next observation uses the term "sublattice", which we have not defined at this point, but we would hope that you could anticipate a definition, and we will leave it as an exercise to do so.

It can be shown that a lattice is nondistributive if and only if it contains a sublattice isomorphic to one of the lattices in Figure 13.2.6, p. 108. The ordering diagram on the right of this figure, produces the **diamond lattice**, which is precisely the one that is defined in Example 13.2.5, p. 108. The lattice based on the left hand poset is called the **pentagon lattice**.

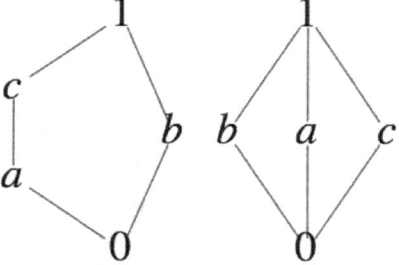

Figure 13.2.6 Nondistributive lattices, the pentagon and diamond lattices

Exercises

1. Let L be the set of all propositions generated by p and q. What are the meet and join operations in this lattice under implication? What are the maximum and minimum elements?
2. Which of the posets in Exercise 13.1.3, p. 106 are lattices? Which of the lattices are distributive?

3.
   (a) State the commutative laws, associative laws, idempotent laws, and absorption laws for lattices.

   (b) Prove laws you stated.

4. Demonstrate that the pentagon lattice is nondistributive.

5. What is a reasonable definition of the term **sublattice**?

6. Let $[L; \vee, \wedge]$ be a lattice based on a partial ordering $\preceq$. Prove that if $a, b, c \in L$,

   (a) $a \preceq a \vee b$.

   (b) $a \wedge b \preceq a$.

   (c) $b \preceq a$ and $c \preceq a \Rightarrow b \vee c \preceq a$.

13.3 Boolean Algebras

In order to define a Boolean algebra, we need the additional concept of complementation. A lattice must have both a greatest element and a least element in order for complementation to take place. The following definition will save us some words in the rest of this section.

Definition 13.3.1 Bounded Lattice. A bounded lattice is a lattice that contains both a least element and a greatest element. ◊

We use the symbols **0** and **1** for the least and greatest elements of a bounded lattice in the remainder of this section.

Definition 13.3.2 The Complement of a Lattice Element. Let $[L; \vee, \wedge]$ be a bounded lattice. If $a \in L$, then a has a complement if there exists $b \in L$ such that

$$a \vee b = \mathbf{1}$$
and
$$a \wedge b = \mathbf{0}$$

◊

Notice that by the commutative laws for lattices, if b complements a, then a complements b.

Definition 13.3.3 Complemented Lattice. Let $\mathcal{L} = [L; \vee, \wedge]$ be a bounded lattice. $\mathcal{L}$ is a complemented lattice if every element of L has a complement in L. ◊

Example 13.3.4 Set Complement is a Complement. In Chapter 1, we defined the complement of a subset of any universe. This turns out to be a concrete example of the general concept we have just defined, but we will reason through why this is the case here. Let $L = \mathcal{P}(A)$, where $A = \{a, b, c\}$. Then $[L; \cup, \cap]$ is a bounded lattice with $\mathbf{0} = \emptyset$ and $\mathbf{1} = A$. To find the complement, if it exists, of $B = \{a, b\} \in L$, for example, we want D such that

$$\{a, b\} \cap D = \emptyset$$
and
$$\{a, b\} \cup D = A$$

It's not too difficult to see that $D = \{c\}$, since we need to include c to make the first condition true and can't include a or b if the second condition is to be true. Of course this is precisely how we defined A^c in Chapter 1. Since it can be shown that each element of L has a complement (see Exercise 1), $[L; \cup, \cap]$ is a complemented lattice. Note that if A is any set and $L = \mathcal{P}(A)$, then $[L; \cup, \cap]$ is a complemented lattice where the complement of $B \in L$ is $B^c = A - B$. □

In Example 13.3.4, p. 109, we observed that the complement of each element of L is unique. Is this always true in a complemented lattice? The answer is no. Consider the following.

Example 13.3.5 A Lattice for which complements are not unique. Let $L = \{1, 2, 3, 5, 30\}$ and consider the lattice $[L; \vee, \wedge]$ (under "divides"). The least element of L is 1 and the greatest element is 30. Let us compute the complement of the element $a = 2$. We want to determine $\bar{a}$ such that $2 \wedge \bar{a} = 1$ and $2 \vee \bar{a} = 30$. Certainly, $\bar{a} = 3$ works, but so does $\bar{a} = 5$, so the complement of $a = 2$ in this lattice is not unique. However, $[L; \vee, \wedge]$ is still a complemented lattice since each element does have at least one complement. □

Definition 13.3.6 Complementation as an operation. If a complemented lattice has the property that the complement of every element is unique, then we consider complementation to be a unary operation. The usual notation for the complement of a is $\bar{a}$. ◇

The following theorem gives us an insight into when uniqueness of complements occurs.

Theorem 13.3.7 One condition for unique complements. *If $[L; \vee, \wedge]$ is a complemented, distributive lattice, then the complement of each element $a \in L$ is unique.*

Proof. Let $a \in L$ and assume to the contrary that a has two complements, namely a_1 and a_2. Then by the definition of complement,

$$a \wedge a_1 = 0 \text{ and } a \vee a_1 = 1,$$
$$\text{and}$$
$$a \wedge a_2 = 0 \text{ and } a \vee a_2 = 1,$$

Then

$$a_1 = a_1 \wedge \mathbf{1} = a_1 \wedge (a \vee a_2)$$
$$= (a_1 \wedge a) \vee (a_1 \wedge a_2)$$
$$= \mathbf{0} \vee (a_1 \wedge a_2)$$
$$= a_1 \wedge a_2$$

On the other hand,

$$a_2 = a_2 \wedge \mathbf{1} = a_2 \wedge (a \vee a_1)$$
$$= (a_2 \wedge a) \vee (a_2 \wedge a_1)$$
$$= \mathbf{0} \vee (a_2 \wedge a_1)$$
$$= a_2 \wedge a_1$$
$$= a_1 \wedge a_2$$

Hence $a_1 = a_2$, which contradicts the assumption that a has two different complements. ■

Definition 13.3.8 Boolean Algebra. A Boolean algebra is a lattice that contains a least element and a greatest element and that is both complemented and distributive. The notation $[B; \vee, \wedge, ^-]$ is used to denote the boolean algebra with operations join, meet and complementation. ◇

13.3. BOOLEAN ALGEBRAS

Since the complement of each element in a Boolean algebra is unique (by Theorem 13.3.7, p. 110), complementation is a valid unary operation over the set under discussion, which is why we will list it together with the other two operations to emphasize that we are discussing a set together with three operations. Also, to help emphasize the distinction between lattices and lattices that are Boolean algebras, we will use the letter B as the generic symbol for the set of a Boolean algebra; that is, $[B; \vee, \wedge, \bar{}\,]$ will stand for a general Boolean algebra.

Example 13.3.9 Boolean Algebra of Sets. Let A be any set, and let $B = \mathcal{P}(A)$. Then $[B; \cup, \cap, {}^c]$ is a Boolean algebra. Here, ^{c} stands for the complement of an element of B with respect to A, $A - B$.

This is a key example for us since all finite Boolean algebras and many infinite Boolean algebras look like this example for some A. In fact, a glance at the basic Boolean algebra laws in Table 13.3.11, p. 111, in comparison with the set laws of Chapter 4 and the basic laws of logic of Chapter 3, indicates that all three systems behave the same; that is, they are isomorphic. □

Example 13.3.10 Divisors of 30. A somewhat less standard example of a boolean algebra is derived from the lattice of divisors of 30 under the relation "divides". If you examine the ordering diagram for this lattice, you see that it is structurally the same as the boolean algebra of subsets of a three element set. Therefore, the join, meet and complementation operations act the same as union, intersection and set complementation. We might conjecture that the lattice of divisors of any integer will produce a boolean algebra, but it is only the case of certain integers. Try out a few integers to see if you can identify what is necessary to produce a boolean algebra. □

Table 13.3.11 Basic Boolean Algebra Laws

Commutative Laws	$a \vee b = b \vee a$	$a \wedge b = b \wedge a$
Associative Laws	$a \vee (b \vee c) = (a \vee b) \vee c$	$a \wedge (b \wedge c) = (a \wedge b) \wedge c$
Distributive Laws	$a \wedge (b \vee c) = (a \wedge b) \vee (a \wedge c)$	$a \vee (b \wedge c) = (a \vee b) \wedge (a \vee c)$
Identity Laws	$a \vee 0 = 0 \vee a = a$	$a \wedge 1 = 1 \wedge a = a$
Complement Laws	$a \vee \bar{a} = 1$	$a \wedge \bar{a} = 0$
Idempotent Laws	$a \vee a = a$	$a \wedge a = a$
Null Laws	$a \vee 1 = 1$	$a \wedge 0 = 0$
Absorption Laws	$a \vee (a \wedge b) = a$	$a \wedge (a \vee b) = a$
DeMorgan's Laws	$\overline{a \vee b} = \bar{a} \wedge \bar{b}$	$\overline{a \wedge b} = \bar{a} \vee \bar{b}$
Involution Law	$\bar{\bar{a}} = a$	

The "pairings" of the boolean algebra laws reminds us of the principle of duality, which we state for a Boolean algebra.

Definition 13.3.12 Principle of Duality for Boolean Algebras. Let $\mathcal{B} = [B; \vee, \wedge, {}^c]$ be a Boolean algebra under $\preceq$, and let S be a true statement for $\mathcal{B}$. If S^* is obtained from S by replacing $\preceq$ with $\succeq$ (this is equivalent to turning the graph upside down), $\vee$ with $\wedge$, $\wedge$ with $\vee$, $\mathbf{0}$ with $\mathbf{1}$, and $\mathbf{1}$ with $\mathbf{0}$, then S^* is also a true statement in $\mathcal{B}$. ◊

Exercises

1. Determine the complement of each element $B \in L$ in Example 13.3.4, p. 109. Is this lattice a Boolean algebra? Why?

2.
    (a) Determine the complement of each element of D_6 in $[D_6; \vee, \wedge]$.

(b) Repeat part a using the lattice in Example 13.2.5, p. 108.

(c) Repeat part a using the lattice in Exercise 13.1.1, p. 105.

(d) Are the lattices in parts a, b, and c Boolean algebras? Why?

3. Determine which of the lattices of Exercise 13.1.3, p. 106 of Section 13.1 are Boolean algebras.

4. Let $A = \{a, b\}$ and $B = \mathcal{P}(A)$.

   (a) Prove that $[B; \cup, \cap, {}^c]$ is a Boolean algebra.

   (b) Write out the operation tables for the Boolean algebra.

5. It can be shown that the following statement, S, holds for any Boolean algebra $[B; \vee, \wedge, -] : (a \wedge b) = a$ if and only if $a \leq b$.

   (a) Write the dual, S^*, of the statement S.

   (b) Write the statement S and its dual, S^*, in the language of sets.

   (c) Are the statements in part b true for all sets?

   (d) Write the statement S and its dual, S^*, in the language of logic.

   (e) Are the statements in part d true for all propositions?

6. State the dual of:

   (a) $a \vee (b \wedge a) = a$.

   (b) $a \vee \left(\overline{(\bar{b} \vee a) \wedge b}\right) = 1$.

   (c) $\left(\overline{a \wedge \bar{b}}\right) \wedge b = a \vee b$.

7. Formulate a definition for isomorphic Boolean algebras.

8. For what positive integers, n, does the lattice $[D_n, |]$ produce a boolean algebra?

13.4 Atoms of a Boolean Algebra

In this section we will look more closely at something we've hinted at, which is that every finite Boolean algebra is isomorphic to an algebra of sets. We will show that every finite Boolean algebra has 2^n elements for some n with precisely n generators, called atoms.

Consider the Boolean algebra $[B; \vee, \wedge, ^-]$, whose ordering diagram is depicted in Figure 13.4.1, p. 113

13.4. ATOMS OF A BOOLEAN ALGEBRA

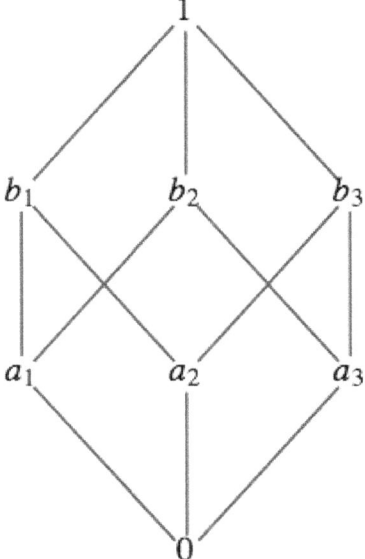

Figure 13.4.1 Illustration of the atom concept

We note that $1 = a_1 \vee a_2 \vee a_3$, $b_1 = a_1 \vee a_2$, $b_2 = a_1 \vee a_3$, and $b_3 = a_2 \vee a_3$; that is, each of the elements above level one can be described completely and uniquely in terms of the elements on level one. The a_i's have uniquely generated the non-least elements of B much like a basis in linear algebra generates the elements in a vector space. We also note that the a_i's are the immediate successors of the minimum element, 0. In any Boolean algebra, the immediate successors of the minimum element are called **atoms**. For example, let A be any nonempty set. In the Boolean algebra $[\mathcal{P}(A); \cup, \cap, {}^c]$ (over $\subseteq$), the singleton sets are the generators, or atoms, of the algebraic structure since each element $\mathcal{P}(A)$ can be described completely and uniquely as the join, or union, of singleton sets.

Definition 13.4.2 Atom. A non-least element a in a Boolean algebra $[B; \vee, \wedge, ^-]$ is called an atom if for every $x \in B$, $x \wedge a = a$ or $x \wedge a = 0$. ◊

The condition that $x \wedge a = a$ tells us that x is a successor of a; that is, $a \preceq x$, as depicted in Figure 13.4.3, p. 114(a)

The condition $x \wedge a = 0$ is true only when x and a are "not connected." This occurs when x is another atom or if x is a successor of atoms different from a, as depicted in Figure 13.4.3, p. 114(b).

Figure 13.4.3 Conditions for an atom

An alternate definition of an atom is based on the concept of "covering."

Definition 13.4.4 The Covering Relation. Given a Boolean algebra $[B; \vee, \wedge, ^-]$, let $x, z \in B$. We say that z **covers** x iff $x \prec z$ and there does not exist $y \in B$ with $x \prec y \prec z$. ◊

It can be proven that the atoms of Boolean algebra are precisely those elements that cover the zero element.

The set of atoms of the Boolean algebra $[D_{30}; \vee, \wedge, ^-]$ is $M = \{2, 3, 5\}$. To see that $a = 2$ is an atom, let x be any non-least element of D_{30} and note that one of the two conditions $x \wedge 2 = 2$ or $x \wedge 2 = 1$ holds. Of course, to apply the definition to this Boolean algebra, we must remind ourselves that in this case the 0-element is 1, the operation $\wedge$ is greatest common divisor, and the poset relation is "divides." So if $x = 10$, we have $10 \wedge 2 = 2$ (or $2 \mid 10$), so Condition 1 holds. If $x = 15$, the first condition is not true. (Why?) However, Condition 2, $15 \wedge 2 = 1$, is true. The reader is encouraged to show that 3 and 5 also satisfy the definition of an atom. Next, if we should compute the join (the least common multiple in this case) of all possible combinations of the atoms 2, 3, and 5 to generate all nonzero (non-1 in this case) elements of D_{30}. For example, $2 \vee 3 \vee 5 = 30$ and $2 \vee 5 = 10$. We state this concept formally in the following theorem, which we give without proof.

Theorem 13.4.5 Let $\mathcal{B} = [B; \vee, \wedge,]$ be any finite Boolean algebra. Let $A = \{a_1, a_2, \ldots, a_n\}$ be the set of all atoms of $\mathcal{B}$. Then every element in B can be expressed uniquely as the join of a subset of A.

The least element in relation to this theorem bears noting. If we consider the empty set of atoms, we would consider the join of elements in the empty set to be the least element. This makes the statement of the theorem above a bit more tidy since we don't need to qualify what elements can be generated from atoms.

We now ask ourselves if we can be more definitive about the structure of different Boolean algebras of a given order. Certainly, the Boolean algebras $[D_{30}; \vee, \wedge, \wedge ^-]$ and $[\mathcal{P}(A); \cup, \cap, \ ^c]$ have the same graph (that of Figure 13.4.1, p. 113), the same number of atoms, and, in all respects, look the same except for the names of the elements and the operations. In fact, when we apply corresponding operations to corresponding elements, we obtain corresponding results. We know from Chapter 11 that this means that the two structures are isomorphic as Boolean algebras. Furthermore, the graphs of these examples are exactly the same as that of Figure 13.4.1, p. 113, which is an arbitrary Boolean algebra of order $8 = 2^3$.

In these examples of a Boolean algebra of order 8, we note that each had 3 atoms and $2^3 = 8$ number of elements, and all were isomorphic to $[\mathcal{P}(A); \cup, \cap, \ ^c]$, where $A = \{a, b, c\}$. This leads us to the following questions:

13.4. ATOMS OF A BOOLEAN ALGEBRA

- Are there any different (nonisomorphic) Boolean algebras of order 8?

- What is the relationship, if any, between finite Boolean algebras and their atoms?

- How many different (nonisomorphic) Boolean algebras are there of order 2? Order 3? Order 4? etc.

The answers to these questions are given in the following theorem and corollaries.

Theorem 13.4.6 *Let $\mathcal{B} = [B; \vee, \wedge, -]$ be any finite Boolean algebra, and let A be the set of all atoms of $\mathcal{B}$. Then $[\mathcal{P}(A); \cup, \cap, {}^c]$ is isomorphic to $[B; \vee, \wedge, -]$*
Proof. An isomorphism that serves to prove this theorem is $T : \mathcal{P}(A) \to B$ defined by $T(S) = \bigvee_{a \in S} a$, where $T(\emptyset)$ is interpreted as the zero of $\mathcal{B}$. We leave it to the reader to prove that this is indeed an isomorphism. ■

Corollary 13.4.7 *Every finite Boolean algebra $\mathcal{B} = [B; \vee, \wedge, {}^-]$ has 2^n elements for some positive integer n.*
Proof. Let A be the set of all atoms of $\mathcal{B}$ and let $|A| = n$. Then there are exactly 2^n elements (subsets) in $\mathcal{P}(A)$, and by Theorem 13.4.6, p. 115, $[B; \vee, \wedge, {}^-]$ is isomorphic to $[\mathcal{P}(A); \cup, \cap {}^c]$ and must also have 2^n elements. ■

Corollary 13.4.8 *All Boolean algebras of order 2^n are isomorphic to one another.*
Proof.

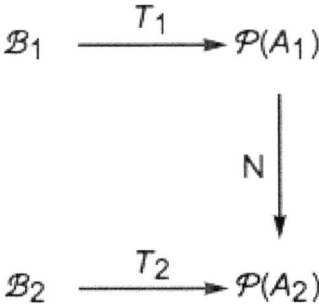

Figure 13.4.9 Isomorphisms to be combined

Every Boolean algebra of order 2^n is isomorphic to $[\mathcal{P}(A); \cup, \cap, {}^c]$ when $|A| = n$. Hence, if $\mathcal{B}_1$ and $\mathcal{B}_2$ each have 2^n elements, they each have n atoms. Suppose their sets of atoms are A_1 and A_2, respectively. We know there are isomorphisms T_1 and T_2, where $T_i : \mathcal{B}_i \to \mathcal{P}(A_i)$, $i = 1, 2$. In addition we have an isomorphism, N from $\mathcal{P}(A_1)$ into $\mathcal{P}(A_2)$, which we ask you to prove in Exercise 13.4.9, p. 116. We can combine these isomorphisms to produce the isomorphism $T_2^{-1} \circ N \circ T_1 : \mathcal{B}_1 \to \mathcal{B}_2$, which proves the corollary. ■

The above theorem and corollaries tell us that we can only have finite Boolean algebras of orders $2^1, 2^2, 2^3, ..., 2^n$, and that all finite Boolean algebras of any given order are isomorphic. These are powerful tools in determining the structure of finite Boolean algebras. In the next section, we will discuss one of the easiest ways of describing a Boolean algebra of any given order.

Exercises

1.
    (a) Show that $a = 2$ is an atom of the Boolean algebra $[D_{30}; \vee, \wedge, -]$.

    (b) Repeat part a for the elements 3 and 5 of D_{30}.

(c) Verify Theorem 13.4.5, p. 114 for the Boolean algebra $[D_{30}; \vee, \wedge, -]$.

2. Let $A = \{a, b, c\}$.

   (a) Rewrite the definition of atom for $[\mathcal{P}(A); \cup, \cap, c]$. What does $a \leq x$ mean in this example?

   (b) Find all atoms of $[\mathcal{P}(A); \cup, \cap, c]$.

   (c) Verify Theorem 13.4.5, p. 114 for $[\mathcal{P}(A); c, \cup, \cap]$.

3. Verify Theorem 13.4.6, p. 115 and its corollaries for the Boolean algebras in Exercises 1 and 2 of this section.

4. Give an example of a Boolean algebra of order 16 whose elements are certain subsets of the set $\{1, 2, 3, 4, 5, 6, 7\}$

5. Corollary 13.4.7, p. 115 implies that there do not exist Boolean algebras of orders 3, 5, 6, 7, 9, etc. (orders different from 2^n). Without this corollary, directly show that we cannot have a Boolean algebra of order 3.

   Hint. Assume that $[B; \vee, \wedge, -]$ is a Boolean algebra of order 3 where $B = \{0, x, 1\}$ and show that this cannot happen by investigating the possibilities for its operation tables.

6.
   (a) There are many different, yet isomorphic, Boolean algebras with two elements. Describe one such Boolean algebra that is derived from a power set, $\mathcal{P}(A)$, under $\subseteq$. Describe a second that is described from D_n, for some $n \in P$, under "divides."

   (b) Since the elements of a two-element Boolean algebra must be the greatest and least elements, 1 and 0, the tables for the operations on $\{0, 1\}$ are determined by the Boolean algebra laws. Write out the operation tables for $[\{0, 1\}; \vee, \wedge, -]$.

7. Find a Boolean algebra with a countably infinite number of elements.

8. Prove that the direct product of two Boolean algebras is a Boolean algebra.

   Hint. "Copy" the corresponding proof for groups in Section 11.6.

9. Prove if two finite sets A_1 and A_2 both have n elements then $[\mathcal{P}(A_1); \cup, \cap, {}^c]$ is isomorphic to $[\mathcal{P}(A_2); \cup, \cap, {}^c]$

10. Prove an element of a Boolean algebra is an atom if and only if it covers the zero element.

13.5 Finite Boolean Algebras as n-tuples of 0's and 1's

From the previous section we know that all finite Boolean algebras are of order 2^n, where n is the number of atoms in the algebra. We can therefore completely describe every finite Boolean algebra by the algebra of power sets. Is there a more convenient, or at least an alternate way, of defining finite Boolean algebras? In Chapter 11 we found that we could produce new groups by taking Cartesian products of previously known groups. We imitate this process for Boolean algebras.

The simplest nontrivial Boolean algebra is the Boolean algebra on the set $B_2 = \{0, 1\}$. The ordering on B_2 is the natural one, $0 \leq 0, 0 \leq 1, 1 \leq 1$. If we treat 0 and 1 as the truth values "false" and "true," respectively, we see that the Boolean operations $\vee$(join) and $\wedge$(meet) are nothing more than the logical operation with the same symbols. The Boolean operation, $-$, (complementation) is the logical ¬ (negation). In fact, this is why these symbols were chosen as the names of the Boolean operations. The operation tables for $[B_2; \vee, \wedge, -]$ are simply those of "or," "and," and "not," which we repeat here.

$\vee$	0	1		$\wedge$	0	1		u	$\bar{u}$
0	0	1		0	0	0		0	1
1	1	1		1	0	1		1	0

By Theorem 13.4.6, p. 115 and its corollaries, all Boolean algebras of order 2 are isomorphic to this one.

We know that if we form $B_2 \times B_2 = B_2^2$ we obtain the set $\{(0,0), (0,1), (1,0), (1,1)\}$, a set of order 4. We define operations on B_2^2 the natural way, namely componentwise, so that $(0,1) \vee (1,1) = (0 \vee 1, 1 \vee 1) = (1,1)$, $(0,1) \wedge (1,1) = (0 \wedge 1, 1 \wedge 1) = (0,1)$ and $\overline{(0,1)} = (\bar{0}, \bar{1}) = (1,0)$. We claim that B_2^2 is a Boolean algebra under the componentwise operations. Hence, $[B_2^2; \vee, \wedge, ^-]$ is a Boolean algebra of order 4. Since all Boolean algebras of order 4 are isomorphic to one other, we have found a simple way of describing all Boolean algebras of order 4.

It is quite clear that we can describe any Boolean algebra of order 8 by considering $B_2 \times B_2 \times B_2 = B_2^3$ and, more generally, any Boolean algebra of order 2^n with $B_2^n = B_2 \times B_2 \times \cdots \times B_2$ (n factors).

Exercises

1.
    (a) Write out the operation tables for $[B_2^2; \vee, \wedge, -]$.

    (b) Draw the Hasse diagram for $[B_2^2; \vee, \wedge, -]$ and compare your results with Figure 6.0.1, p. 12.

    (c) Find the atoms of this Boolean algebra.

2.
    (a) Write out the operation tables for $[B_2^3; \vee, \wedge, -]$.

    (b) Draw the Hasse diagram for $[B_2^3; \vee, \wedge, -]$

3.
    (a) List all atoms of B_2^4.

    (b) Describe the atoms of $B_2^n, n \geq 1$.

4. Theorem 13.4.6, p. 115 tells us we can think of any finite Boolean algebra in terms of sets. In Chapter 4, we defined minsets 4.0.1, p. 7 and minset normal form 4.0.2, p. 7. Rephrase these definitions in the language of Boolean algebra. The generalization of minsets are called **minterms**.

13.6 Boolean Expressions

In this section, we will use our background from the previous sections and set theory to develop a procedure for simplifying Boolean expressions. This procedure has considerable application to the simplification of circuits in switching theory or logical design.

Definition 13.6.1 Boolean Expression. Let $[B; \vee, \wedge, -]$ be any Boolean algebra, and let $x_1, x_2, \ldots, x_k$ be variables in B; that is, variables that can assume values from B. A Boolean expression generated by $x_1, x_2, \ldots, x_k$ is any valid combination of the x_i and the elements of B with the operations of meet, join, and complementation. ◊

This definition is the analog of the definition of a proposition generated by a set of propositions, presented in Section 3.2.

Each Boolean expression generated by k variables, $e(x_1, \ldots, x_k)$, defines a function $f : B^k \to B$ where $f(a_1, \ldots, a_k) = e(a_1, \ldots, a_k)$. If B is a finite Boolean algebra, then there are a finite number of functions from B^k into B. Those functions that are defined in terms of Boolean expressions are called Boolean functions. As we will see, there is an infinite number of Boolean expressions that define each Boolean function. Naturally, the "shortest" of these expressions will be preferred. Since electronic circuits can be described as Boolean functions with $B = B_2$, this economization is quite useful.

In what follows, we make use of Exercise 7.7, p. 13 in Section 7.1 for counting number of functions.

Example 13.6.2 Two variables over B_2. Consider any Boolean algebra of order 2, $[B; \vee, \wedge, -]$. How many functions $f : B^2 \to B$ are there? First, all Boolean algebras of order 2 are isomorphic to $[B_2; \vee, \wedge, -]$ so we want to determine the number of functions $f : B_2^2 \to B_2$. If we consider a Boolean function of two variables, x_1 and x_2, we note that each variable has two possible values 0 and 1, so there are 2^2 ways of assigning these two values to the $k = 2$ variables. Hence, the table below has $2^2 = 4$ rows. So far we have a table such as this one:

x_1	x_2	$f(x_1, x_2)$
0	0	?
0	1	?
1	0	?
1	1	?

How many possible different functions can there be? To list a few: $f_1(x_1, x_2) = x_1$, $f_2(x_1, x_2) = x_2$, $f_3(x_1, x_2) = x_1 \vee x_2$, $f_4(x_1, x_2) = (x_1 \wedge \overline{x_2}) \vee x_2$, $f_5(x_1, x_2) = x_1 \wedge x_2 \vee \overline{x_2}$, etc. Each of these will fill in the question marks in the table above. The tables for f_1 and f_3 are

x_1	x_2	$f_1(x_1, x_2)$	x_1	x_2	$f_3(x_1, x_2)$
0	0	0	0	0	0
0	1	0	0	1	1
1	0	1	1	0	1
1	1	1	1	1	1

Two functions are different if and only if their tables are different for at least one row. Of course by using the basic laws of Boolean algebra we can see that $f_3 = f_4$. Why? So if we simply list by brute force all "combinations" of x_1 and x_2 we will obtain unnecessary duplication. However, we note that for any combination of the variables x_1, and x_2 there are only two possible values for $f(x_1, x_2)$, namely 0 or 1. Thus, we could write $2^4 = 16$ different functions on 2 variables. □

Now, let's count the number of different Boolean functions in a more general setting. We will consider two cases: first, when $B = B_2$, and second, when B is any finite Boolean algebra with 2^n elements.

Let $B = B_2$. Each function $f : B^k \to B$ is defined in terms of a table having 2^k rows. Therefore, since there are two possible images for each element of B^k, there are 2 raised to the 2^k, or 2^{2^k} different functions. We will show that every one of these functions is a Boolean function.

Now suppose that $|B| = 2^n > 2$. A function from B^k into B can still be defined in terms of a table. There are $|B|^k$ rows to each table and $|B|$ possible

13.6. BOOLEAN EXPRESSIONS

images for each row. Therefore, there are 2^n raised to the power 2^{nk} different functions. We will show that if $n > 1$, not every one of these functions is a Boolean function.

Since all Boolean algebras are isomorphic to a Boolean algebra of sets, the analogues of statements in sets are useful in Boolean algebras.

Definition 13.6.3 Minterm. A Boolean expression generated by $x_1, x_2, \ldots, x_k$ that has the form
$$\bigwedge_{i=1}^{k} y_i,$$
where each y_i may be either x_i or $\bar{x_i}$ is called a minterm generated by $x_1, x_2, \ldots, x_k$. We use the notation $M_{\delta_1 \delta_2 \cdots \delta_k}$ for the minterm generated by $x_1, x_2, \ldots, x_k$, where $y_i = x_i$ if $\delta_i = 1$ and $y_i = \bar{x}_i$ if $\delta_i = 0$ ◇

An example of the notation is that $M_{110} = x_1 \wedge x_2 \wedge \bar{x}_3$.

By a direct application of the Rule of Products we see that there are 2^k different minterms generated by $x_1, \ldots, x_k$.

Definition 13.6.4 Minterm Normal Form. A Boolean expression generated by $x_1, \ldots, x_k$ is in minterm normal form if it is the join of expressions of the form $a \wedge m$, where $a \in B$ and m is a minterm generated by $x_1, \ldots, x_k$. That is, it is of the form
$$\bigvee_{j=1}^{p} (a_j \wedge m_j) \tag{13.6.1}$$
where $p = 2^k$, and $m_1, m_2, \ldots, m_p$ are the minterms generated by $x_1, \ldots, x_k$.
◇

Note 13.6.5

- We seem to require every minterm generated by $x_1, \ldots, x_k$, in (13.6.1), and we really do. However, some of the values of a_j can be **0**, which effectively makes the corresponding minterm disappear.

- If $B = B_2$, then each a_j in a minterm normal form is either 0 or 1. Therefore, $a_j \wedge m_j$ is either 0 or m_j.

Theorem 13.6.6 Uniqueness of Minterm Normal Form. *Let $e(x_1, \ldots, x_k)$ be a Boolean expression over B. There exists a unique minterm normal form $M(x_1, \ldots, x_k)$ that is equivalent to $e(x_1, \ldots, x_k)$ in the sense that e and M define the same function from B^k into B.*

The uniqueness in this theorem does not include the possible ordering of the minterms in M (commonly referred to as "uniqueness up to the order of minterms"). The proof of this theorem would be quite lengthy, and not very instructive, so we will leave it to the interested reader to attempt. The implications of the theorem are very interesting, however.

If $|B| = 2^n$, then there are 2^n raised to the 2^k different minterm normal forms. Since each different minterm normal form defines a different function, there are a like number of Boolean functions from B^k into B. If $B = B_2$, there are as many Boolean functions (2 raised to the 2^k) as there are functions from B^k into B, since there are 2 raised to the 2^n functions from B^k into B. The significance of this result is that any desired function can be realized using electronic circuits having 0 or 1 (off or on, positive or negative) values.

More complex, multivalued circuits corresponding to boolean algebras with more than two values would not have this flexibility because of the number of minterm normal forms, and hence the number of boolean functions, is strictly less than the number of functions.

We will close this section by examining minterm normal forms for expressions over B_2, since they are a starting point for circuit economization.

Example 13.6.7 Consider the Boolean expression $f(x_1, x_2) = x_1 \vee \overline{x_2}$. One method of determining the minterm normal form of f is to think in terms of sets. Consider the diagram with the usual translation of notation in Figure 13.6.8, p. 120. Then

$$f(x_1, x_2) = (\overline{x_1} \wedge \overline{x_2}) \vee (x_1 \wedge \overline{x_2}) \vee (x_1 \wedge x_2)$$
$$= M_{00} \vee M_{10} \vee M_{11}$$

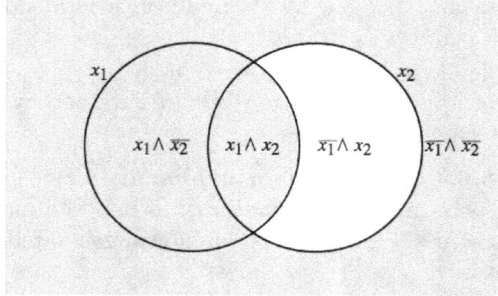

Figure 13.6.8 Visualization of minterms for $x_1 \vee \bar{x}_2$

□

Table 13.6.9 Definition of the boolean function g

x_1	x_2	x_3	$g(x_1, x_2, x_3)$
0	0	0	1
0	0	1	0
0	1	0	0
0	1	1	1
1	0	0	0
1	0	1	0
1	1	0	1
1	1	1	0

Example 13.6.10 Consider the function $g : B_2^3 \to B_2$ defined by Table 13.6.9, p. 120.

The minterm normal form for g can be obtained by taking the join of minterms that correspond to rows that have an image value of 1. If $g(a_1, a_2, a_3) = 1$, then include the minterm $y_1 \wedge y_2 \wedge y_3$ where

$$y_j = \begin{cases} x_j & \text{if } a_j = 1 \\ \bar{x}_j & \text{if } a_j = 0 \end{cases}$$

Or, to use alternate notation, include $M_{a_1 a_2 a_3}$ in the expression if and only if $g(a_1, a_2, a_3) = 1$

Therefore,

$$g(x_1, x_2, x_3) = (\overline{x_1} \wedge \overline{x_2} \wedge \overline{x_3}) \vee (\overline{x_1} \wedge x_2 \wedge x_3) \vee (x_1 \wedge x_2 \wedge \overline{x_3}).$$

□

The minterm normal form is a first step in obtaining an economical way of expressing a given Boolean function. For functions of more than three variables,

Exercises

1.
   (a) Write the 16 possible functions of Example 13.6.2, p. 118.

   (b) Write out the tables of several of the above Boolean functions to show that they are indeed different.

   (c) Determine the minterm normal forms of

   (i) $g_1(x_1, x_2) = x_1 \vee x_2$,

   (ii) $g_2(x_1, x_2) = \overline{x_1} \vee \overline{x_2}$

   (iii) $g_3(x_1, x_2) = \overline{x_1 \wedge x_2}$

   (iv) $g_4(x_1, x_2) = 1$

2. Consider the Boolean expression $f(x_1, x_2, x_3) = (\overline{x_3} \wedge x_2) \vee (\overline{x_1} \wedge x_3) \vee (x_2 \wedge x_3)$ on $\left[B_2^3; \vee, \wedge, -\right]$.

   (a) Simplify this expression using basic Boolean algebra laws.

   (b) Write this expression in minterm normal form.

   (c) Write out the table for the given function defined by f and compare it to the tables of the functions in parts a and b.

   (d) How many possible different functions in three variables on $[B_2; \vee, \wedge, -]$ are there?

3. Let $[B; \vee, \wedge, -]$ be a Boolean algebra of order 4, and let f be a Boolean function of two variables on B.

   (a) How many elements are there in the domain of f?

   (b) How many different Boolean functions are there of two, variables? Three variables?

   (c) Determine the minterm normal form of $f(x_1, x_2) = x_1 \vee x_2$.

   (d) If $B = \{0, a, b, 1\}$, define a function from B^2 into B that is not a Boolean function.

13.7 A Brief Introduction to Switching Theory and Logic Design

Disclaimer: I'm still looking for a good application for drawing logic gates. The figures here are quite rough.

Early computers relied on many switches to perform the logical operations needed for computation. This was true as late as the 1970's when early personal computers such as the Altair (Figure 13.7.1, p. 122) started to appear. Pioneering computer scientists such as Claude Shannon realized that the operation of these computers could be simplified by making use of an isomorphism between computer circuits and boolean algebra. The term **Switching Theory** was used at the time. Logical gates realized through increasingly smaller and smaller integrated circuits still perform the same functions as in early computers, but using purely electronic means. In this section, we give examples of some switching circuits. Soon afterward, we will transition to the more modern form of circuits that are studied in **Logic Design**, where gates replace switches. Our main goal is to give you an overview of how boolean functions corresponds to any such circuit. We will introduce the common system notation used in logic design and show how it corresponds with the mathematical notation of Boolean algebras. Any computer scientist should be familiar with both systems.

Figure 13.7.1 The Altair Computer, an early PC, by Todd Dailey, Creative Commons

The simplest switching device is the on-off switch. If the switch is closed/ON, current will pass through it; if it is open/OFF, current will not pass through it. If we designate ON by 1, and OFF by 0, we can describe electrical circuits containing switches by Boolean expressions with the variables representing the variable states of switches or the variable bits passing through gates.

The electronics involved in these switches take into account whether we are negating a switch or not. For electromagnetic switches, a magnet is used to control whether the switch is open or closed. The magnets themselves may be controlled by simple ON/OFF switches. There are two types of electromagnetic switches. One is normally open (OFF) when the magnet is not activated, but activating the magnet will close the circuit and the switch is then ON. A separate type of switch corresponds with a negated switch. For that type, the switch is closed when the magnet is not activated, and when the magnet is activated, the switch opens. We won't be overly concerned with the details of these switches or the electronics corresponding to logical gates. We will simply assume they are available to plug into a circuit. For simplicity, we use the inversion symbol on a variable that labels a switch to indicate that it is a switch of the second type, as in Figure 13.7.3, p. 123.

Standby power generators that many people have in their homes use a transfer switch to connect the generator to the home power system. This switch is open (OFF) if there is power coming from the normal municipal power supply. It stays OFF because a magnet is keeping it open. When power is lost, the magnet is no longer activated, and the switch closes and is ON. So the transfer switch is a normally ON switch.

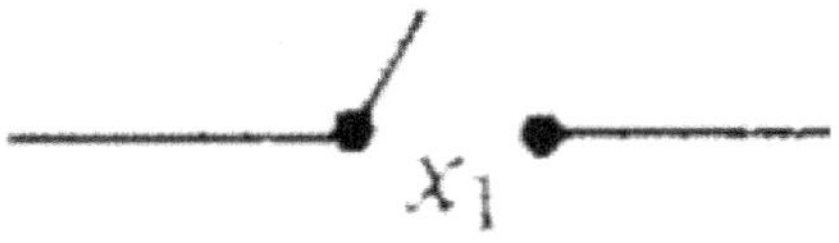

13.7. A BRIEF INTRODUCTION TO SWITCHING THEORY AND LOGIC DESIGN

Figure 13.7.2 Representation of a normally OFF switch controlled by variable x_1

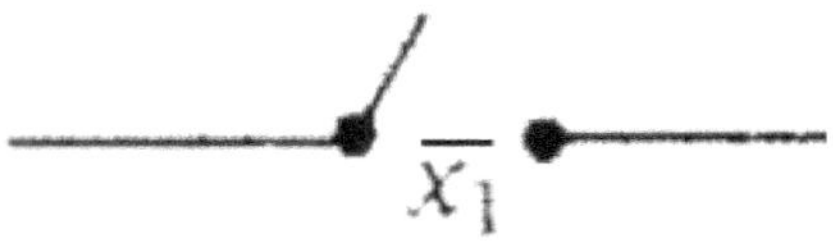

Figure 13.7.3 Representation of a normally ON switch controlled by variable x_1

The standard notation used for Boolean algebra operations in switching theory and logic design is $+$ for join, instead of $\vee$; and $\cdot$ for meet, instead of $\wedge$. Complementation is the same in both notational systems, denoted with an overline.

The expression $x_1 \cdot x_2$ represents the situation in which a series of two switches appears in sequence as in Figure 13.7.4, p. 123. In order for current to flow through the circuit, both switches must be ON; that is, they must both have the value 1. Similarly, a pair of parallel switches, as in Figure 13.7.5, p. 123, is described algebraically by $x_1 + x_2$. Here, current flows through this part of the circuit as long as at least on of the switches is ON.

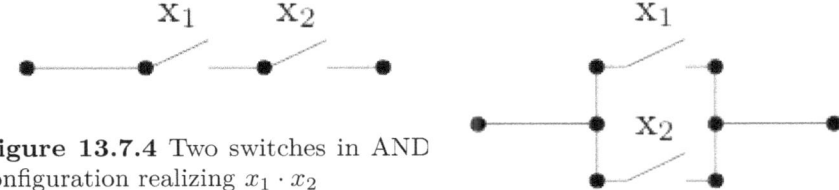

Figure 13.7.4 Two switches in AND configuration realizing $x_1 \cdot x_2$

Figure 13.7.5 Two switches in OR configuration realizing $x_1 + x_2$

All laws and concepts developed previously for Boolean algebras hold. The only change is purely notational. We make the change in this section solely to introduce the reader to another frequently used system of notation.

Many of the laws of Boolean algebra can be visualized thought switching theory. For example, the distributive law of meet over join is expressed as

$$x_1 \cdot (x_2 + x_3) = x_1 \cdot x_2 + x_1 \cdot x_3.$$

The switching circuit analogue of the above statement is that the circuits in the two images below are equivalent. In circuit (b), the presence of two x_1's might represent two electromagnetic switches controlled by the same magnet.

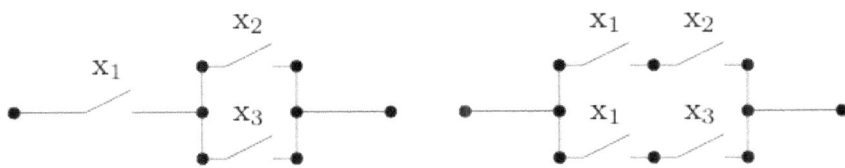

Figure 13.7.6 (a)

Figure 13.7.7 (b)

The circuits in a computer are now composed of large quantities of gates, which serve the same purpose as switches, but can be miniaturized to a great degree. For example, the OR gate, usually drawn as in Figure 13.7.8, p. 124 implements the logical OR function. This happens electronically, but is equivalent to Figure 13.7.5, p. 123. The AND gate, which is equivalent to two sequential switches is shown in Figure 13.7.8, p. 124.

Figure 13.7.8 An OR gate 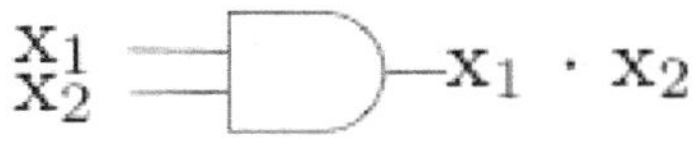 **Figure 13.7.9** An AND gate

The complementation process is represented in a gate diagram by an inverter as pictured in Figure 13.7.10, p. 124.

Figure 13.7.10 Inverter, or NOT gate

When drawing more complex circuits, multiple AND's or OR's are sometimes depicted using a more general gate drawing. For example if we want to depict an OR gate with three inputs that is ON as long as at least one input is ON, we would draw it as in Figure 13.7.11, p. 124, although this would really be two binary gates, as in Figure 13.7.12, p. 124. Both diagrams are realizing the boolean expression $x_1 + x_2 + x_3$. Strictly speaking, the gates in Figure 13.7.12, p. 124 represent $(x_1 + x_2) + x_3$, but the associative law for join tells us that the grouping doesn't matter.

Figure 13.7.11 Simple version of a ternary OR gate

Figure 13.7.12 A ternary OR gate created with binary OR gates

In Figure 13.7.13, p. 124, we show a few other commonly used gates, XOR NAND, and NOR, which correspond to the boolean exressions $x_1 \oplus x_2$, $\overline{x_1 \cdot x_2}$ and $\overline{x_1 + x_2}$, respectively.

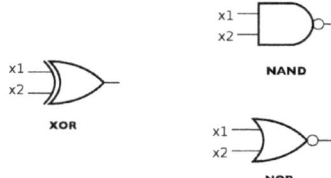

Figure 13.7.13 Other common gates

Let's start with a logic circuit and see how the laws of boolean algebra can help us simplify it.

Example 13.7.14 Simplification of a circuit. Consider the circuit in Figure 13.7.15, p. 125. As usual, we assume that three inputs enter on the left and the output exits on the right.

13.7. A BRIEF INTRODUCTION TO SWITCHING THEORY AND LOGIC DESIGN

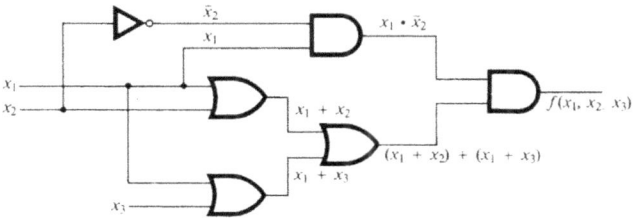

Figure 13.7.15 Initial gate diagram

If we trace the inputs through the gates we see that this circuit realizes the boolean function

$$f(x_1, x_2, x_3) = x_1 \cdot \overline{x_2} \cdot ((x_1 + x_2) + (x_1 + x_3)).$$

We simplify the boolean expression that defines f, simplifying the circuit in so doing. You should be able to identify the laws of Boolean algebra that are used in each of the steps. See Exercise 13.7.1, p. 126.

$$\begin{aligned}
x_1 \cdot \overline{x_2} \cdot ((x_1 + x_2) + (x_1 + x_3)) &= x_1 \cdot \overline{x_2} \cdot (x_1 + x_2 + x_3) \\
&= x_1 \cdot \overline{x_2} \cdot x_1 + x_1 \cdot \overline{x_2} \cdot x_2 + x_1 \cdot \overline{x_2} \cdot x_3 \\
&= x_1 \cdot \overline{x_2} + 0 \cdot x_1 + x_3 \cdot x_1 \cdot \overline{x_2} \\
&= x_1 \cdot \overline{x_2} + x_3 \cdot x_1 \cdot \overline{x_2} \\
&= x_1 \cdot \overline{x_2} \cdot (1 + x_3) \\
&= x_1 \cdot \overline{x_2}
\end{aligned}$$

Therefore, $f(x_1, x_2, x_3) = x_1 \cdot \overline{x_2}$, which can be realized with the much simpler circuit in Figure 13.7.16, p. 125, without using the input x_3.

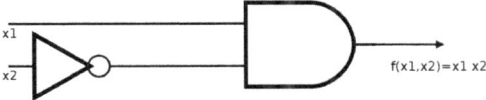

Figure 13.7.16 Simplified gate diagram

□

Next, we start with a table of desired outputs based on three bits of input and design an efficient circuit to realize this output.

Example 13.7.17 Consider the following table of desired outputs for the three input bits x_1, x_2, x_3.

Table 13.7.18 Desired output table

x_1	x_2	x_3	$f(x_1,x_2,x_3)$
0	0	0	0
0	0	1	1
0	1	0	0
0	1	1	0
1	0	0	1
1	0	1	1
1	1	0	0
1	1	1	0

The first step is to write the Minterm Normal Form, p. 119 of f. Since we are working with the two value Boolean algebra, B_2, the constants in each minterm are either 0 or 1, and we simply list the minterms that have a 1. These correspond with the rows of the table above that have an output of 1. We will then attempt to simplify the expression as much as possible.

$$f(x_1,x_2,x_3) = (\overline{x_1} \cdot \overline{x_2} \cdot x_3) + (x_1 \cdot \overline{x_2} \cdot \overline{x_3}) + (x_1 \cdot \overline{x_2} \cdot x_3)$$
$$= \overline{x_2} \cdot ((\overline{x_1} \cdot x_3) + (x_1 \cdot \overline{x_3}) + (x_1 \cdot x_3))$$
$$= \overline{x_2} \cdot ((\overline{x_1} \cdot x_3) + x_1 \cdot (\overline{x_3} + x_3))$$
$$= \overline{x_2} \cdot ((\overline{x_1} \cdot x_3) + x_1)$$

Therefore we can realize our table with the boolean function $f(x_1,x_2,x_3) = \overline{x_2} \cdot ((\overline{x_1} \cdot x_3) + x_1)$. A circuit diagram for this function is Figure 13.7.19, p. 126. But is this the simplest circuit that realizes the table? See Exercise 13.7.3, p. 126.

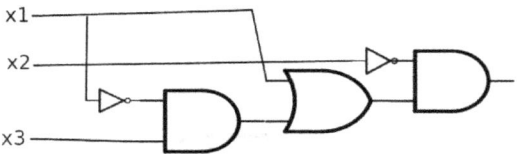

Figure 13.7.19 A realization of the table of desired outputs.

Exercises

1. List the laws of boolean algebra that justify the steps in the simplification of the boolean function $f(x_1,x_2,x_3)$ in Example 13.7.14, p. 124. Some steps use more than one law.

2. Write the following Boolean expression in the notation of logic design.

$$(x_1 \wedge \overline{x_2}) \vee (x_1 \wedge x_2) \vee (\overline{x_1} \wedge x_2).$$

3. Find a further simplification of the boolean function in Example 13.7.17, p. 125, and draw the corresponding gate diagram for the circuit that it realizes.

4. Consider the switching circuit in Figure 13.7.20, p. 127.

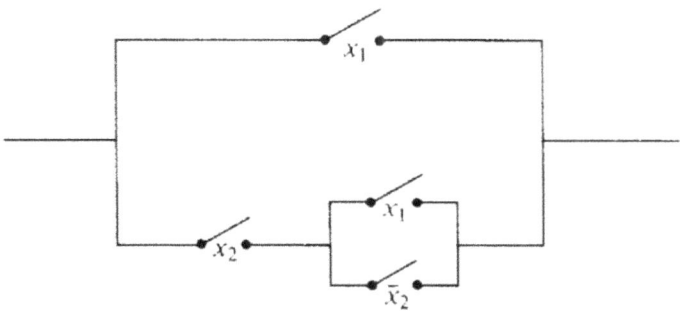

Figure 13.7.20 Can this circuit be simplifed?

(a) Draw the corresponding gate diagram for this circuit.

(b) Construct a table of outputs for each of the eight inputs to this circuit.

(c) Determine the minterm normal of the Boolean function based on the table.

(d) Simplify the circuit as much as possible.

5. Consider the circuit in Figure 13.7.21, p. 127.

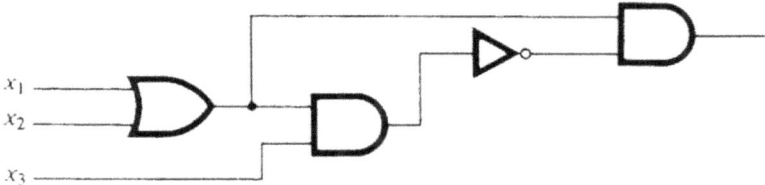

Figure 13.7.21 Can this circuit be simplifed?

(a) Trace the inputs though this circuit and determine the Boolean function that it realizes.

(b) Construct a table of outputs for each of the eight inputs to this circuit.

(c) Find the minterm normal form of f.

(d) Draw the circuit based on the minterm normal form.

(e) Simplify the circuit algebraically and draw the resulting circuit.

6. Consider the Boolean function $f(x_1, x_2, x_3, x_4) = x_1 + (x_2 \cdot (\overline{x_1} + x_4) + x_3 \cdot (\overline{x_2} + \overline{x_4}))$.

(a) Simplify f algebraically.

(b) Draw the gate diagram based on the simplified version of f.

7. Draw a logic circuit using only AND, OR and NOT gates that realizes an XOR gate.

8. Draw a logic circuit using only AND, OR and NOT gates that realizes the Boolean function on three variables that returns 1 if the majority of inputs are 1 and 0 otherwise.

Chapter 14

Monoids and Automata

At first glance, the two topics that we will discuss in this chapter seem totally unrelated. The first is monoid theory, which we touched upon in Chapter 11. The second is automata theory, in which computers and other machines are described in abstract terms. After short independent discussions of these topics, we will describe how the two are related in the sense that each monoid can be viewed as a machine and each machine has a monoid associated with it.

14.1 Monoids

Recall that in Section 11.2, p. 25 we introduced systems called monoids. Here is the formal definition.

Definition 14.1.1 Monoid. A monoid is a set M together with a binary operation $*$ with the properties

- $*$ is associative: $\forall a, b, c \in M$, $(a * b) * c = a * (b * c)$ and

- $*$ has an identity in M: $\exists e \in M$ such that $\forall a \in M$, $a * e = e * a = a$

$\Diamond$

Note 14.1.2 Since the requirements for a group contain the requirements for a monoid, every group is a monoid.

Example 14.1.3 Some Monoids.

(a) The power set of any set together with any one of the operations intersection, union, or symmetric difference is a monoid.

(b) The set of integers, $\mathbb{Z}$, with multiplication, is a monoid. With addition, $\mathbb{Z}$ is also a monoid.

(c) The set of $n \times n$ matrices over the integers, $M_n(\mathbb{Z})$, $n \geq 2$, with matrix multiplication, is a monoid. This follows from the fact that matrix multiplication is associative and has an identity, I_n. This is an example of a noncommutative monoid since there are matrices, A and B, for which $AB \neq BA$.

(d) $[\mathbb{Z}_n; \times_n]$, $n \geqslant 2$, is a monoid with identity 1.

(e) Let X be a nonempty set. The set of all functions from X into X, often denoted X^X, is a monoid over function composition. In Chapter 7, we saw that function composition is associative. The function $i : X \to X$ defined by $i(a) = a$ is the identity element for this system. If $|X|$ is greater than 1 then it is a noncommutative monoid. If X is finite, $|X^X| = |X|^{|X|}$. For example, if $B = \{0,1\}$, $|B^B| = 4$. The functions $z, u, i,$ and t, defined by the graphs in Figure 14.1.4, p. 130, are the elements of B^B. This monoid is not a group. Do you know why?

One reason why B^B is noncommutative is that $t \circ z \neq z \circ t$ because $(t \circ z)(0) = t(z(0)) = t(0) = 1$ while $(z \circ t)(0) = z(t(0)) = z(1) = 0$.

$\square$

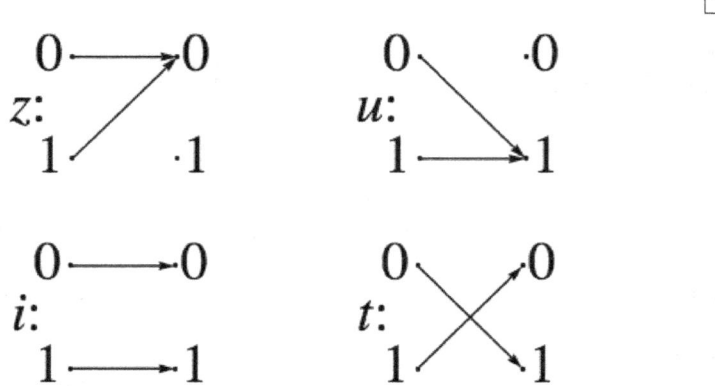

Figure 14.1.4 The functions on B_2

Virtually all of the group concepts that were discussed in Chapter 11 are applicable to monoids. When we introduced subsystems, we saw that a submonoid of monoid M is a subset of M; that is, it is a monoid with the operation of M. To prove that a subset is a submonoid, you can apply the following theorem.

Theorem 14.1.5 Submonoid Test. *Assume $[M;*]$ is a monoid and K is a nonempty subset of M. Then K is a submonoid of M if and only if the following two conditions are met.*

- *If $a, b \in K$, then. $a * b \in K$; i. e., K is closed with under $*$.*

- *The identity of M belongs to K.*

Often we will want to discuss the smallest submonoid that includes a certain subset S of a monoid M. This submonoid can be defined recursively by the following definition.

Definition 14.1.6 Submonoid Generated by a Set. If S is a subset of monoid $[M;*]$, the submonoid generated by S, $\langle S \rangle$, is defined by:.

(a) (Basis) The identity of M belongs to $\langle S \rangle$; and $a \in S \Rightarrow a \in \langle S \rangle$.

(b) (Recursion) $a, b \in \langle S \rangle \Rightarrow a * b \in \langle S \rangle$.

$\diamond$

Note 14.1.7 If $S = \{a_1, a_2, \ldots, a_n\}$, we write $\langle a_1, a_2, \ldots, a_n \rangle$ in place of $\langle \{a_1, a_2, \ldots, a_n\} \rangle$.

Example 14.1.8 Some Submonoids.

(a) One example of a submonoid of $[\mathbb{Z}; +]$ is $\langle 2 \rangle = \{0, 2, 4, 6, 8, \ldots\}$.

(b) The power set of $\mathbb{Z}$, $\mathcal{P}(\mathbb{Z})$, over union is a monoid with identity $\emptyset$. If $S = \{\{1\}, \{2\}, \{3\}\}$, then $\langle S \rangle$ is the power set of $\{1, 2, 3\}$. If $S = \{\{n\} : n \in \mathbb{Z}\}$, then $\langle S \rangle$ is the set of finite subsets of the integers.

14.1. MONOIDS

As you might expect, two monoids are isomorphic if and only if there exists a translation rule between them so that any true proposition in one monoid is translated to a true proposition in the other.

Example 14.1.9 $M = [\mathcal{P}\{1,2,3\}; \cap]$ is isomorphic to $M_2 = [\mathbb{Z}_2^3; \cdot]$, where the operation in M_2 is componentwise mod 2 multiplication. A translation rule is that if $A \subseteq \{1,2,3\}$, then it is translated to (d_1, d_2, d_3) where

$$d_i = \begin{cases} 1 & \text{if } i \in A \\ 0 & \text{if } i \notin A \end{cases}$$

Two cases of how this translation rule works are:

$\{1,2,3\}$ is the identity for M_1 $\{1,2\} \cap \{2,3\} = \{2\}$
$\updownarrow$ $\updownarrow$
$(1,1,1)$ is the identity for M_2 $(1,1,0) \cdot (0,1,1) = (0,1,0)$

A more precise definition of a monoid isomorphism is identical to the definition of a group isomorphism, Definition 11.7.9, p. 57.

Exercises

1. For each of the subsets of the indicated monoid, determine whether the subset is a submonoid.

   (a) $S_1 = \{0, 2, 4, 6\}$ and $S_2 = \{1, 3, 5, 7\}$ in $[\mathbb{Z}_8; \times_8]$.

   (b) $\{f \in \mathbb{N}^\mathbb{N} : f(n) \leq n, \forall n \in \mathbb{N}\}$ and $\{f \in \mathbb{N}^\mathbb{N} : f(1) = 2\}$ in the monoid $[\mathbb{N}^\mathbb{N}; \circ]$.

   (c) $\{A \subseteq \mathbb{Z} \mid A \text{ is finite}\}$ and $\{A \subseteq \mathbb{Z} \mid A^c \text{ is finite}\}$ in $[\mathcal{P}(\mathbb{Z}); \cup]$.

2. For each subset, describe the submonoid that it generates.

   (a) $\{3\}$ in $[\mathbb{Z}_{12}; \times_{12}]$

   (b) $\{5\}$ in $[\mathbb{Z}_{25}; \times_{25}]$

   (c) the set of prime numbers in $[\mathbb{P}; \cdot]$

   (d) $\{3, 5\}$ in $[\mathbb{N}; +]$

3. An $n \times n$ matrix of real numbers is called **stochastic** if and only if each entry is nonnegative and the sum of entries in each column is 1. Prove that the set of stochastic matrices is a monoid over matrix multiplication.

4. A **semigroup** is an algebraic system $[S; *]$ with the only axiom that $*$ be associative on S. Prove that if S is a finite set, then there must exist an idempotent element, that is, an $a \in S$ such that $a * a = a$.

5. Let B be a Boolean algebra and M the set of all Boolean functions on B. Let $*$ be defined on M by $(f * g)(a) = f(a) \wedge g(a)$. Prove that $[M; *]$ is a monoid. Construct the operation table of $[M; *]$ for the case of $B = B_2$.

14.2 Free Monoids and Languages

In this section, we will introduce the concept of a language. Languages are subsets of a certain type of monoid, the free monoid over an alphabet. After defining a free monoid, we will discuss languages and some of the basic problems relating to them. We will also discuss the common ways in which languages are defined.

Let A be a nonempty set, which we will call an alphabet. Our primary interest will be in the case where A is finite; however, A could be infinite for most of the situations that we will describe. The elements of A are called letters or symbols. Among the alphabets that we will use are $B = \{0, 1\}$, and the set of ASCII (American Standard Code for Information Interchange) characters, which we symbolize as $ASCII$.

Definition 14.2.1 Strings over an Alphabet. A string of length n, $n \geqslant 1$ over alphabet A is a sequence of n letters from A: $a_1 a_2 \ldots a_n$. The null string, λ, is defined as the string of length zero containing no letters. The set of strings of length n over A is denoted by A^n. The set of all strings over A is denoted A^*. ◇

Note 14.2.2

(a) If the length of string s is n, we write $|s| = n$.

(b) The null string is not the same as the empty set, although they are similar in many ways. $A^0 = \{\lambda\}$.

(c) $A^* = A^0 \cup A^1 \cup A^2 \cup A^3 \cup \cdots$ and if $i \neq j, A^i \cap A^j = \emptyset$; that is, $\{A^0, A^1, A^2, A^3, \ldots\}$ is a partition of A^*.

(d) An element of A can appear any number of times in a string.

Theorem 14.2.3 *If A is countable, then A^* is countable.*

Proof. Case 1. Given the alphabet $B = \{0, 1\}$, we can define a bijection from the positive integers into B^*. Each positive integer has a binary expansion $d_k d_{k-1} \cdots d_1 d_0$, where each d_j is 0 or 1 and $d_k = 1$. If n has such a binary expansion, then $2^k \leq n \leq 2^{k+1}$. We define $f : \mathbb{P} \to B^*$ by $f(n) = f(d_k d_{k-1} \cdots d_1 d_0) = d_{k-1} \cdots d_1 d_0$, where $f(1) = \lambda$. Every one of the 2^k strings of length k are the images of exactly one of the integers between 2^k and $2^{k+1} - 1$. From its definition, f is clearly a bijection; therefore, B^* is countable.

Case 2: A is Finite. We will describe how this case is handled with an example first and then give the general proof. If $A = \{a, b, c, d, e\}$, then we can code the letters in A into strings from B^3. One of the coding schemes (there are many) is $a \leftrightarrow 000, b \leftrightarrow 001, c \leftrightarrow 010, d \leftrightarrow 011$, and $e \leftrightarrow 100$. Now every string in A^* corresponds to a different string in B^*; for example, ace. would correspond with 000010100. The cardinality of A^* is equal to the cardinality of the set of strings that can be obtained from this encoding system. The possible coded strings must be countable, since they are a subset of a countable set, B^*. Therefore, A^* is countable.

If $|A| = m$, then the letters in A can be coded using a set of fixed-length strings from B^*. If $2^{k-1} < m \leq 2^k$, then there are at least as many strings of length k in B^k as there are letters in A. Now we can associate each letter in A with with a different element of B^k. Then any string in A^*. corresponds to a string in B^*. By the same reasoning as in the example above, A^* is countable.

Case 3: A is Countably Infinite. We will leave this case as an exercise. ∎

Definition 14.2.4 Concatenation. Let $a = a_1 a_2 \cdots a_m$ and $b = b_1 b_2 \cdots b_n$ be strings of length m and n, respectively. The concatenation of a with b, $a + b$, is the string $a_1 a_2 \cdots a_m b_1 b_2 \cdots b_n$ of length $m + n$. ◇

There are several symbols that are used for concatenation. We chose to use the one that is also used in Python and SageMath.

```
'good'+'bye'
```

`'goodbye'`

The set of strings over any alphabet is a monoid under concatenation.

Note 14.2.5

(a) The null string is the identity element of $[A^*; +]$. Henceforth, we will denote the monoid of strings over A by A^*.

(b) Concatenation is noncommutative, provided $|A| > 1$.

(c) If $|A_1| = |A_2|$, then the monoids A_1^* and A_2^* are isomorphic. An isomorphism can be defined using any bijection $f : A_1 \to A_2$. If $a = a_1 a_2 \cdots a_n \in A_1^*$, $f^*(a) = (f(a_1) f(a_2) \cdots f(a_n))$ defines a bijection from A_1^* into A_2^*. We will leave it to the reader to prove that for all $a, b, \in A_1^*$, $f^*(a + b) = f^*(a) + f^*(b)$.

The languages of the world, English, German, Russian, Chinese, and so forth, are called natural languages. In order to communicate in writing in any one of them, you must first know the letters of the alphabet and then know how to combine the letters in meaningful ways. A formal language is an abstraction of this situation.

Definition 14.2.6 Formal Language. If A is an alphabet, a formal language over A is a subset of A^*. ◇

Example 14.2.7 Some Formal Languages.

(a) English can be thought of as a language over of letters $A, B, \cdots Z$, both upper and lower case, and other special symbols, such as punctuation marks and the blank. Exactly what subset of the strings over this alphabet defines the English language is difficult to pin down exactly. This is a characteristic of natural languages that we try to avoid with formal languages.

(b) The set of all ASCII stream files can be defined in terms of a language over ASCII. An ASCII stream file is a sequence of zero or more lines followed by an end-of-file symbol. A line is defined as a sequence of ASCII characters that ends with the a "new line" character. The end-of-file symbol is system-dependent.

(c) The set of all syntactically correct expressions in any computer language is a language over the set of ASCII strings.

(d) A few languages over B are

- $L_1 = \{s \in B^* \mid s \text{ has exactly as many 1's as it has 0's}\}$
- $L_2 = \{1 + s + 0 \mid s \in B^*\}$
- $L_3 = \langle 0, 01 \rangle =$ the submonoid of B^* generated by $\{0, 01\}$.

□

Investigation 14.2.1 Two Fundamental Problems: Recognition and Generation. The generation and recognition problems are basic to computer programming. Given a language, L, the programmer must know how to write (or generate) a syntactically correct program that solves a problem. On the other hand, the compiler must be written to recognize whether a program contains any syntax errors.

Problem 14.2.8 The Recognition Problem. Given a formal language over alphabet A, the Recognition Problem is to design an algorithm that determines the truth of $s \in L$ in a finite number of steps for all $s \in A^*$. Any such algorithm is called a recognition algorithm. □

Definition 14.2.9 Recursive Language. A language is recursive if there exists a recognition algorithm for it. ◊

Example 14.2.10 Some Recursive Languages.

(a) The language of syntactically correct propositions over set of propositional variables expressions is recursive.

(b) The three languages in 7, p. 133(d) are all recursive. Recognition algorithms for L_1 and L_2 should be easy for you to imagine. The reason a recognition algorithm for L_3 might not be obvious is that the definition of L_3 is more cryptic. It doesn't tell us what belongs to L_3, just what can be used to create strings in L_3. This is how many languages are defined. With a second description of L_3, we can easily design a recognition algorithm. You can prove that

$$L_3 = \{s \in B^* \mid s = \lambda \text{ or } s \text{ starts with a 0 and has no consecutive 1's}\}.$$

□

Problem 14.2.11 The Generation Problem. Design an algorithm that generates or produces any string in L. Here we presume that A is either finite or countably infinite; hence, A^* is countable by Theorem 14.2.3, p. 132, and $L \subseteq A^*$ must be countable. Therefore, the generation of L amounts to creating a list of strings in L. The list may be either finite or infinite, and you must be able to show that every string in L appears somewhere in the list. □

Theorem 14.2.12 Recursive implies Generating.

(a) If A is countable, then there exists a generating algorithm for A^*.

(b) If L is a recursive language over a countable alphabet, then there exists a generating algorithm for L.

Proof. Part (a) follows from the fact that A^* is countable; therefore, there exists a complete list of strings in A^*.

To generate all strings of L, start with a list of all strings in A^* and an empty list, W, of strings in L. For each string s, use a recognition algorithm (one exists since L is recursive) to determine whether $s \in L$. If $s \in L$, add it to W; otherwise "throw it out." Then go to the next string in the list of A^*. ■

Example 14.2.13 Since all of the languages in 7, p. 133(d) are recursive, they must have generating algorithms. The one given in the proof of Theorem 14.2.12, p. 134 is not usually the most efficient. You could probably design more efficient generating algorithms for L_2 and L_3; however, a better generating algorithm for L_1 is not quite so obvious. □

The recognition and generation problems can vary in difficulty depending on how a language is defined and what sort of algorithms we allow ourselves to use. This is not to say that the means by which a language is defined determines

14.2. FREE MONOIDS AND LANGUAGES

whether it is recursive. It just means that the truth of the statement "L is recursive" may be more difficult to determine with one definition than with another. We will close this section with a discussion of grammars, which are standard forms of definition for a language. When we restrict ourselves to only certain types of algorithms, we can affect our ability to determine whether $s \in L$ is true. In defining a recursive language, we do not restrict ourselves in any way in regard to the type of algorithm that will be used. In the next section, we will consider machines called finite automata, which can only perform simple algorithms.

One common way of defining a language is by means of a **phrase structure grammar** (or grammar, for short). The set of strings that can be produced using set of grammar rules is called a phrase structure language.

Example 14.2.14 Zeros before Ones. We can define the set of all strings over B for which all 0's precede all 1's as follows. Define the starting symbol S and establish rules that S can be replaced with any of the following: λ, $0S$, or $S1$. These replacement rules are usually called production rules. They are usually written in the format $S \to \lambda$, $S \to 0S$, and $S \to S1$. Now define L to be the set of all strings that can be produced by starting with S and applying the production rules until S no longer appears. The strings in L are exactly the ones that are described above. □

Definition 14.2.15 Phrase Structure Grammar. A phrase structure grammar consists of four components:

(1) A nonempty finite set of terminal characters, T. If the grammar is defining a language over A, T is a subset of A^*.

(2) A finite set of nonterminal characters, N.

(3) A starting symbol, $S \in N$.

(4) A finite set of production rules, each of the form $X \to Y$, where X and Y are strings over $A \cup N$ such that $X \neq Y$ and X contains at least one nonterminal symbol.

If G is a phrase structure grammar, $L(G)$ is the set of strings that can be produced by starting with S and applying the production rules a finite number of times until no nonterminal characters remain. If a language can be defined by a phrase structure grammar, then it is called a phrase structure language.
◇

Example 14.2.16 Alternating bits language. The language over B consisting of strings of alternating 0's and 1's is a phrase structure language. It can be defined by the following grammar:

(1) Terminal characters: λ, 0, and 1

(2) Nonterminal characters: S, T, and U

(3) Starting symbol: S

(4) Production rules:

$$\begin{array}{lll} S \to T & S \to U & S \to \lambda \\ S \to 0 & & S \to 1 \\ S \to 0T & & S \to 1U \\ T \to 10T & & T \to 10 \\ U \to 01U & & U \to 01 \end{array}$$

These rules can be visualized with a graph:

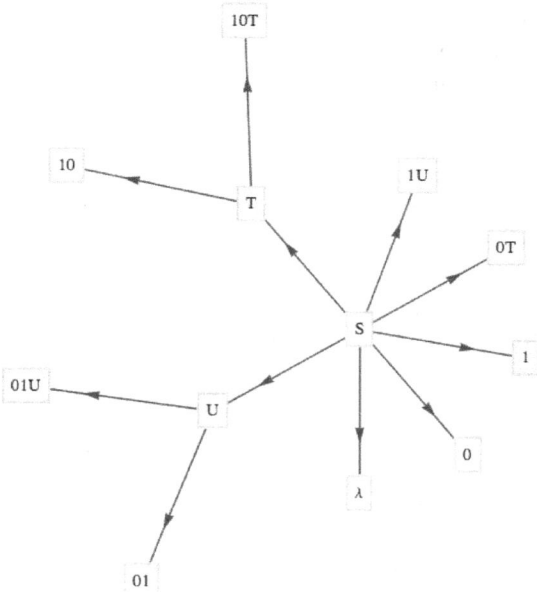

Figure 14.2.17 Production rules for the language of alternating 0's and 1's

We can verify that a string such as 10101 belongs to the language by starting with S and producing 10101 using the production rules a finite number of times: $S \to 1U \to 101U \to 10101$. □

Example 14.2.18 Valid SageMath Variables. Let G be the grammar with components:

(1) Terminal symbols = all letters of the alphabet (both upper and lower case), digits 0 through 9, and underscore

(2) Nonterminal symbols: $\{I, X\}$,

(3) Starting symbol: I

(4) Production rules: $I \to \alpha$, where α is any letter, $I \to \alpha + X$ for any letter α, $X \to X + \beta$ for any letter, digit or underscore, β, and $X \to \beta$ for any letter, digit or underscore, β. There are a total of $52 + 52 + 63 + 63 = 230$ production rules for this grammar. The language $L(G)$ consists of all valid SageMath variable names.

□

Example 14.2.19 Backus-Naur Form. Backus-Naur form (BNF) is a popular alternate form of defining the production rules in a grammar. If the production rules $A \to B_1, A \to B_2, \ldots A \to B_n$ are part of a grammar, they would be written in BNF as $A ::= B_1 \mid B_2 \mid \cdots \mid B_n$. The symbol | in BNF is read as "or" while the ::= is read as "is defined as." Additional notations of BNF are that $\{x\}$, represents zero or more repetitions of x and $[y]$ means that y is optional.

A BNF version of the production rules for a SageMath variable, I, is

$$letter ::= a \mid b \mid c \mid \cdots \mid z \mid A \mid B \mid \cdots \mid Z$$
$$digit ::= 0 \mid 1 \mid \cdots \mid 9$$
$$I ::= letter\{letter \mid digit \mid \_\}$$

□

Example 14.2.20 The language of simple arithmetic expressions. An arithmetic expression can be defined in BNF. For simplicity, we will consider only expressions obtained using addition and multiplication of integers. The terminal symbols are (,),+,*, -, and the digits 0 through 9. The nonterminal symbols are E (for expression), T (term), F (factor), and N (number). The starting symbol is E. Production rules are

$$E ::= E + T \mid T$$
$$T ::= T * F \mid F$$
$$F ::= (E) \mid N$$
$$N ::= [-]digit\{digit\}$$

□

One particularly simple type of phrase structure grammar is the regular grammar.

Definition 14.2.21 Regular Grammar. A regular (right-hand form) grammar is a grammar whose production rules are all of the form $A \to t$ and $A \to tB$, where A and B are nonterminal and t is terminal. A left-hand form grammar allows only $A \to t$ and $A \to Bt$. A language that has a regular phrase structure language is called a regular language. ◇

Example 14.2.22

(a) The set of Sage variable names is a regular language since the grammar by which we defined the set is a regular grammar.

(b) The language of all strings for which all 0's precede all 1's (Example 14.2.14, p. 135) is regular; however, the grammar by which we defined this set is not regular. Can you define these strings with a regular grammar?

(c) The language of arithmetic expressions is not regular.

□

Exercises

1.

 (a) If a computer is being designed to operate with a character set of 350 symbols, how many bits must be reserved for each character? Assume each character will use the same number of bits.

 (b) Do the same for 3,500 symbols.

2. It was pointed out in the text that the null string and the null set are different. The former is a string and the latter is a set, two different kinds of objects. Discuss how the two are similar.

3. What sets of strings are defined by the following grammar?
   (a) Terminal symbols: λ, 0 and 1
   (b) Nonterminal symbols: S and E
   (c) Starting symbol: S
   (d) Production rules: $S \to 0S0, S \to 1S1, S \to E, E \to \lambda, E \to 0, E \to 1$
4. What sets of strings are defined by the following grammar?
   (a) Terminal symbols: λ, a, b, and c
   (b) Nonterminal symbols: S, T, U and E
   (c) Starting symbol: S
   (d) Production rules:

$$S \to aS \quad S \to T \quad T \to bT$$
$$T \to U \quad U \to cU \quad U \to E$$
$$E \to \lambda$$

5. Define the following languages over B with phrase structure grammars. Which of these languages are regular?
   (a) The strings with an odd number of characters.
   (b) The strings of length 4 or less.
   (c) The palindromes, strings that are the same backwards as forwards.
6. Define the following languages over B with phrase structure grammars. Which of these languages are regular?
   (a) The strings with more 0's than 1's.
   (b) The strings with an even number of 1's.
   (c) The strings for which all 0's precede all 1's.
7. Prove that if a language over A is recursive, then its complement is also recursive.
8. Use BNF to define the grammars in Exercises 3 and 4.
9.
   (a) Prove that if $X_1, X_2, \ldots$ is a countable sequence of countable sets, the union of these sets, $\bigcup_{i=1}^{\infty} X_i$ is countable.
   (b) Using the fact that the countable union of countable sets is countable, prove that if A is countable, then A^* is countable.

14.3 Automata, Finite-State Machines

In this section, we will introduce the concept of an abstract machine. The machines we will examine will (in theory) be capable of performing many of the tasks associated with digital computers. One such task is solving the recognition problem for a language. We will concentrate on one class of machines, finite-state machines (finite automata). And we will see that they are precisely the machines that are capable of recognizing strings in a regular grammar.

14.3. AUTOMATA, FINITE-STATE MACHINES

Given an alphabet X, we will imagine a string in X^* to be encoded on a tape that we will call an input tape. When we refer to a tape, we might imagine a strip of material that is divided into segments, each of which can contain either a letter or a blank.

The typical abstract machine includes an input device, the read head, which is capable of reading the symbol from the segment of the input tape that is currently in the read head. Some more advanced machines have a read/write head that can also write symbols onto the tape. The movement of the input tape after reading a symbol depends on the machine. With a finite-state machine, the next segment of the input tape is always moved into the read head after a symbol has been read. Most machines (including finite-state machines) also have a separate output tape that is written on with a write head. The output symbols come from an output alphabet, Z, that may or may not be equal to the input alphabet. The most significant component of an abstract machine is its memory structure. This structure can range from a finite number of bits of memory (as in a finite-state machine) to an infinite amount of memory that can be stored in the form of a tape that can be read from and written on (as in a Turing machine).

Definition 14.3.1 Finite-State Machine. A finite-state machine is defined by a quintet (S, X, Z, w, t) where

(1) $S = \{s_1, s_2, \ldots, s_r\}$ is the state set, a finite set that corresponds to the set of memory configurations that the machine can have at any time.

(2) $X = \{x_1, x_2, \ldots, x_m\}$ is the input alphabet.

(3) $Z = \{z_1, z_2, \ldots, z_n\}$ is the output alphabet.

(4) $w : X \times S \to Z$ is the output function, which specifies which output symbol $w(x,s) \in Z$ is written onto the output tape when the machine is in state s and the input symbol x is read.

(5) $t : X \times S \to S$ is the next-state (or transition) function, which specifies which state $t(x,s) \in S$ the machine should enter when it is in state s and it reads the symbol x.

$\diamond$

Example 14.3.2 Vending Machine as a Finite-State Machine. Many mechanical devices, such as simple vending machines, can be thought of as finite-state machines. For simplicity, assume that a vending machine dispenses packets of gum, spearmint (S), peppermint (P), and bubble (B), for 25 cents each. We can define the input alphabet to be

$$\{\text{deposit } 25 \text{ cents}, \text{press S}, \text{press P}, \text{press B}\}$$

and the state set to be {Locked, Select}, where the deposit of a quarter unlocks the release mechanism of the machine and allows you to select a flavor of gum. We will leave it to the reader to imagine what the output alphabet, output function, and next-state function would be. You are also invited to let your imagination run wild and include such features as a coin-return lever and change maker. □

Example 14.3.3 A Parity Checking Machine. The following machine is called a parity checker. It recognizes whether or not a string in B^* contains an even number of 1s. The memory structure of this machine reflects the fact that in order to check the parity of a string, we need only keep track of whether an odd or even number of 1's has been detected.

The input alphabet is $B = \{0, 1\}$ and the output alphabet is also B. The state set is $\{even, odd\}$. The following table defines the output and next-state functions.

x	s	$w(x,s)$	$t(x,s)$
0	even	0	even
0	odd	1	odd
1	even	1	odd
1	odd	0	even

Note how the value of the most recent output at any time is an indication of the current state of the machine. Therefore, if we start in the even state and read any finite input tape, the last output corresponds to the final state of the parity checker and tells us the parity of the string on the input tape. For example, if the string 11001010 is read from left to right, the output tape, also from left to right, will be 10001100. Since the last character is a 0, we know that the input string has even parity. □

An alternate method for defining a finite-state machine is with a transition diagram. A transition diagram is a directed graph that contains a node for each state and edges that indicate the transition and output functions. An edge (s_i, s_j) that is labeled x/z indicates that in state s_i the input x results in an output of z and the next state is s_j. That is, $w(x, s_i) = z$ and $t(x, s_i) = s_j$. The transition diagram for the parity checker appears in Figure 14.3.4, p. 140. In later examples, we will see that if there are different inputs, x_i and x_j, while in the same state resulting in the same transitions and outputs, we label a single edge $x_i, x_j/z$ instead of drawing two edges with labels x_i/z and x_j/z.

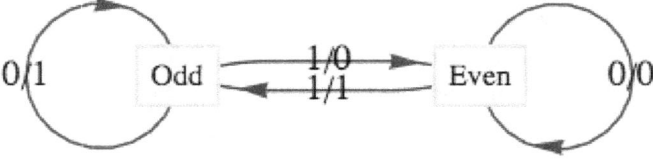

Figure 14.3.4 Transition Diagram for a Parity Checker

One of the most significant features of a finite-state machine is that it retains no information about its past states that can be accessed by the machine itself. For example, after we input a tape encoded with the symbols 01101010 into the parity checker, the current state will be even, but we have no indication within the machine whether or not it has always been in even state. Note how the output tape is not considered part of the machine's memory. In this case, the output tape does contain a "history" of the parity checker's past states. We assume that the finite-state machine has no way of recovering the output sequence for later use.

Example 14.3.5 A Baseball Machine. Consider the following simplified version of the game of baseball. To be precise, this machine describes one half-inning of a simplified baseball game. Suppose that in addition to home plate, there is only one base instead of the usual three bases. Also, assume that there are only two outs per inning instead of the usual three. Our input alphabet will consist of the types of hits that the batter could have: out (O), double play (DP), single (S), and home run (HR). The input DP is meant to represent a batted ball that would result in a double play (two outs), if possible. The input DP can then occur at any time. The output alphabet is the numbers 0, 1, and 2 for the number of runs that can be scored as a result of any input.

14.3. AUTOMATA, FINITE-STATE MACHINES

The state set contains the current situation in the inning, the number of outs, and whether a base runner is currently on the base. The list of possible states is then 00 (for 0 outs and 0 runners), 01, 10, 11, and end (when the half-inning is over). The transition diagram for this machine appears in Figure 14.3.6, p. 141

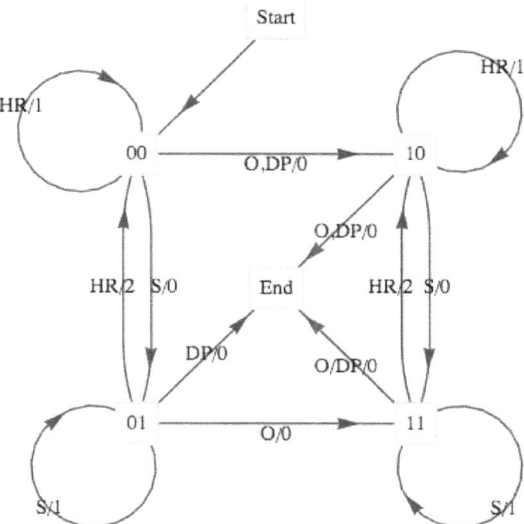

Figure 14.3.6 Transition Diagram for a simplified game of baseball

Let's concentrate on one state. If the current state is 01, 0 outs and 1 runner on base, each input results in a different combination of output and next-state. If the batter hits the ball poorly (a double play) the output is zero runs and the inning is over (the limit of two outs has been made). A simple out also results in an output of 0 runs and the next state is 11, one out and one runner on base. If the batter hits a single, one run scores (output = 1) while the state remains 01. If a home run is hit, two runs are scored (output = 2) and the next state is 00. If we had allowed three outs per inning, this graph would only be marginally more complicated. The usual game with three bases would be quite a bit more complicated, however. □

Example 14.3.7 Recognition in Regular Languages. As we mentioned at the outset of this section, finite-state machines can recognize strings in a regular language. Consider the language L over $\{a, b, c\}$ that contains the strings of positive length in which each a is followed by b and each b is followed by c. One such string is $bccabcbc$. This language is regular. A grammar for the language would be nonterminal symbols $\{A, B, C\}$ with starting symbol C and production rules $A \to bB$, $B \to cC$, $C \to aA$, $C \to bB$, $C \to cC$, $C \to c$. A finite-state machine (Figure 14.3.8, p. 142) that recognizes this language can be constructed with one state for each nonterminal symbol and an additional state (Reject) that is entered if any invalid production takes place. At the end of an input tape that encodes a string in $\{a, b, c\}^*$, we will know when the string belongs to L based on the final output. If the final output is 1, the string belongs to L and if it is 0, the string does not belong to L. In addition, recognition can be accomplished by examining the final state of the machine. The input string belongs to the language if and only if the final state is C.

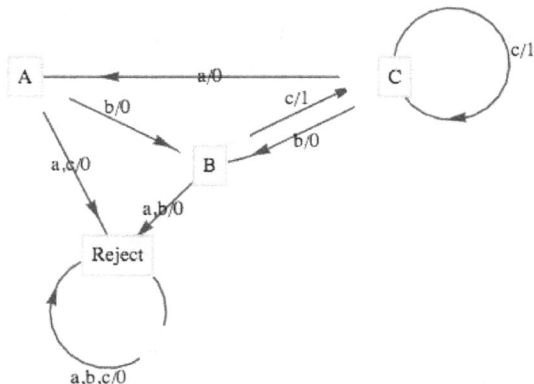

Figure 14.3.8

The construction of this machine is quite easy: note how each production rule translates into an edge between states other than Reject. For example, $C \to bB$ indicates that in State C, an input of b places the machine into State B. Not all sets of production rules can be as easily translated to a finite-state machine. Another set of production rules for L is $A \to aB$, $B \to bC$, $C \to cA$, $C \to cB$, $C \to cC$ and $C \to c$. Techniques for constructing finite-state machines from production rules is not our objective here. Hence we will only expect you to experiment with production rules until appropriate ones are found. $\square$

Example 14.3.9 A Binary Adder. A finite-state machine can be designed to add positive integers of any size. Given two integers in binary form, $a = a_n a_{n-1} \cdots a_1 a_0$ and $b = b_n b_{n-1} \cdots b_1 b_0$, the machine take as its input sequence the corresponding bits of a and b reading from right to left with a "parity bit" added

$$a_0 b_0 (a_0 +_2 b_0), a_1 b_1 (a_1 +_2 b_1) \ldots, a_n b_n (a_n +_2 b_n), 111$$

Notice the special input 111 at the end. All possible inputs except the last one must even parity (contain an even number of ones). The output sequence is the sum of a and b, starting with the units digit, and comes from the set $\{0, 1, \lambda\}$. The transition diagram for this machine appears in Figure 14.3.10, p. 142.

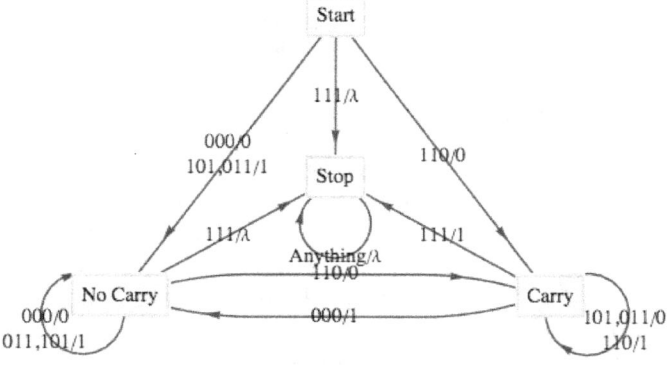

Figure 14.3.10 Transition Diagram for a binary adder

$\square$

Exercises

1. Draw a transition diagram for the vending machine described in Example 14.3.2, p. 139.
2. Construct finite-state machines that recognize the regular languages that you identified in Section 14.2.
3. What is the input set for the binary adding machine in Example 14.3.9, p. 142?
4. What input sequence would be used to compute the sum of 1101 and 0111 (binary integers)? What would the output sequence be?
5. The Gray Code Decoder. The finite-state machine defined by the following figure has an interesting connection with the Gray Code.

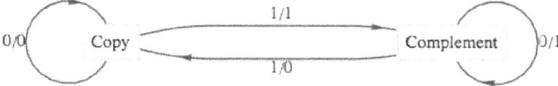

Figure 14.3.11 Gray Code Decoder

Given a string $x = x_1 x_2 \cdots x_n \in B^n$, we may ask where x appears in G_n. Starting in Copy state, the input string x will result in an output string $z \in B^n$, which is the binary form of the position of x in G_n. Recall that positions are numbered from 0 to $2^n - 1$.

(a) In what positions (0 − 31) do 10110, 00100, and 11111 appear in G_5?

(b) Prove that the Gray Code Decoder always works.

14.4 The Monoid of a Finite-State Machine

In this section, we will see how every finite-state machine has a monoid associated with it. For any finite-state machine, the elements of its associated monoid correspond to certain input sequences. Because only a finite number of combinations of states and inputs is possible for a finite-state machine there is only a finite number of input sequences that summarize the machine. This idea is illustrated best with a few examples.

Consider the parity checker. The following table summarizes the effect on the parity checker of strings in B^1 and B^2. The row labeled "Even" contains the final state and final output as a result of each input string in B^1 and B^2 when the machine starts in the even state. Similarly, the row labeled "Odd" contains the same information for input sequences when the machine starts in the odd state.

Input String	0	1	00	01	10	11
Even	(Even, 0)	(Odd, 1)	(Even, 0)	(Odd, 1)	(Odd, 1)	(Even, 0)
Odd	(Odd, 1)	(Even, 1)	(Odd, 1)	(Even, 1)	(Even, 0)	(Odd, 1)
Same Effect as			0	1	1	0

Note how, as indicated in the last row, the strings in B^2 have the same effect as certain strings in B^1. For this reason, we can summarize the machine in terms of how it is affected by strings of length 1. The actual monoid that we will now describe consists of a set of functions, and the operation on the functions will be based on the concatenation operation.

Let T_0 be the final effect (state and output) on the parity checker of the input 0. Similarly, T_1 is defined as the final effect on the parity checker of the input 1. More precisely,

$$T_0(\text{ even}) = (\text{ even}, 0) \quad \text{and} \quad T_0(\text{ odd}) = (\text{ odd}, 1),$$

while

$$T_1(\text{ even}) = (\text{ odd}, 1) \quad \text{and} \quad T_1(\text{ odd}) = (\text{ even}, 0).$$

In general, we define the operation on a set of such functions as follows: if s, t are input sequences and T_s and T_t, are functions as above, then $T_s * T_t = T_{st}$, that is, the result of the function that summarizes the effect on the machine by the concatenation of s with t. Since, for example, 01 has the same effect on the parity checker as 1, $T_0 * T_1 = T_{01} = T_1$. We don't stop our calculation at T_{01} because we want to use the shortest string of inputs to describe the final result.

A complete table for the monoid of the parity checker is

$*$	T_0	T_1
T_0	T_0	T_1
T_1	T_1	T_0

What is the identity of this monoid? The monoid of the parity checker is isomorphic to the monoid $[\mathbb{Z}_2; +_2]$.

This operation may remind you of the composition operation on functions, but there are two principal differences. The domain of T_s is not the codomain of T_t and the functions are read from left to right unlike in composition, where they are normally read from right to left.

You may have noticed that the output of the parity checker echoes the state of the machine and that we could have looked only at the effect on the machine as the final state. The following example has the same property, hence we will only consider the final state.

Example 14.4.1 The transition diagram for the machine that recognizes strings in $B*$ that have no consecutive 1's appears in Figure 14.4.2, p. 144. Note how it is similar to the graph in Figure 9.0.1, p. 17. Only a "reject state" has been added, for the case when an input of 1 occurs while in State a. We construct a similar table to the one in the previous example to study the effect of certain strings on this machine. This time, we must include strings of length 3 before we recognize that no "new effects" can be found.

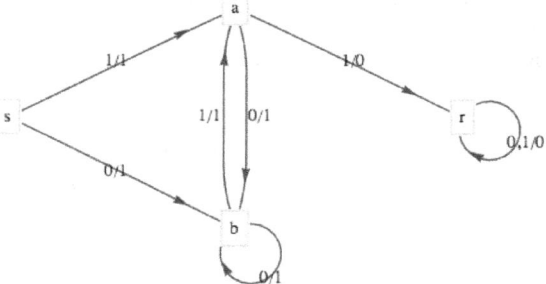

Figure 14.4.2 No Consecutive Ones Monoid

Inputs	0	1	00	01	10	11	000	001	010	011	100	101	110	111
s	b	a	b	a	b	r	b	a	b	r	b	a	r	r
a	b	r	b	a	r	r	b	a	b	r	r	r	r	r
b	b	a	b	a	b	r	b	a	b	r	b	a	r	r
r	r	r	r	r	r	r	r	r	r	r	r	r	r	r
Same as			0				0	01	0	11	10	1	11	11

The following table summarizes how combinations of the strings 0, 1, 01, 10, and 11

14.4. THE MONOID OF A FINITE-STATE MACHINE

affect this machine.

*	T_0	T_1	T_{01}	T_{10}	T_{11}
T_0	T_0	T_1	T_{01}	T_{10}	T_{11}
T_1	T_{10}	T_{11}	T_1	T_{11}	T_{11}
T_{01}	T_0	T_{11}	T_{01}	T_{11}	T_{11}
T_{10}	T_{10}	T_1	T_1	T_{10}	T_{11}
T_{11}	T_{11}	T_{11}	T_{11}	T_{11}	T_{11}

All the results in this table can be obtained using the previous table. For example,

$$T_{10} * T_{01} = T_{1001} = T_{100} * T_1 = T_{10} * T_1 = T_{101} = T_1$$
and
$$T_{01} * T_{01} = T_{0101} = T_{010}T_1 = T_0T_1 = T_{01}$$

Note that none of the elements that we have listed in this table serves as the identity for our operation. This problem can always be remedied by including the function that corresponds to the input of the null string, T_λ. Since the null string is the identity for concatenation of strings, $T_s T_\lambda = T_\lambda T_s = T_s$ for all input strings s. □

Example 14.4.3 The Unit-time Delay Machine. A finite-state machine called the unit-time delay machine does not echo its current state, but prints its previous state. For this reason, when we find the monoid of the unit-time delay machine, we must consider both state and output. The transition diagram of this machine appears in Figure 14.4.4, p. 145.

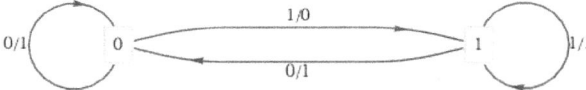

Figure 14.4.4

Input	0	1	00	01	10	11	100 or 000	101 or 001	110 or 101	111 or 011
0	(0,0)	(1,0)	(0,0)	(1,0)	(0,1)	(1,1)	(0,0)	(1,0)	(0,1)	(1,1)
1	(0,1)	(1,1)	(0,0)	(1,0)	(0,1)	(1,1)	(0,0)	(1,0)	(0,1)	(1,1)
Same as							00	01	10	11

Again, since no new outcomes were obtained from strings of length 3, only strings of length 2 or less contribute to the monoid of the machine. The table for the strings of positive length shows that we must add T_λ to obtain a monoid.

*	T_0	T_1	T_{00}	T_{01}	T_{10}	T_{11}
T_0	T_{00}	T_{01}	T_{00}	T_{01}	T_{10}	T_{11}
T_1	T_{10}	T_{11}	T_{00}	T_{01}	T_{10}	T_{11}
T_{00}	T_{00}	T_{01}	T_{00}	T_{01}	T_{10}	T_{11}
T_{01}	T_{10}	T_{11}	T_{00}	T_{01}	T_{10}	T_{11}
T_{10}	T_{00}	T_{01}	T_{00}	T_{01}	T_{10}	T_{11}
T_{11}	T_{10}	T_{11}	T_{00}	T_{01}	T_{10}	T_{11}

□

Exercises

1. For each of the transition diagrams in Figure 5, p. 146, write out tables for their associated monoids. Identify the identity in terms of a string of positive length, if possible.

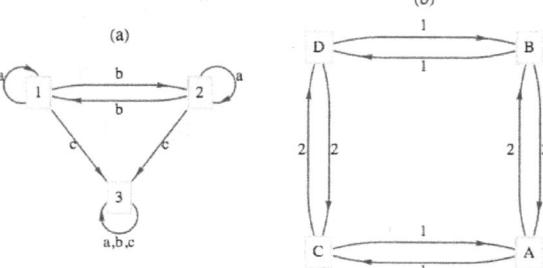

Figure 14.4.5 Exercise 1

Hint. Where the output echoes the current state, the output can be ignored.

2. What common monoids are isomorphic to the monoids obtained in the previous exercise?

3. Can two finite-state machines with nonisomorphic transition diagrams have isomorphic monoids?

14.5 The Machine of a Monoid

Any finite monoid $[M; *]$ can be represented in the form of a finite-state machine with input and state sets equal to M. The output of the machine will be ignored here, since it would echo the current state of the machine. Machines of this type are called **state machines**. It can be shown that whatever can be done with a finite-state machine can be done with a state machine; however, there is a trade-off. Usually, state machines that perform a specific function are more complex than general finite-state machines.

Definition 14.5.1 Machine of a Monoid. If $[M; *]$ is a finite monoid, then the machine of M, denoted $m(M)$, is the state machine with state set M, input set M, and next-state function $t : M \times M \to M$ defined by $t(s,x) = s * x$. ◊

Example 14.5.2 We will construct the machine of the monoid $[\mathbb{Z}_2; +_2]$. As mentioned above, the state set and the input set are both $\mathbb{Z}_2$. The next state function is defined by $t(s, x) = s +_2 x$. The transition diagram for $m(\mathbb{Z}_2)$ appears in Figure 14.5.3, p. 146. Note how it is identical to the transition diagram of the parity checker, which has an associated monoid that was isomorphic to $[\mathbb{Z}_2; +_2]$.

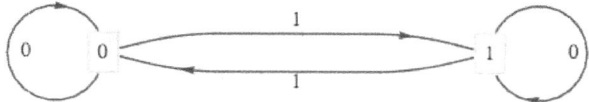

Figure 14.5.3 The machine of $[\mathbb{Z}_2; +_2]$

□

Example 14.5.4 The transition diagram of the monoids $[\mathbb{Z}_2; \times_2]$ and $[\mathbb{Z}_3; \times_3]$ appear in Figure 14.5.5, p. 147.

14.5. THE MACHINE OF A MONOID

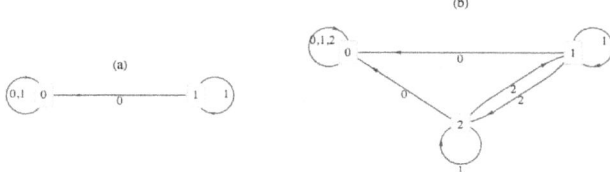

Figure 14.5.5 The machines of $[\mathbb{Z}_2; \times_2]$ and $[\mathbb{Z}_3; \times_3]$

□

Example 14.5.6 Let U be the monoid that we obtained from the unit-time delay machine (Example 14.4.3, p. 145). We have seen that the machine of the monoid of the parity checker is essentially the parity checker. Will we obtain a unit-time delay machine when we construct the machine of U? We can't expect to get exactly the same machine because the unit-time delay machine is not a state machine and the machine of a monoid is a state machine. However, we will see that our new machine is capable of telling us what input was received in the previous time period. The operation table for the monoid serves as a table to define the transition function for the machine. The row headings are the state values, while the column headings are the inputs. If we were to draw a transition diagram with all possible inputs, the diagram would be too difficult to read. Since U is generated by the two elements, T_0 and T_1, we will include only those inputs. Suppose that we wanted to read the transition function for the input T_{01}. Since $T_{01} = T_0 T_1$, in any state s, $t(s, T_{01}) = t(t(s, T_0), T_1)$. The transition diagram appears in Figure 14.5.7, p. 147.

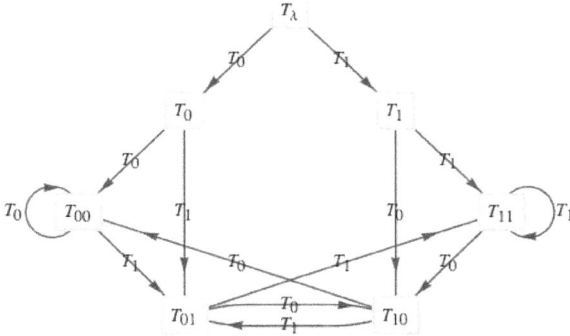

Figure 14.5.7 Unit time delay machine

If we start reading a string of 0's and 1's while in state T_λ and are in state T_{ab} at any one time, the input from the previous time period (not the input that sent us into T_{ab}, the one before that) is a. In states T_λ, T_0 and T_1, no previous input exists. □

Exercises

1. Draw the transition diagrams for the machines of the following monoids:

    (a) $[\mathbb{Z}_4; +_4]$

    (b) The direct product of $[\mathbb{Z}_2; \times_2]$ with itself.

2. Even though a monoid may be infinite, we can visualize it as an infinite-state machine provided that it is generated by a finite number of elements. For example, the monoid B^* is generated by 0 and 1. A section of its transition diagram can be obtained by allowing input only from the

generating set. The monoid of integers under addition is generated by the set $\{-1, 1\}$. The transition diagram for this monoid can be visualized by drawing a small portion of it, as in Figure 10, p. 148. The same is true for the additive monoid of integers, as seen in Figure 11, p. 148.

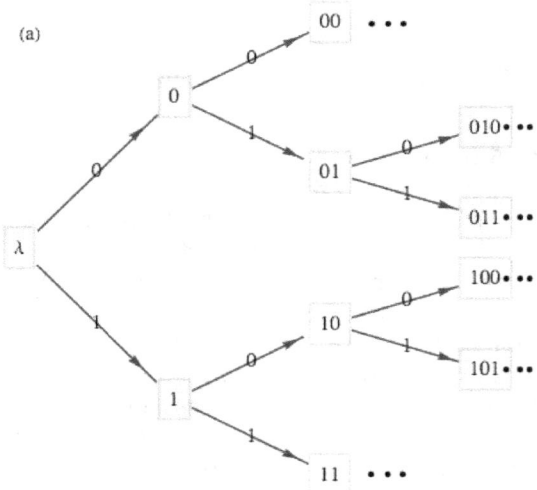

Figure 14.5.8 An infinite machine B^*

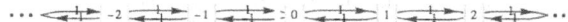

Figure 14.5.9 An infinite machine $[\mathbb{Z}; +]$

(a) Draw a transition diagram for $\{a, b, c\}$

(b) Draw a transition diagram for $[\mathbb{Z} \times \mathbb{Z}; \text{componentwise addition}]$.

(c) Draw a transition diagram for $[\mathbb{Z}; +]$ with generating set $\{5, -2\}$.

Chapter 15

Group Theory and Applications

alternating group

N objects are ordered, and you
Switch consecutive pairs two by two.
All reorders you get
Will comprise a new set
Called an **alternating group** when you're through.

Chris Doyle, The Omnificent English Dictionary In Limerick Form

In Chapter 11, we introduced groups as a typical algebraic system. The associated concepts of subgroup, group isomorphism, and direct products of groups were also introduced. Groups were chosen for that chapter because they are among the simplest types of algebraic systems. Despite this simplicity, group theory abounds with interesting applications. In this chapter we will introduce some more important concepts in elementary group theory, and some of their applications.

15.1 Cyclic Groups

Groups are classified according to their size and structure. A group's structure is revealed by a study of its subgroups and other properties (e.g., whether it is abelian) that might give an overview of it. Cyclic groups have the simplest structure of all groups.

Definition 15.1.1 Cyclic Group. Group G is cyclic if there exists $a \in G$ such that the cyclic subgroup generated by a, $\langle a \rangle$, equals all of G. That is, $G = \{na | n \in \mathbb{Z}\}$, in which case a is called a generator of G. The reader should note that additive notation is used for G. ◊

Example 15.1.2 A Finite Cyclic Group. $\mathbb{Z}_{12} = [\mathbb{Z}_{12}; +_{12}]$, where $+_{12}$ is addition modulo 12, is a cyclic group. To verify this statement, all we need to do is demonstrate that some element of $\mathbb{Z}_{12}$ is a generator. One such element is 5; that is, $\langle 5 \rangle = \mathbb{Z}_{12}$. One more obvious generator is 1. In fact, 1 is a generator of every $[\mathbb{Z}_n; +_n]$. The reader is asked to prove that if an element is a generator, then its inverse is also a generator. Thus, $-5 = 7$ and $-1 = 11$ are the other generators of $\mathbb{Z}_{12}$. The remaining eight elements of the group are not generators.

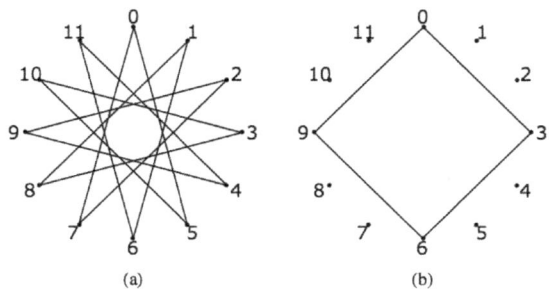

Figure 15.1.3 Examples of "string art"

Figure 15.1.3, p. 150(a) is an example of "string art" that illustrates how 5 generates $\mathbb{Z}_{12}$. Twelve tacks are placed evenly around a circle and numbered 0 through 11. A string is tied to tack 0, and is then looped around every fifth tack. As a result, the numbers of the tacks that are reached are exactly the ordered multiples of 5 modulo 12: 5, 10, 3, ... , 7, 0. Note that if every seventh tack were used, the same artwork would be produced. If every third tack were connected, as in Figure 15.1.3, p. 150(b), the resulting loop would only use four tacks; thus 3 does not generate $\mathbb{Z}_{12}$. □

Example 15.1.4 The Group of Integers is Cyclic. The additive group of integers, $[\mathbb{Z}; +]$, is cyclic:

$$\mathbb{Z} = \langle 1 \rangle = \{n \cdot 1 | n \in \mathbb{Z}\}$$

This observation does not mean that every integer is the product of an integer times 1. It means that

$$\mathbb{Z} = \{0\} \cup \{\overbrace{1+1+\cdots+1}^{n \text{ terms}} \mid n \in \mathbb{P}\} \cup \{\overbrace{(-1)+(-1)+\cdots+(-1)}^{n \text{ terms}} \mid n \in \mathbb{P}\}$$

□

Theorem 15.1.5 Cyclic Implies Abelian. *If $[G; *]$ is cyclic, then it is abelian.*
Proof. Let a be any generator of G and let $b, c \in G$. By the definition of the generator of a group, there exist integers m and n such that $b = ma$ and $c = na$. Thus, using Theorem 11.3.14, p. 33,

$$\begin{aligned} b * c &= (ma) * (na) \\ &= (m+n)a \\ &= (n+m)a \\ &= (na) * (ma) \\ &= c * b \end{aligned}$$

■

One of the first steps in proving a property of cyclic groups is to use the fact that there exists a generator. Then every element of the group can be expressed as some multiple of the generator. Take special note of how this is used in theorems of this section.

Up to now we have used only additive notation to discuss cyclic groups. Theorem 15.1.5, p. 150 actually justifies this practice since it is customary to use additive notation when discussing abelian groups. Of course, some concrete groups for which we employ multiplicative notation are cyclic. If one of its

15.1. CYCLIC GROUPS

elements, a, is a generator,
$$\langle a \rangle = \{a^n \mid n \in \mathbb{Z}\}$$

Example 15.1.6 A Cyclic Multiplicative Group. The group of positive integers modulo 11 with modulo 11 multiplication, $[\mathbb{U}_{11}; \times_{11}]$, is cyclic. One of its generators is 6: $6^1 = 6$, $6^2 = 3$, $6^3 = 7, \ldots, 6^9 = 2$, and $6^{10} = 1$, the identity of the group. □

Example 15.1.7 A Non-cyclic Group. The real numbers with addition, $[\mathbb{R}; +]$ is a noncyclic group. The proof of this statement requires a bit more generality since we are saying that for all $r \in \mathbb{R}$, $\langle r \rangle$ is a proper subset of $\mathbb{R}$. If r is nonzero, the multiples of r are distributed over the real line, as in Figure 15.1.8, p. 151. It is clear then that there are many real numbers, like $r/2$, that are not in $\langle r \rangle$.

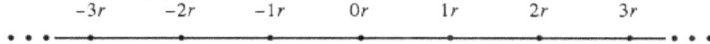

Figure 15.1.8 Elements of $\langle r \rangle, r > 0$

□

The next two proofs make use of the Theorem 11.4.1, p. 35.

The following theorem shows that a cyclic group can never be very complicated.

Theorem 15.1.9 Possible Cyclic Group Structures. *If G is a cyclic group, then G is either finite or countably infinite. If G is finite and $|G| = n$, it is isomorphic to $[\mathbb{Z}_n; +_n]$. If G is infinite, it is isomorphic to $[\mathbb{Z}; +]$.*

Proof. Case 1: $|G| < \infty$. If a is a generator of G and $|G| = n$, define $\phi : \mathbb{Z}_n \to G$ by $\phi(k) = ka$ for all $k \in \mathbb{Z}_n$.

Since $\langle a \rangle$ is finite, we can use the fact that the elements of $\langle a \rangle$ are the first n nonnegative multiples of a. From this observation, we see that ϕ is a surjection. A surjection between finite sets of the same cardinality must be a bijection. Finally, if $p, q \in \mathbb{Z}_n$,

$$\begin{aligned}
\phi(p) + \phi(q) &= pa + qa \\
&= (p+q)a \\
&= (p +_n q)a \quad \text{see exercise 10} \\
&= \phi(p +_n q)
\end{aligned}$$

Therefore ϕ is an isomorphism.

Case 2: $|G| = \infty$. We will leave this case as an exercise. ∎

Theorem 15.1.10 Subgroups of Cyclic Groups. *Every subgroup of a cyclic group is cyclic.*

Proof. Let G be cyclic with generator a and let $H \leq G$. If $H = \{e\}$, H has e as a generator. We may now assume that $|H| \geq 2$ and $a \neq e$. Let m be the least positive integer such that ma belongs to H. This is the key step. It lets us get our hands on a generator of H. We will now show that $c = ma$ generates H. Certainly, $\langle c \rangle \subseteq H$, but suppose that $\langle c \rangle \neq H$. Then there exists $b \in H$ such that $b \notin \langle c \rangle$. Now, since b is in G, there exists $n \in \mathbb{Z}$ such that $b = na$. We now apply the division property and divide n by m. $b = na = (qm+r)a = (qm)a + ra$, where $0 \leq r < m$. We note that r cannot be zero for otherwise we would have $b = na = q(ma) = qc \in \langle c \rangle$. Therefore, $ra = na - (qm)a \in H$. This contradicts our choice of m because $0 < r < m$. ∎

Example 15.1.11 All subgroups of $\mathbb{Z}_{10}$. The only proper subgroups of $\mathbb{Z}_{10}$ are $H_1 = \{0, 5\}$ and $H_2 = \{0, 2, 4, 6, 8\}$. They are both cyclic: $H_1 = \langle 5 \rangle$, while

$H_2 = \langle 2 \rangle = \langle 4 \rangle = \langle 6 \rangle = \langle 8 \rangle$. The generators of $\mathbb{Z}_{10}$ are 1, 3, 7, and 9. □

Example 15.1.12 All subgroups of $\mathbb{Z}$. With the exception of $\{0\}$, all subgroups of $\mathbb{Z}$ are isomorphic to $\mathbb{Z}$. If $H \leq \mathbb{Z}$, then H is the cyclic subgroup generated by the least positive element of H. It is infinite and so by Theorem 15.1.10, p. 151 it is isomorphic to $\mathbb{Z}$. □

We now cite a useful theorem for computing the order of cyclic subgroups of a cyclic group:

Theorem 15.1.13 The order of elements of a finite cyclic group. *If G is a cyclic group of order n and a is a generator of G, the order of ka is n/d, where d is the greatest common divisor of n and k.*
Proof. The proof of this theorem is left to the reader. ∎

Example 15.1.14 Computation of an order in a cyclic group. To compute the order of $\langle 18 \rangle$ in $\mathbb{Z}_{30}$, we first observe that 1 is a generator of $\mathbb{Z}_{30}$ and $18 = 18(1)$. The greatest common divisor of 18 and 30 is 6. Hence, the order of $\langle 18 \rangle$ is 30/6, or 5. □

At this point, we will introduce the idea of a fast adder, a relatively modern application (Winograd, 1965) of an ancient theorem, Sun Tzu's Theorem. We will present only an overview of the theory and rely primarily on examples.

Out of necessity, integer addition with a computer is addition modulo n, for n some larger number. Consider the case where n is small, like 64. Then addition involves the addition of six-digit binary numbers. Consider the process of adding 31 and 1. Assume the computer's adder takes as input two bit strings $a = \{a_0, a_1, a_2, a_3, a_4, a_5\}$ and $b = \{b_0, b_1, b_2, b_3, b_4, b_5\}$ and outputs $s = \{s_0, s_1, s_2, s_3, s_4, s_5\}$, the sum of a and b. Then, if $a = 31 = (1, 1, 1, 1, 1, 0)$ and $b = 1 = (1, 0, 0, 0, 0, 0)$, s will be $(0, 0, 0, 0, 0, 1)$, or 32. The output $s = 1$ cannot be determined until all other outputs have been determined. If addition is done with a finite-state machine, as in Example 14.3.9, p. 142, the time required to get s will be six time units, where one time unit is the time it takes to get one output from the machine. In general, the time required to obtain s will be proportional to the number of bits. Theoretically, this time can be decreased, but the explanation would require a long digression and our relative results would not change that much. We will use the rule that the number of time units needed to perform addition modulo n is proportional to $\lceil \log_2 n \rceil$.

Now we will introduce a hypothetical problem that we will use to illustrate the idea of a fast adder. Suppose that we had to add 1,000 numbers modulo $27720 = 8 \cdot 9 \cdot 5 \cdot 7 \cdot 11$. By the rule above, since $2^{14} < 27720 < 2^{15}$, each addition would take 15 time units. If the sum is initialized to zero, 1,000 additions would be needed; thus, 15,000 time units would be needed to do the additions. We can improve this time dramatically by applying Sun Tzu's Theorem. Recall that $k\%n$ is the remainder upon division of k by n.

Theorem 15.1.15 Sun Tzu's Theorem. *Let $n_1, n_2, \ldots, n_p$ be integers that have no common factor greater than one between any pair of them; i. e., they are relatively prime. Let $n = n_1 n_2 \cdots n_p$. Define*

$$\theta : \mathbb{Z}_n \to \mathbb{Z}_{n_1} \times \mathbb{Z}_{n_2} \times \cdots \times \mathbb{Z}_{n_p}$$

by

$$\theta(k) = (k_1, k_2, \ldots, k_p) = (k\%n_1, k\%n_2, \ldots, k\%n_p)$$

where for $1 \leq i \leq p$, $0 \leq k_i < n_i$ and $k \equiv k_i \,(mod\, n_i)$. Then θ is an isomorphism from $\mathbb{Z}_n$ into $\mathbb{Z}_{n_1} \times \mathbb{Z}_{n_2} \times \cdots \times \mathbb{Z}_{n_p}$.

15.1. CYCLIC GROUPS

Sun Tzu's Theorem can be stated in several different forms, and its proof can be found in many abstract algebra texts. Older texts most likely will refer to the theorem as the Chinese Remainder Theorem.

As we saw in Chapter 11, $\mathbb{Z}_6$ is isomorphic to $\mathbb{Z}_2 \times \mathbb{Z}_3$. This is the smallest case to which Sun Tzu's Theorem can be applied. An isomorphism between $\mathbb{Z}_6$ and $\mathbb{Z}_2 \times \mathbb{Z}_3$ is

$$\theta(0) = (0,0) \quad \theta(3) = (1,0)$$
$$\theta(1) = (1,1) \quad \theta(4) = (0,1)$$
$$\theta(2) = (0,2) \quad \theta(5) = (1,2)$$

Let's consider a somewhat larger case. We start by selecting a modulus that can be factored into a product of relatively prime integers: $n = 21,600 = 2^5 3^3 5^2$. In this case the factors are $2^5 = 32$, $3^3 = 27$, and $5^2 = 25$. They need not be powers of primes, but it is easy to break the factors into this form to assure relatively prime numbers. To add in $\mathbb{Z}_n$, we need $\lceil \log_2 n \rceil = 15$ time units. Let $G = \mathbb{Z}_{32} \times \mathbb{Z}_{27} \times \mathbb{Z}_{25}$. Sun Tzu's Theorem gives us an isomorphism between $\mathbb{Z}_{21600}$ and G. The basic idea behind the fast adder, illustrated in Figure 15.1.16, p. 153, is to make use of this isomorphism. The notation x += a is interpreted as the instruction to add the value of a to the variable x.

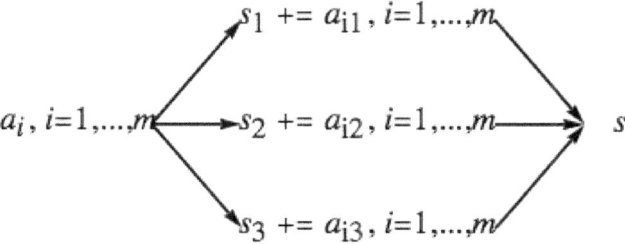

Figure 15.1.16 Fast Adder Scheme

Assume we have several integers $a_1, \ldots, a_m$ to be added. Here, we assume $m = 20$. We compute the sum s to compare our result with this true sum.

```
a=[1878,1384,84,2021,784,1509,1740,1201,2363,1774,
   1865,33,1477,894,690,520,198,1349,1278,650]
s =0
for t in a:
    s+=t
s
```

23692

Although our sum is an integer calculation, we will put our calculation in the context of the integers modulo 21600. The isomophism from $\mathbb{Z}_{21600}$ into $G = \mathbb{Z}_{32} \times \mathbb{Z}_{27} \times \mathbb{Z}_{25}$ is defined in Sage as theta. In addition we demonstrate that the operations in these groups are preserved by theta.

```
G=cartesian_product([Integers(32),Integers(27),Integers(25)])
def theta(x):
    return G((x%32,x%27,x%25))
[theta(1878)+theta(1384),theta(1878+1384)]
```

[(30, 22, 12), (30, 22, 12)]

We initialize the sums in each factor of the range of theta to zero and decompose each summand t into a triple $\theta(t) = (t_1, t_2, t_3) \in G$.

```
sum=G((0,0,0))
for t in a:
    sum+=theta(t)
sum
```

(12, 13, 17)

Addition in G can be done in parallel so that each new subtotal in the form of the triple (s_1, s_2, s_3) takes only as long to compute as it takes to add in the largest modulus, $\log_2 32 = 5$ time units, if calculations are done in parallel. By the time rule that we have established, the addition of 20 numbers can be done in $20 \cdot 5 = 100$ time units, as opposed to $20 \cdot 15 = 300$ time units if we do the calculations in $\mathbb{Z}_{21600}$. However the result is a triple in G. The function that performs the inverse of theta is built into most mathematics programs, including Sage. In Sage the function is crt, short for Chinese Remainder Theorem, the other common name of Sun Tzu's Theorem. We use this function to compute the inverse of our triple, which is an element of $\mathbb{Z}_{21600}$. The result isn't the true sum because the modulus 21600 is not large enough. However, we verify that our result is congruent to the true sum modulo 21600.

```
isum=crt([12,13,17],[32,27,25])
[isum,(s-isum)%(21600)]
```

[2092, 0]

In order to get the true sum from our scheme, the modulus would need to be increased by moving from 21600 to, for example, $21600 * 23 = 496800$. Mapping into the new group, $G = \mathbb{Z}_{32} \times \mathbb{Z}_{27} \times \mathbb{Z}_{25} \times \mathbb{Z}_{23}$ will take slightly longer, as will the inversion process with crt, but adding the summands that are in the form of quadruples can be done with no additional time.

The computation of $\theta^{-1}(s_1, s_2, s_3)$ that is done by the Sage function crt can be accomplished in a variety of ways. All of them ultimately are simplified by the fact that θ^{-1} is also an isomorphism. One approach is to use the isomorphism property to realize that the value of $\theta^{-1}(s_1, s_2, s_3)$ is $s_1\theta^{-1}(1,0,0) + s_2\theta^{-1}(0,1,0) + s_3\theta^{-1}(0,0,1)$. The arithmetic in this expression is in the domain of θ and is more time consuming, but it need only be done once. This is why the fast adder is only practical in situations where many additions must be performed to get a single sum.

The inverse images of the "unit vectors" can be computed ahead of time.

```
u=[crt([1,0,0],[32,27,25]),
   crt([0,1,0],[32,27,25]),crt([0,0,1],[32,27,25])]
u
```

[7425, 6400, 7776]

The result we computed earlier can be computed directly by in the larger modulus.

```
(7425*12 + 6400*13+ 7776* 17)%21600
```

2092

To further illustrate the potential of fast adders, consider increasing the modulus to $n = 2^5 3^3 5^2 7^2 11 \cdot 13 \cdot 17 \cdot 19 \cdot 23 \cdot 29 \cdot 31 \cdot 37 \cdot 41 \cdot 43 \cdot 47 \approx 3.1 \times 10^{21}$. Each addition using the usual modulo n addition with full adders would take

72 time units. By decomposing each summand into 15-tuples according to Sun Tzu's Theorem, the time is reduced to $\lceil \log_2 49 \rceil = 6$ time units per addition.

Exercises

1. What generators besides 1 does $[\mathbb{Z}; +]$ have?

2. Suppose $[G; *]$ is a cyclic group with generator g. If you build a graph of with vertices from the elements of G and edge set $E = \{(a, g * a) \mid a \in G\}$, what would the graph look like? If G is a group of even order, what would a graph with edge set $E' = \{(a, g^2 * a) \mid a \in G\}$ look like?

3. Prove that if $|G| > 2$ and G is cyclic, G has at least two generators.

4. If you wanted to list the generators of $\mathbb{Z}_n$ you would only have to test the first $n/2$ positive integers. Why?

5. Which of the following groups are cyclic? Explain.

    (a) $[\mathbb{Q}; +]$

    (b) $[\mathbb{R}^+; \cdot]$

    (c) $[6\mathbb{Z}; +]$ where $6\mathbb{Z} = \{6n \mid n \in \mathbb{Z}\}$

    (d) $\mathbb{Z} \times \mathbb{Z}$

    (e) $\mathbb{Z}_2 \times \mathbb{Z}_3 \times \mathbb{Z}_{25}$

6. For each group and element, determine the order of the cyclic subgroup generated by the element:

    (a) $\mathbb{Z}_{25}$, 15

    (b) $\mathbb{Z}_4 \times \mathbb{Z}_9$, $(2, 6)$ (apply Exercise 8)

    (c) $\mathbb{Z}_{64}$, 2

7. How can Theorem 15.1.13, p. 152 be applied to list the generators of $\mathbb{Z}_n$? What are the generators of $\mathbb{Z}_{25}$? Of $\mathbb{Z}_{256}$?

8. Prove that if the greatest common divisor of n and m is 1, then $(1, 1)$ is a generator of $\mathbb{Z}_n \times \mathbb{Z}_m$, and hence, $\mathbb{Z}_n \times \mathbb{Z}_m$ is isomorphic to $\mathbb{Z}_{nm}$.

9.
    (a) Illustrate how the fast adder can be used to add the numbers 21, 5, 7, and 15 using the isomorphism between $\mathbb{Z}_{77}$ and $\mathbb{Z}_7 \times \mathbb{Z}_{11}$.

    (b) If the same isomorphism is used to add the numbers 25, 26, and 40, what would the result be, why would it be incorrect, and how would the answer differ from the answer in part a?

10. Prove that if G is a cyclic group of order n with generator a, and $p, q \in \{0, 1, \ldots, n-1\}$, then $(p+q)a = (p +_n q) a$.

15.2 Cosets and Factor Groups

Consider the group $[\mathbb{Z}_{12}; +_{12}]$. As we saw in the previous section, we can picture its cyclic properties with the string art of Figure 15.1.3, p. 150. Here we will be interested in the non-generators, like 3. The solid lines in Figure 15.2.1, p. 156 show that only one-third of the tacks have been reached by starting at zero and jumping to every third tack. The numbers of these tacks correspond to $\langle 3 \rangle = \{0, 3, 6, 9\}$.

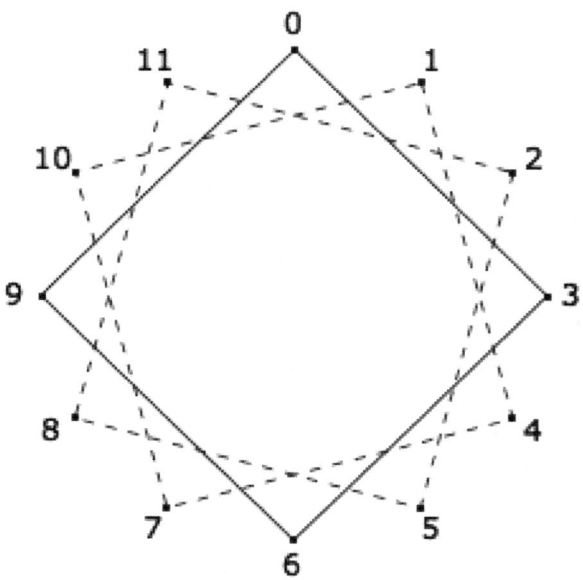

Figure 15.2.1 "String art" cosets

What happens if you start at one of the unused tacks and again jump to every third tack? The two broken paths on Figure 15.2.1, p. 156 show that identical squares are produced. The tacks are thus partitioned into very similar subsets. The subsets of $\mathbb{Z}_{12}$ that they correspond to are $\{0, 3, 6, 9\}$, $\{1, 4, 7, 10\}$, and $\{2, 5, 8, 11\}$. These subsets are called cosets. In particular, they are called cosets of the subgroup $\{0, 3, 6, 9\}$. We will see that under certain conditions, cosets of a subgroup can form a group of their own. Before pursuing this example any further we will examine the general situation.

Definition 15.2.2 Coset. If $[G; *]$ is a group, $H \leq G$ and $a \in G$, the left coset of H generated by a is

$$a * H = \{a * h | h \in H\}$$

and the right coset of H generated by a is

$$H * a = \{h * a | h \in H\}.$$

$\Diamond$

Note 15.2.3

(a) H itself is both a left and right coset since $e * H = H * e = H$.

(b) If G is abelian, $a * H = H * a$ and the left-right distinction for cosets can be dropped. We will normally use left coset notation in that situation.

Definition 15.2.4 Coset Representative. Any element of a coset is called a representative of that coset. $\Diamond$

One might wonder whether a is in any way a special representative of $a * H$ since it seems to define the coset. It is not, as we shall see.

Remark 15.2.5 A Duality Principle. A duality principle can be formulated concerning cosets because left and right cosets are defined in such similar ways. Any theorem about left and right cosets will yield a second theorem when "left" and "right" are exchanged for "right" and "left."

15.2. COSETS AND FACTOR GROUPS

Theorem 15.2.6 *If $b \in a * H$, then $a * H = b * H$, and if $b \in H * a$, then $H * a = H * b$.*

Proof. In light of the remark above, we need only prove the first part of this theorem. Suppose that $x \in a * H$. We need only find a way of expressing x as "b times an element of H." Then we will have proven that $a * H \subseteq b * H$. By the definition of $a * H$, since b and x are in $a * H$, there exist h_1 and h_2 in H such that $b = a * h_1$ and $x = a * h_2$. Given these two equations, $a = b h_1^{-1}$ and

$$x = a * h_2 = (b * h_1^{-1}) * h_2 = b * (h_1^{-1} * h_2)$$

Since $h_1, h_2 \in H$, $h_1^{-1} * h_2 \in H$, and we are done with this part of the proof. In order to show that $b * H \subseteq a * H$, one can follow essentially the same steps, which we will let the reader fill in. ∎

Example 15.2.7 In Figure 15.2.1, p. 156, you can start at either 1 or 7 and obtain the same path by taking jumps of three tacks in each step. Thus,

$$1 +_{12} \{0, 3, 6, 9\} = 7 +_{12} \{0, 3, 6, 9\} = \{1, 4, 7, 10\}.$$

□

The set of left (or right) cosets of a subgroup partition a group in a special way:

Theorem 15.2.8 Cosets Partition a Group. *If $[G; *]$ is a group and $H \leq G$, the set of left cosets of H is a partition of G. In addition, all of the left cosets of H have the same cardinality. The same is true for right cosets.*

Proof. That every element of G belongs to a left coset is clear because $a \in a * H$ for all $a \in G$. If $a * H$ and $b * H$ are left cosets, we will prove that they are either equal or disjoint. If $a * H$ and $b * H$ are not disjoint, $a * H \cap b * H$ is nonempty and some element $c \in G$ belongs to the intersection. Then by Theorem 15.2.6, p. 157, $c \in a * H \Rightarrow a * H = c * H$ and $c \in b * H \Rightarrow b * H = c * H$. Hence $a * H = b * H$.

We complete the proof by showing that each left coset has the same cardinality as H. To do this, we simply observe that if $a \in G$, $\rho: H \to a * H$ defined by $\rho(h) = a * h$ is a bijection and hence $|H| = |a * H|$. We will leave the proof of this statement to the reader. ∎

The function ρ has a nice interpretation in terms of our opening example. If $a \in \mathbb{Z}_{12}$, the graph of $\{0, 3, 6, 9\}$ is rotated $(30a)°$ to coincide with one of the three cosets of $\{0, 3, 6, 9\}$.

Corollary 15.2.9 A Coset Counting Formula. *If $|G| < \infty$ and $H \leq G$, the number of distinct left cosets of H equals $\frac{|G|}{|H|}$. For this reason we use G/H to denote the set of left cosets of H in G*

Proof. This follows from the partitioning of G into equal sized sets, one of which is H. ∎

Example 15.2.10 The set of integer multiples of four, $4\mathbb{Z}$, is a subgroup of $[\mathbb{Z}; +]$. Four distinct cosets of $4\mathbb{Z}$ partition the integers. They are $4\mathbb{Z}$, $1 + 4\mathbb{Z}$, $2 + 4\mathbb{Z}$, and $3 + 4\mathbb{Z}$, where, for example, $1 + 4\mathbb{Z} = \{1 + 4k | k \in \mathbb{Z}\}$. $4\mathbb{Z}$ can also be written $0 + 4\mathbb{Z}$. □

Convention 15.2.11 Distinguished Representatives. Although we have seen that any representative can describe a coset, it is often convenient to select a distinguished representative from each coset. The advantage to doing this is that there is a unique name for each coset in terms of its distinguished representative. In numeric examples such as the one above, the distinguished representative is usually the smallest nonnegative representative. Remember, this is purely a convenience and there is absolutely nothing wrong in writing

$-203 + 4\mathbb{Z}$, $5 + 4\mathbb{Z}$, or $621 + 4\mathbb{Z}$ in place of $1 + 4\mathbb{Z}$ because $-203, 5, 621 \in 1 + 4\mathbb{Z}$.

Before completing the main thrust of this section, we will make note of a significant implication of Theorem 15.2.8, p. 157. Since a finite group is divided into cosets of a common size by any subgroup, we can conclude:

Theorem 15.2.12 Lagrange's Theorem. *The order of a subgroup of a finite group must divide the order of the group.*

One immediate implication of Lagrange's Theorem is that if p is prime, $\mathbb{Z}_p$ has no proper subgroups.

We will now describe the operation on cosets which will, under certain circumstances, result in a group. For most of this section, we will assume that G is an abelian group. This is one sufficient (but not necessary) condition that guarantees that the set of left cosets will form a group.

Definition 15.2.13 Operation on Cosets. Let C and D be left cosets of H, a subgroup of G with representatives c and d, respectively. Then

$$C \otimes D = (c * H) \otimes (d * H) = (c * d) * H$$

The operation $\otimes$ is called the operation induced on left cosets by $*$. ◇

In Theorem 15.2.18, p. 159, later in this section, we will prove that if G is an abelian group, $\otimes$ is indeed an operation. In practice, if the group G is an additive group, the symbol $\otimes$ is replaced by $+$, as in the following example.

Example 15.2.14 Computing with cosets of $4\mathbb{Z}$. Consider the cosets described in Example 15.2.10, p. 157. For brevity, we rename $0 + 4\mathbb{Z}$, $1 + 4\mathbb{Z}$, $2 + 4\mathbb{Z}$, and $3 + 4\mathbb{Z}$ with the symbols $\bar{0}$, $\bar{1}$, $\bar{2}$, and $\bar{3}$. Let's do a typical calculation, $\bar{1} + \bar{3}$. We will see that the result is always going to be $\bar{0}$, no matter what representatives we select. For example, $9 \in \bar{1}$, $7 \in \bar{3}$, and $9 + 7 = 16 \in \bar{0}$. Our choice of the representatives $\bar{1}$ and $\bar{3}$ were completely arbitrary. □

In general, $C \otimes D$ can be computed in many ways, and so it is necessary to show that the choice of representatives does not affect the result. When the result we get for $C \otimes D$ is always independent of our choice of representatives, we say that "$\otimes$ is well defined." Addition of cosets is a well-defined operation on the left cosets of $4\mathbb{Z}$ and is summarized in the following table. Do you notice anything familiar?

$\otimes$	$\bar{0}$	$\bar{1}$	$\bar{2}$	$\bar{3}$
$\bar{0}$	$\bar{0}$	$\bar{1}$	$\bar{2}$	$\bar{3}$
$\bar{1}$	$\bar{1}$	$\bar{2}$	$\bar{3}$	$\bar{0}$
$\bar{2}$	$\bar{2}$	$\bar{3}$	$\bar{0}$	$\bar{1}$
$\bar{3}$	$\bar{3}$	$\bar{0}$	$\bar{1}$	$\bar{2}$

Example 15.2.15 Cosets of the integers in the group of Real numbers. Consider the group of real numbers, $[\mathbb{R}; +]$, and its subgroup of integers, $\mathbb{Z}$. Every element of $\mathbb{R}/\mathbb{Z}$ has the same cardinality as $\mathbb{Z}$. Let $s, t \in \mathbb{R}$. $s \in t + \mathbb{Z}$ if s can be written $t + n$ for some $n \in \mathbb{Z}$. Hence s and t belong to the same coset if they differ by an integer. (See Exercise 15.2.6, p. 160 for a generalization of this fact.)

Now consider the coset $0.25 + \mathbb{Z}$. Real numbers that differ by an integer from 0.25 are $1.25, 2.25, 3.25, \ldots$ and $-0.75, -1.75, -2.75, \ldots$. If any real number is selected, there exists a representative of its coset that is greater than or equal to 0 and less than 1. We will call that representative the distinguished representative of the coset. For example, 43.125 belongs to the coset represented by 0.125; $-6.382 + \mathbb{Z}$ has 0.618 as its distinguished representative. The operation on $\mathbb{R}/\mathbb{Z}$ is commonly called addition modulo 1. A few typical calculations in

15.2. COSETS AND FACTOR GROUPS

$\mathbb{R}/\mathbb{Z}$ are
$$(0.1 + \mathbb{Z}) + (0.48 + \mathbb{Z}) = 0.58 + \mathbb{Z}$$
$$(0.7 + \mathbb{Z}) + (0.31 + \mathbb{Z}) = 0.01 + \mathbb{Z}$$
$$-(0.41 + \mathbb{Z}) = -0.41 + \mathbb{Z} = 0.59 + \mathbb{Z}$$
$$\text{and in general,} \; -(a + \mathbb{Z}) = (1 - a) + \mathbb{Z}$$

□

Example 15.2.16 Cosets in a Direct Product. Consider $F = (\mathbb{Z}_4 \times \mathbb{Z}_2)/H$, where $H = \{(0,0), (0,1)\}$. Since $\mathbb{Z}_4 \times \mathbb{Z}_2$ is of order 8, each element of F is a coset containing two ordered pairs. We will leave it to the reader to verify that the four distinct cosets are $(0,0) + H$, $(1,0) + H$, $(2,0) + H$ and $(3,0) + H$. The reader can also verify that F is isomorphic to $\mathbb{Z}_4$, since F is cyclic. An educated guess should give you a generator. □

Example 15.2.17 Consider the group $\mathbb{Z}_2{}^4 = \mathbb{Z}_2 \times \mathbb{Z}_2 \times \mathbb{Z}_2 \times \mathbb{Z}_2$. Let H be $\langle (1,0,1,0) \rangle$, the cyclic subgroup of $\mathbb{Z}_2{}^4$ generate by (1,0,1,0). Since

$$(1,0,1,0) + (1,0,1,0) = (1 +_2 1, 0 +_2 0, 1 +_2 1, 0 +_2 0) = (0,0,0,0)$$

the order of H is 2 and, $\mathbb{Z}_2{}^4/H$ has $|\mathbb{Z}_2^4/H| = \frac{|\mathbb{Z}_2^4|}{|H|} = \frac{16}{2} = 8$ elements. A typical coset is

$$C = (0,1,1,1) + H = \{(0,1,1,1), (1,1,0,1)\}$$

Note that since $2(0,1,1,1) = (0,0,0,0)$, $2C = C \otimes C = H$, the identity for the operation on $\mathbb{Z}_2{}^4/H$. The orders of non-identity elements of this factor group are all 2, and it can be shown that the factor group is isomorphic to $\mathbb{Z}_2{}^3$. □

Theorem 15.2.18 Coset operation is well-defined (Abelian Case). *If G is an abelian group, and $H \leq G$, the operation induced on cosets of H by the operation of G is well defined.*

Proof. Suppose that a, b, and a', b'. are two choices for representatives of cosets C and D. That is to say that $a, a' \in C$, $b, b' \in D$. We will show that $a * b$ and $a' * b'$ are representatives of the same coset. Theorem 15.2.6, p. 1571 implies that $C = a * H$ and $D = b * H$, thus we have $a' \in a * H$ and $b' \in b * H$. Then there exists $h_1, h_2 \in H$ such that $a' = a * h_1$ and $b' = b * h_2$ and so

$$a' * b' = (a * h_1) * (b * h_2) = (a * b) * (h_1 * h_2)$$

by various group properties and the assumption that G is abelian, which lets us reverse the order in which b and h_1 appear in the chain of equalities. This last expression for $a' * b'$ implies that $a' * b' \in (a * b) * H$ since $h_1 * h_2 \in H$ because H is a subgroup of G. Thus, we get the same coset for both pairs of representatives. ∎

Theorem 15.2.19 *Let G be a group and $H \leq G$. If the operation induced on left cosets of H by the operation of G is well defined, then the set of left cosets forms a group under that operation.*

Proof. Let C_1, C_2, and C_3 be the left cosets with representatives r_1, r_2, and r_3, respectively. The values of $C_1 \otimes (C_2 \otimes C_3)$ and $(C_1 \otimes C_2) \otimes C_3$ are determined by $r_1 * (r_2 * r_3)$ and $(r_1 * r_2) * r_3$, respectively. By the associativity of $*$ in G, these two group elements are equal and so the two coset expressions must be equal. Therefore, the induced operation is associative. As for the identity and inverse properties, there is no surprise. The identity coset is H, or $e * H$, the coset that contains G's identity. If C is a coset with representative a; that is, if

$C = a * H$, then C^{-1} is $a^{-1} * H$.

$$(a * H) \otimes (a^{-1} * H) = (a * a^{-1}) * H = e * H = \text{ identity coset.}$$

∎

Definition 15.2.20 Factor Group. Let G be a group and $H \leq G$. If the set of left cosets of H forms a group, then that group is called the factor group of "G modulo H." It is denoted G/H. ◊

Note 15.2.21 If G is abelian, then every subgroup of G yields a factor group. We will delay further consideration of the non-abelian case to Section 15.4.

Remark 15.2.22 On Notation. It is customary to use the same symbol for the operation of G/H as for the operation on G. The reason we used distinct symbols in this section was to make the distinction clear between the two operations.

Exercises

1. Consider $\mathbb{Z}_{10}$ and the subsets of $\mathbb{Z}_{10}$, $\{0, 1, 2, 3, 4\}$ and $\{5, 6, 7, 8, 9\}$. Why is the operation induced on these subsets by modulo 10 addition not well defined?

2. Can you think of a group G, with a subgroup H such that $|H| = 6$ and $|G/H| = 6$? Is your answer unique?

3. For each group and subgroup, what is G/H isomorphic to?

    (a) $G = \mathbb{Z}_4 \times \mathbb{Z}_2$ and $H = \langle (2, 0) \rangle$. Compare to Example 15.2.16, p. 159.

    (b) $G = [\mathbb{C}; +]$ and $H = \mathbb{R}$.

    (c) $G = \mathbb{Z}_{20}$ and $H = \langle 8 \rangle$.

4. For each group and subgroup, what is G/H isomorphic to?

    (a) $G = \mathbb{Z} \times \mathbb{Z}$ and $H = \{(a, a) | a \in \mathbb{Z}\}$.

    (b) $G = [\mathbb{R}^*; \cdot]$ and $H = \{1, -1\}$.

    (c) $G = \mathbb{Z}_2^5$ and $H = \langle (1, 1, 1, 1, 1) \rangle$.

5. Assume that G is a group, $H \leq G$, and $a, b \in G$. Prove that $a * H = b * H$ if and only if $b^{-1} * a \in H$.

6.
    (a) Real addition modulo r, $r > 0$, can be described as the operation induced on cosets of $\langle r \rangle$ by ordinary addition. Describe a system of distinguished representatives for the elements of $\mathbb{R}/\langle r \rangle$.

    (b) Consider the trigonometric function sine. Given that $\sin(x + 2\pi k) = \sin x$ for all $x \in \mathbb{R}$ and $k \in \mathbb{Z}$, show how the distinguished representatives of $\mathbb{R}/\langle 2\pi \rangle$ can be useful in developing an algorithm for calculating the sine of a number.

7. Complete the proof of Theorem 15.2.8, p. 157 by proving that if $a \in G$, $\rho : H \to a * H$ defined by $\rho(h) = a * h$ is a bijection.

15.3 Permutation Groups

15.3.1 The Symmetric Groups

At the risk of boggling the reader's mind, we will now examine groups whose elements are functions. Recall that a permutation on a set A is a bijection from A into A. Suppose that $A = \{1, 2, 3\}$. There are $3! = 6$ different permutations on A. We will call the set of all 6 permutations S_3. They are listed in the following table. The matrix form for describing a function on a finite set is to list the domain across the top row and the image of each element directly below it. For example $r_1(1) = 2$.

Table 15.3.1 Elements of S_3

$$i = \begin{pmatrix} 1 & 2 & 3 \\ 1 & 2 & 3 \end{pmatrix} \quad r_1 = \begin{pmatrix} 1 & 2 & 3 \\ 2 & 3 & 1 \end{pmatrix} \quad r_2 = \begin{pmatrix} 1 & 2 & 3 \\ 3 & 1 & 2 \end{pmatrix}$$

$$f_1 = \begin{pmatrix} 1 & 2 & 3 \\ 1 & 3 & 2 \end{pmatrix} \quad f_2 = \begin{pmatrix} 1 & 2 & 3 \\ 3 & 2 & 1 \end{pmatrix} \quad f_3 = \begin{pmatrix} 1 & 2 & 3 \\ 2 & 1 & 3 \end{pmatrix}$$

The operation that will give $\{i, r_1, r_2, f_1, f_2, f_3\}$ a group structure is function composition. Consider the "product" $r_1 \circ f_3$:

$$r_1 \circ f_3(1) = r_1(f_3(1)) = r_1(2) = 3$$
$$r_1 \circ f_3(2) = r_1(f_3(2)) = r_1(1) = 2 \ .$$
$$r_1 \circ f_3(3) = r_1(f_3(3)) = r_1(3) = 1$$

The images of 1, 2, and 3 under $r_1 \circ f_3$ and f_2 are identical. Thus, by the definition of equality for functions, we can say $r_1 \circ f_3 = f_2$. The complete table for the operation of function composition is given in Table 15.3.2, p. 161.

Table 15.3.2 Operation Table for S_3

$\circ$	i	r_1	r_2	f_1	f_2	f_3
i	i	r_1	r_2	f_1	f_2	f_3
r_1	r_1	r_2	i	f_3	f_1	f_2
r_2	r_2	i	r_1	f_2	f_3	f_1
f_1	f_1	f_2	f_3	i	r_1	r_2
f_2	f_2	f_3	f_1	r_2	i	r_1
f_3	f_3	f_1	f_2	r_1	r_2	i

List 15.3.3

We don't even need the table to verify that we have a group. Based on the following observations, the set of all permutations on any finite set will be a group.

(1) Function composition is always associative.

(2) The identity for the group is i. If g is any one of the permutations on A and $x \in A$,

$$(g \circ i)(x) = g(i(x)) = g(x) \quad (i \circ g)(x) = i(g(x)) = g(x)$$

Therefore $g \circ i = i \circ g = g$.

(3) A permutation, by definition, is a bijection. In Chapter 7 we proved that this implies that it must have an inverse and the inverse itself is a bijection and hence a permutation. Hence all

elements of S_3 have an inverse in S_3. If a permutation is displayed in matrix form, its inverse can be obtained by exchanging the two rows and rearranging the columns so that the top row is in order. The first step is actually sufficient to obtain the inverse, but the sorting of the top row makes it easier to recognize the inverse.

For example, let's consider a typical permutation on $\{1, 2, 3, 4, 5\}$,
$$f = \begin{pmatrix} 1 & 2 & 3 & 4 & 5 \\ 5 & 3 & 2 & 1 & 4 \end{pmatrix}.$$

$$f^{-1} = \begin{pmatrix} 5 & 3 & 2 & 1 & 4 \\ 1 & 2 & 3 & 4 & 5 \end{pmatrix} = \begin{pmatrix} 1 & 2 & 3 & 4 & 5 \\ 4 & 3 & 2 & 5 & 1 \end{pmatrix}.$$

Note 15.3.4 From Table 15.3.2, p. 161, we can see that S_3 is non-abelian. Remember, non-abelian is the negation of abelian. The existence of two elements that don't commute is sufficient to make a group non-abelian. In this group, r_1 and f_3 is one such pair: $r_1 \circ f_3 = f_2$ while $f_3 \circ r_1 = f_1$, so $r_1 \circ f_3 \neq f_3 \circ r_1$. Caution: Don't take this to mean that every pair of elements has to have this property. There are several pairs of elements in S_3 that do commute. In fact, the identity, i, must commute with everything. Also every element must commute with its inverse.

Definition 15.3.5 Symmetric Group. Let A be a nonempty set. The set of all permutations on A with the operation of function composition is called the symmetric group on A, denoted S_A.

The cardinality of a finite set A is more significant than the elements, and we will denote by S_n the symmetric group on any set of cardinality n, $n \geq 1$.
◊

Example 15.3.6 The significance of S_3. Our opening example, S_3, is the smallest non-abelian group. For that reason, all of its proper subgroups are abelian: in fact, they are all cyclic. Figure 15.3.7, p. 162 shows the Hasse diagram for the subgroups of S_3.

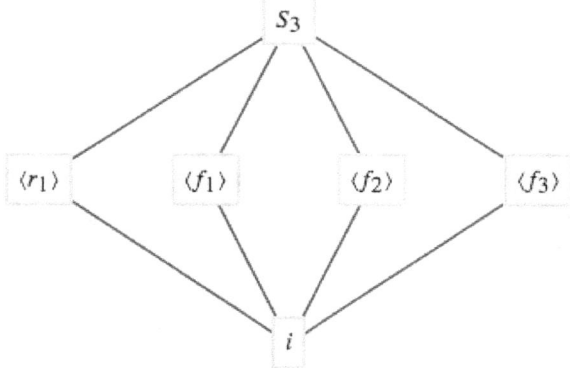

Figure 15.3.7 Lattice diagram of subgroups of S_3

□

Example 15.3.8 Smallest Symmetric Groups. The only abelian symmetric groups are S_1 and S_2, with 1 and 2 elements, respectively. The elements of S_2 are $i = \begin{pmatrix} 1 & 2 \\ 1 & 2 \end{pmatrix}$ and $\alpha = \begin{pmatrix} 1 & 2 \\ 2 & 1 \end{pmatrix}$. S_2 is isomorphic to $\mathbb{Z}_2$. □

Theorem 15.3.9 *For $n \geq 1$, $|S_n| = n!$ and for $n \geq 3$, S_n is non-abelian.*

15.3. PERMUTATION GROUPS

Proof. The first part of the theorem follows from the extended rule of products (see Chapter 2). We leave the details of proof of the second part to the reader after the following hint. Consider f in S_n where $f(1) = 2$, $f(2) = 3$, $f(3) = 1$, and $f(j) = j$ for $3 < j \leq n$. Therefore the cycle representation of f is $(1, 2, 3)$. Now define g in a similar manner so that when you compare $f(g(1))$ and $g(f(1))$ you get different results. ∎

15.3.2 Cycle Notation

A second way of describing a permutation is by means of cycles, which we will introduce first with an example. Consider $f \in S_8$ defined using the now-familiar matrix notation:
$$f = \begin{pmatrix} 1 & 2 & 3 & 4 & 5 & 6 & 7 & 8 \\ 8 & 2 & 7 & 6 & 5 & 4 & 1 & 3 \end{pmatrix}$$

Consider the images of 1 when f is applied repeatedly. The images $f(1)$, $f(f(1))$, $f(f(f(1)))$,... are $8, 3, 7, 1, 8, 3, 7, \ldots$. In Figure 15.3.10, p. 163(a), this situation is represented by a graph with vertices 1, 8, 3, and 7 and shows that the values that you get by repeatedly applying f cycle through those values. This is why we refer to this part of f as a cycle of length 4. Of course starting at 8, 3, or 7 also produces the same cycle with only the starting value changing.

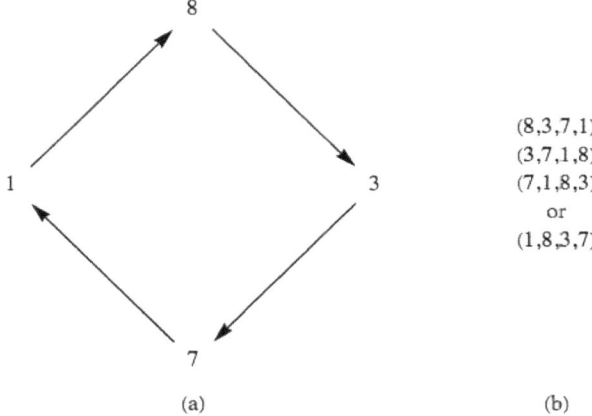

Figure 15.3.10 Representations of a cycle of length 4

Figure 15.3.10, p. 163(a) illustrates how the cycle can be represented in a visual manner, but it is a bit awkward to write. Part (b) of the figure presents a more universally recognized way to write a cycle. In (b), a cycle is represented by a list where the image of any number in the list is its successor. In addition, the last number in the list has as its image the first number.

The other elements of the domain of f are never reached if you start in the cycle $(1, 8, 3, 7)$, and so looking at the images of these other numbers will produce numbers that are disjoint from the set $\{1, 8, 3, 7\}$. The other disjoint cycles of f are (2), $(4, 6)$, and (5). We can express f as a product of disjoint cycles: $f = (1, 8, 3, 7)(2)(4, 6)(5)$ or $f = (1, 8, 3, 7)(4, 6)$, where the absence of 2 and 5 implies that $f(2) = 2$ and $f(5) = 5$.

Note 15.3.11 Disjoint Cycles. We say that two cycles are disjoint if no number appears in both cycles, as is the case in our expressions for f above. Disjoint cycles can be written in any order. Thus, we could also say that $f = (4, 6)(1, 8, 3, 7)$.

Note 15.3.12 Composition of Permutations. We will now consider the composition of permutations written in cyclic form by an example. Suppose that $f = (1, 8, 3, 7)(4, 6)$ and $g = (1, 5, 6)(8, 3, 7, 4)$ are elements of S_8. To calculate $f \circ g$, we start with simple concatenation:

$$f \circ g = (1, 8, 3, 7)(4, 6)(1, 5, 6)(8, 3, 7, 4) \tag{15.3.1}$$

Although this is a valid expression for $f \circ g$, our goal is to express the composition as a product of disjoint cycles as f and g were individually written. We will start by determining the cycle that contains 1. When combining any number of cycles, they are always read from right to left, as with all functions. The first cycle in (15.3.1) does not contain 1; thus we move on to the second. The image of 1 under that cycle is 5. Now we move on to the next cycle, looking for 5, which doesn't appear. The fourth cycle does not contain a 5 either; so $f \circ g(1) = 5$.

At this point, we would have written "$f \circ g = (1, 5$" on paper. We repeat the steps to determine $f \circ g(5)$. This time the second cycle of (15.3.1) moves 5 to 6 and then the third cycle moves 6 to 4. Therefore, $f \circ g(5) = 4$. We continue until the cycle $(1, 5, 4, 3)$ is completed by determining that $f \circ g(3) = 1$. The process is then repeated starting with any number that does not appear in the cycle(s) that have already been completed.

The final result for our example is $f \circ g = (1, 5, 4, 3)(6, 8, 7)$. Since $f(2) = 2$ and $g(2) = 2$, $f \circ g(2) = 2$ and we need not include the one-cycle (2) in the final result, although it can be included.

Example 15.3.13 Some Compositions.

(a) $(1, 2, 3, 4)(1, 2, 3, 4) = (1, 3)(2, 4)$

(b) $(1, 4)(1, 3)(1, 2) = (1, 2, 3, 4)$.

Notice that cyclic notation does not indicate the set which is being permuted. The examples above could be in S_5, where the image of 5 is 5. This ambiguity is usually overcome by making the context clear at the start of a discussion. □

Definition 15.3.14 Transposition. A transposition is a cycle of length 2.
◊

Observation 15.3.15 About transpositions. $f = (1, 4)$ and $g = (4, 5)$ are transpositions in S_5. However, $f \circ g = (1, 4, 5)$ and $g \circ f = (1, 5, 4)$ are not transpositions; thus, the set of transpositions is not closed under composition. Since $f^2 = f \circ f$ and $g^2 = g \circ g$ are both equal to the identity permutation, f and g are their own inverses. In fact, every transposition is its own inverse.

Theorem 15.3.16 Decomposition into Cycles. *Every cycle of length greater than 2 can be expressed as a product of transpositions.*
Proof. We need only indicate how the product of transpositions can be obtained. It is easy to verify that a cycle of length k, $(a_1, a_2, a_3, \ldots, a_k)$, is equal to the following product of $k - 1$ transpositions:

$$(a_1, a_k) \cdots (a_1, a_3)(a_1, a_2)$$

■

Of course, a product of cycles can be written as a product of transpositions just as easily by applying the rule above to each cycle. For example,

$$(1, 3, 5, 7)(2, 4, 6) = (1, 7)(1, 5)(1, 3)(2, 6)(2, 4)$$

Unlike the situation with disjoint cycles, we are not free to change the order of these transpositions.

15.3.3 Parity of Permutations and the Alternating Group

A decomposition of permutations into transpositions makes it possible to classify them and identify an important family of groups.

The proofs of the following theorem appears in many abstract algebra texts.

Theorem 15.3.17 *Every permutation on a finite set can be expressed as the product of an even number of transpositions or an odd number of transpositions, but not both.*

Theorem 15.3.17, p. 165 suggests that S_n can be partitioned into its "even" and "odd" elements. For example, the even permutations of S_3 are i, $r_1 = (1,2,3) = (1,3)(1,2)$ and $r_2 = (1,3,2) = (1,2)(1,3)$. They form a subgroup, $\{i, r_1, r_2\}$ of S_3.

In general:

Definition 15.3.18 The Alternating Group. Let $n \geq 2$. The set of even permutations in S_n is a proper subgroup of S_n called the alternating group on $\{1, 2, \ldots, n\}$, denoted A_n. $\diamond$

We justify our statement that A_n is a group:

Theorem 15.3.19 *Let $n \geq 2$. The alternating group is indeed a group and has order $\frac{n!}{2}$.*

Proof. In this proof, the symbols s_i and t_i stand for transpositions and p, q are even nonnegative integers. If $f, g \in A_n$, we can write the two permutations as products of even numbers of transpositions, $f = s_1 s_2 \cdots s_p$ and $g = t_1 t_2 \cdots t_q$. Then

$$f \circ g = s_1 s_2 \cdots s_p t_1 t_2 \cdots t_q$$

Since $p + q$ is even, $f \circ g \in A_n$, and A_n is closed with respect to function composition. With this, we have proven that A_n is a subgroup of S_n by Theorem 11.5.5, p. 46.

To prove the final assertion, let B_n be the set of odd permutations and let $\tau = (1,2)$. Define $\theta : A_n \to B_n$ by $\theta(f) = f \circ \tau$. Suppose that $\theta(f) = \theta(g)$. Then $f \circ \tau = g \circ \tau$ and by the right cancellation law, $f = g$. Hence, θ is an injection. Next we show that θ is also a surjection. If $h \in B_n$, h is the image of an element of A_n. Specifically, h is the image of $h \circ \tau$.

$$\begin{aligned}\theta(h \circ \tau) &= (h \circ \tau) \circ \tau \\ &= h \circ (\tau \circ \tau) \quad \text{Why?} \\ &= h \circ i \quad \text{Why?} \\ &= h\end{aligned}$$

Since θ is a bijection, $|A_n| = |B_n| = \frac{n!}{2}$. ∎

Example 15.3.20 The Sliding Tile Puzzle. Consider the sliding-tile puzzles pictured in Figure 15.3.21, p. 166. Each numbered square is a tile and the dark square is a gap. Any tile that is adjacent to the gap can slide into the gap. In most versions of this puzzle, the tiles are locked into a frame so that they can be moved only in the manner described above. The object of the puzzle is to arrange the tiles as they appear in Configuration (a). Configurations (b) and (c) are typical starting points. We propose to show why the puzzle can be solved starting with (b), but not with (c).

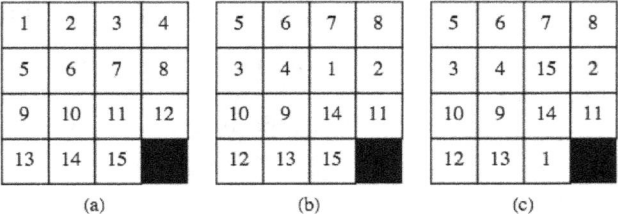

Figure 15.3.21 Configurations of the sliding tile puzzle

We will associate a change in the configuration of the puzzle with an element of S_{16}. Imagine that a tile numbered 16 fills in the gap. For any configuration of the puzzle, the identity i, is the function that leave the configurate "as is." In general, if $f \in S_{16}$, and $1 \leq k \leq 16$, $f(k)$ is the position to which the tile in position k is moved by f that appears in the position of k in configuration (a). If we call the functions that, starting with configuration (a), result in configurations (b) and (c) by the names f_1 and f_2, respectively,

$$f_1 = (1,5,3,7)(2,6,4,8)(9,10)(11,14,13,12)(15)(16)$$

and
$$f_2 = (1,5,3,7,15)(2,6,4,8)(9,10)(11,14,13,12)(16).$$

How can we interpret the movement of one tile as a permutation? Consider what happens when the 12 tile of i slides into the gap. The result is a configuration that we would interpret as $(12, 16)$, a single transposition. Now if we slide the 8 tile into the 12 position, the result is or $(8, 16, 12)$. Hence, by "exchanging" the tiles 8 and 16, we have implemented the function $(8, 12)(12, 16) = (8, 12, 16)$.

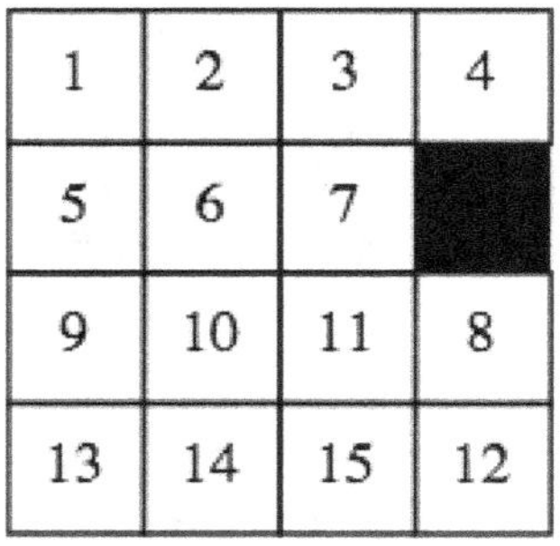

Figure 15.3.22 The configuration $(8, 12, 16)$

Every time you slide a tile into the gap, the new permutation is a transposition composed with the old permutation. Now observe that to start with initial configuration and terminate after a finite number of moves with the gap in its original position, you must make an even number of moves. Thus, configuration corresponding any permutation that leaves 16 fixed cannot be solved if the permutation is odd. Note that f_2 is an odd permutation; thus, Puzzle (c) can't be solved. The proof that all even permutations, such as f_1, can be solved is left to the interested reader to pursue. □

15.3.4 Dihedral Groups

Observation 15.3.23 Realizations of Groups. By now we've seen several instances where a group can appear through an isomorphic copy of itself in various settings. The simplest such example is the cyclic group of order 2. When this group is mentioned, we might naturally think of the group $[\mathbb{Z}_2; +_2]$,

15.3. PERMUTATION GROUPS

but the groups $[\{-1, 1\}; \cdot]$ and $[S_2; \circ]$ are isomorphic to it. None of these groups are necessarily more natural or important than the others. Which one you use depends on the situation you are in and all are referred to as **realizations** of the cyclic group of order 2. The next family of groups we will study, the dihedral groups, has two natural realizations, first as permutations and second as geometric symmetries.

The family of dihedral groups is indexed by the positive integers greater than or equal to 3. For $k \geq 3$, $\mathcal{D}_k$ will have $2k$ elements. We first describe the elements and the operation on them using geometry.

We can describe $\mathcal{D}_n$ in terms of symmetries of a regular n-gon ($n = 3$: equilateral triangle, $n = 4$: square, $n = 5$: regular pentagon,...). Here we will only concentrate on the case of $\mathcal{D}_4$. If a square is fixed in space, there are several motions of the square that will, at the end of the motion, not change the apparent position of the square. The actual changes in position can be seen if the corners of the square are labeled. In Figure 15.3.24, p. 167, the initial labeling scheme is shown, along with the four axes of symmetry of the square.

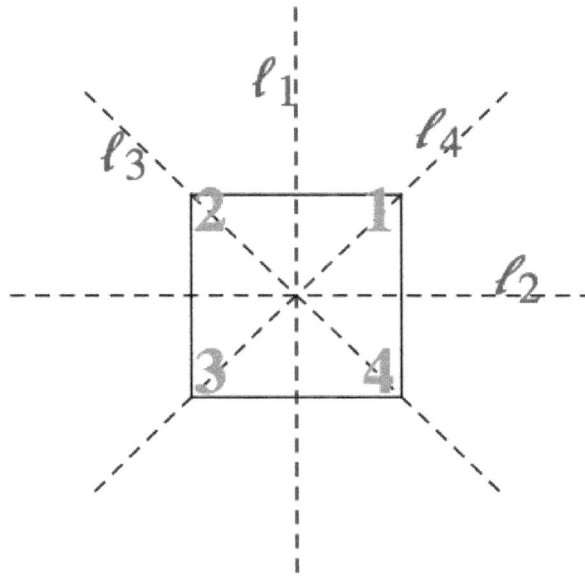

Figure 15.3.24 Axes of symmetry of the square

It might be worthwhile making a square like this with a sheet of paper. Be careful to label the back so that the numbers match the front. Two motions of the square will be considered equivalent if the square is in the same position after performing either motion. There are eight distinct motions. The first four are $0°$, $90°$, $180°$, and $270°$ clockwise rotations of the square, and the other four are the $180°$ flips along the axes l_1, l_2, l_3, and l_4. We will call the rotations i, r_1, r_2, and r_3, respectively, and the flips f_1, f_2, f_3, and f_4, respectively. Figure 15.3.25, p. 168 illustrates r_1 and f_1. For future reference, we also include the permutations to which they correspond.

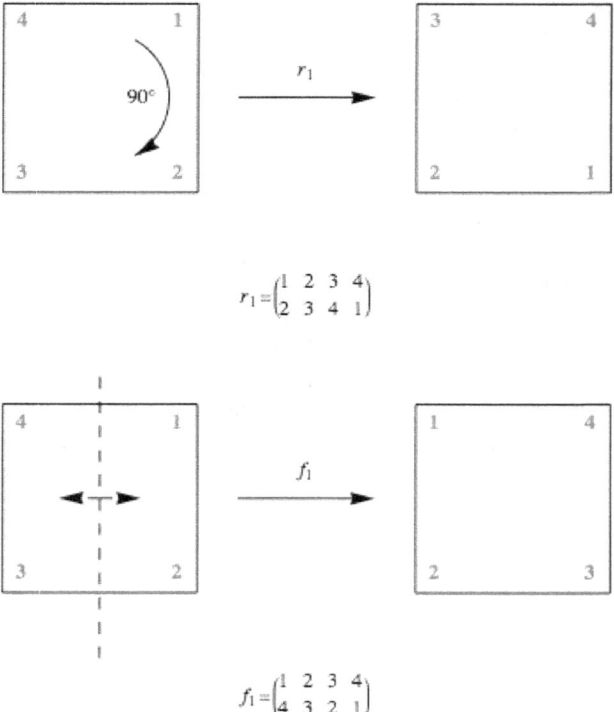

Figure 15.3.25 Two elements of $\mathcal{D}_4$

What is the operation on this set of symmetries? We will call the operation "followed by" and use the symbol $*$ to represent it. The operation will be to combine motions, applying motions from right to left, as with functions. We will illustrate how $*$ is computed by finding $r_1 * f_1$. Starting with the initial configuration, if you perform the f_1 motion, and then immediately perform r_1 on the result, we get the same configuration as if we just performed f_4, which is to flip the square along the line l_4. Therefore, $r_1 * f_1 = f_4$. An important observation is that $f_1 * r_1 \neq f_4$, meaning that this group is nonabelian. The reader is encouraged to verify this on their own.

We can also realize the dihedral groups as permutations. For any symmetric motion of the square we can associate with it a permutation. In the case of $\mathcal{D}_4$, the images of each of the numbers 1 through 4 are the positions on the square that each of the corners 1 through 4 are moved to. For example, since corner 4 moves to position 1 when you perform r_1, the corresponding function will map 4 to 1. In addition, 1 gets mapped to 2, 2 to 3 and 3 to 4. Therefore, r_1 is the cycle $(1,2,3,4)$. The flip f_1 transposes two pairs of corners and corresponds to $(1,4)(2,3)$. If we want to combine these two permutations, using the same names as with motions, we get

$$r_1 \circ f_1 = (1,2,3,4) \circ (1,4)(2,3) = (1)(2,4)(3) = (2,4)$$

Notice that this permutation corresponds with the flip f_4.

Although $\mathcal{D}_4$ isn't cyclic (since it isn't abelian), it can be generated from the two elements r_1 and f_1:

$$\mathcal{D}_4 = \langle r_1, f_1 \rangle = \{i, r_1, r_1{}^2, r_1{}^3, f_1, r_1 \circ f_1, r_1{}^2 \circ f_1, r_1{}^3 \circ f_1\}$$

It is quite easy to describe any of the dihedral groups in a similar fashion. Here is the formal definition

15.3. PERMUTATION GROUPS

Definition 15.3.26 Dihedral Group. Let n be a positive integer greater than or equal to 3. If $r = (1, 2, \ldots, n)$, an n-cycle, and $f = (1, n)(2, n-1)\ldots$ Then

$$\mathcal{D}_n = \langle r, f \rangle = \{i, r, r^2, \ldots, r^{n-1}, f, r \circ f, r^2 \circ f, \ldots, r^{n-1} \circ f\}$$

is the nth dihedral group. ◇

Note 15.3.27 Caution. You might notice that we use a script D, $\mathcal{D}$, for the dihedral groups. Occasionally you might see an ordinary D in other sources for the dihedral groups. Don't confuse it with the set of divisors of n, which we denote by D_n. Normally the context of the discussion should make the meaning of D_n clear.

Example 15.3.28 A Letter-facing Machine. An application of $\mathcal{D}_4$ is in the design of a letter-facing machine. Imagine letters entering a conveyor belt to be postmarked. They are placed on the conveyor belt at random so that two sides are parallel to the belt. Suppose that a postmarker can recognize a stamp in the top right corner of the envelope, on the side facing up. In Figure 15.3.29, p. 169, a sequence of machines is shown that will recognize a stamp on any letter, no matter what position in which the letter starts. The letter P stands for a postmarker. The letters R and F stand for rotating and flipping machines that perform the motions of r_1 and f_1.

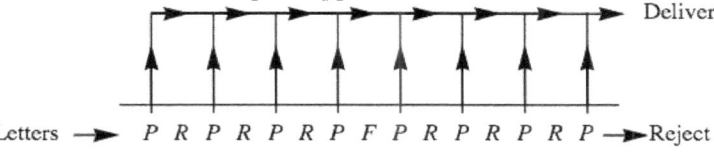

Figure 15.3.29 A letter facer

The arrows pointing up indicate that if a letter is postmarked, it is taken off the conveyor belt for delivery. If a letter reaches the end, it must not have a stamp. Letter-facing machines like this have been designed (see [16]). One economic consideration is that R-machines tend to cost more than F-machines. R-machines also tend to damage more letters. Taking these facts into consideration, the reader is invited to design a better letter-facing machine. Assume that R-machines cost \$800 and F-machines cost \$500. Be sure that all corners of incoming letters will be examined as they go down the conveyor belt. □

15.3.5 Exercises

1. Given $f = \begin{pmatrix} 1 & 2 & 3 & 4 \\ 2 & 1 & 4 & 3 \end{pmatrix}$, $g = \begin{pmatrix} 1 & 2 & 3 & 4 \\ 2 & 3 & 4 & 1 \end{pmatrix}$, and $h = \begin{pmatrix} 1 & 2 & 3 & 4 \\ 3 & 2 & 4 & 1 \end{pmatrix}$, compute
    (a) $f \circ g$
    (b) $g \circ h$
    (c) $(f \circ g) \circ h$
    (d) $f \circ (g \circ h)$
    (e) h^{-1}
    (f) $h^{-1} \circ g \circ h$
    (g) f^{-1}

2. Write f, g, and h from Exercise 1 as products of disjoint cycles and determine whether each is odd or even.

3. Do the left cosets of $A_3 = \{i, r_1, r_2\}$ over S_3 form a group under the induced operation on left cosets of A_3? What about the left cosets of $\langle f_1 \rangle$?

4. In its realization as permutations, the dihedral group $\mathcal{D}_3$ is equal to S_3. Can you give a geometric explanation why? Why isn't $\mathcal{D}_4$ equal to S_4?

5.
   (a) Complete the list of elements of $\mathcal{D}_4$ and write out a table for the group in its realization as symmetries.

   (b) List the subgroups of $\mathcal{D}_4$ in a lattice diagram. Are they all cyclic? To what simpler groups are the subgroups of $\mathcal{D}_4$ isomorphic?

6. Design a better letter-facing machine (see Example 15.3.28, p. 169). How can you verify that a letter-facing machine does indeed check every corner of a letter? Can it be done on paper without actually sending letters through it?

7. Prove by induction that if $r \geq 1$ and each t_i, is a transposition, then $(t_1 \circ t_2 \circ \cdots \circ t_r)^{-1} = t_r \circ \cdots \circ t_2 \circ t_1$

8. How many elements are there in $\mathcal{D}_5$? Describe them geometrically.

9. Complete the proof of Theorem 15.3.9, p. 162.

10. How many left cosets does A_n, $n \geq 2$ have?

11. Prove that $f \circ r = r^{n-1} \circ f$ in $\mathcal{D}_n$.

12.
   (a) Prove that the tile puzzles corresponding to $A_{16} \cap \{f \in S_{16} | f(16) = 16\}$ are solvable.

   (b) If $f(16) \neq 16$, how can you determine whether f's puzzle is solvable?

13.
   (a) Prove that S_3 is isomorphic to R_3, the group of 3×3 rook matrices (see Section 11.2 exercises).

   (b) Prove that for each $n \geq 2$, R_n is isomorphic to S_n.

15.4 Normal Subgroups and Group Homomorphisms

Our goal in this section is to answer an open question from earlier in this chapter and introduce a related concept. The question is: When are left cosets of a subgroup a group under the induced operation? This question is open for non-abelian groups. Now that we have some examples to work with, we can try a few experiments.

15.4.1 Normal Subgroups

Example 15.4.1 Cosets of A_3. We have seen that $A_3 = \{i, r_1, r_2\}$ is a subgroup of S_3, and its left cosets are A_3 itself and $B_3 = \{f_1, f_2, f_3\}$. Whether $\{A_3, B_3\}$ is a group boils down to determining whether the induced operation is well defined. Consider the operation table for S_3 in Figure 15.4.2, p. 171.

15.4. NORMAL SUBGROUPS AND GROUP HOMOMORPHISMS

$\circ$	i	r_1	r_2	f_1	f_2	f_3
i	i	r_1	r_2	f_1	f_2	f_3
r_1	r_1	r_2	i	f_3	f_1	f_2
r_2	r_2	i	r_1	f_2	f_3	f_1
f_1	f_1	f_2	f_3	i	r_1	r_2
f_2	f_2	f_3	f_1	r_2	i	r_1
f_3	f_3	f_1	f_2	r_1	r_2	i

Figure 15.4.2 Operation table for S_3

We have shaded in all occurrences of the elements of B_3 in gray. We will call these elements the gray elements and the elements of A_3 the white ones.

Now consider the process of computing the coset product $A_3 \circ B_3$. The "product" is obtained by selecting one white element and one gray element. Note that white "times" gray is always gray. Thus, $A_3 \circ B_3$ is well defined. Similarly, the other three possible products are well defined. The table for the factor group S_3/A_3 is

$\circ$	A_3	B_3
A_3	A_3	B_3
B_3	B_3	A_3

Clearly, S_3/A_3 is isomorphic to $\mathbb{Z}_2$. Notice that A_3 and B_3 are also the right cosets of A_3. This is significant. $\square$

Example 15.4.3 *Cosets of another subgroup of S_3.* Now let's try the left cosets of $\langle f_1 \rangle$ in S_3. There are three of them. Will we get a complicated version of $\mathbb{Z}_3$? The left cosets are $C_0 = \langle f_1 \rangle$, $C_1 = r_1 \langle f_1 \rangle = \{r_1, f_3\}$, and $C_2 = r_2 \langle f_1 \rangle = \{r_2, f_2\}$.

The reader might be expecting something to go wrong eventually, and here it is. To determine $C_1 \circ C_2$ we can choose from four pairs of representatives:

$$r_1 \in C_1, r_2 \in C_2 \longrightarrow r_1 \circ r_2 = i \in C_0$$
$$r_1 \in C_1, f_2 \in C_2 \longrightarrow r_1 \circ f_2 = f \in C_0$$
$$f_3 \in C_1, r_2 \in C_2 \longrightarrow f_3 \circ r_2 = f_2 \in C_2$$
$$f_3 \in C_1, f_2 \in C_2 \longrightarrow f_3 \circ f_2 = r_2 \in C_2$$

This time, we don't get the same coset for each pair of representatives. Therefore, the induced operation is not well defined and no factor group is produced. $\square$

Observation 15.4.4 This last example changes our course of action. If we had gotten a factor group from $\{C_0, C_1, C_2\}$, we might have hoped to prove that every collection of left cosets forms a group. Now our question is: How can we determine whether we will get a factor group? Of course, this question is equivalent to: When is the induced operation well defined? There was only one step in the proof of Theorem 15.2.18, p. 159, where we used the fact that G

was abelian. We repeat the equations here:

$$a' * b' = (a * h_1) * (b * h_2) = (a * b) * (h_1 * h_2)$$

since G was abelian.

The last step was made possible by the fact that $h_1 * b = b * h_1$. As the proof continued, we used the fact that $h_1 * h_2$ was in H and so $a' * b'$ is $(a * b) * h$ for some h in H. All that we really needed in the "abelian step" was that $h_1 * b = b * (\text{something in } H) = b * h_3$. Then, since H is closed under G's operation, $h_3 * h_2$ is an element of H. The consequence of this observation is that we define a certain kind of subgroup that guarantees that the induced operation is well-defined.

Definition 15.4.5 Normal Subgroup. If G is a group, $H \leq G$, then H is a normal subgroup of G, denoted $H \triangleleft G$, if and only if every left coset of H is a right coset of H; i. e. $a * H = H * a \quad \forall a \in G$ ◇

Theorem 15.4.6 *If $H \leq G$, then the operation induced on left cosets of H by the operation of G is well defined if and only if any one of the following conditions is true:*

(a) H is a normal subgroup of G.

(b) If $h \in H$, $a \in G$, then there exists $h' \in H$ such that $h * a = a * h'$.

(c) If $h \in H$, $a \in G$, then $a^{-1} * h * a \in H$.

Proof. We leave the proof of this theorem to the reader. ■

Be careful, the following corollary is not an "...if and only if..." statement.

Corollary 15.4.7 *If $H \leq G$, then the operation induced on left cosets of H by the operation of G is well defined if either of the following two conditions is true.*

(a) G is abelian.

(b) $|H| = \frac{|G|}{2}$.

Example 15.4.8 A non-normal subgroup. The right cosets of $\langle f_1 \rangle \leq S_3$ are $\{i, f_1\}$, $\{r_1 f_2\}$, and $\{r_2, f_3\}$. These are not the same as the left cosets of $\langle f_1 \rangle$. In addition, $f_2^{-1} f_1 f_2 = f_2 f_1 f_2 = f_3 \notin \langle f_1 \rangle$. Thus, $\langle f_1 \rangle$ is not normal. □

The improper subgroups $\{e\}$ and G of any group G are normal subgroups. $G/\{e\}$ is isomorphic to G. All other normal subgroups of a group, if they exist, are called **proper normal subgroups**.

Example 15.4.9 By Condition b of Corollary 15.4.7, p. 172, A_n is a normal subgroup of S_n and S_n/A_n is isomorphic to $\mathbb{Z}_2$. □

Example 15.4.10 Subgroups of A_5. A_5, a group in its own right with 60 elements, has many proper subgroups, but none are normal. Although this could be done by brute force, the number of elements in the group would make the process tedious. A far more elegant way to approach the verification of this statement is to use the following fact about the cycle structure of permutations. If $f \in S_n$ is a permutation with a certain cycle structure, $\sigma_1 \sigma_2 \cdots \sigma_k$, where the length of σ_i is ℓ_i, then for any $g \in S_n$, $g^{-1} \circ f \circ g$, which is the conjugate of f by g, will have a cycle structure with exactly the same cycle lengths. For example if we take $f = (1,2,3,4)(5,6)(7,8,9) \in S_9$ and conjugate by $g = (1,3,5,7,9)$,

$$g^{-1} \circ f \circ g = (1,9,7,5,3) \circ (1,2,3,4)(5,6)(7,8,9) \circ (1,3,5,7,9)$$
$$= (1,4,9,2)(3,6)(5,8,7)$$

Notice that the condition for normality of a subgroup H of G is that the conjugate of any element of H by an element of G must be remain in H.

To verify that A_5 has no proper normal subgroups, you can start by cataloging the different cycle structures that occur in A_5 and how many elements have those structures. Then consider what happens when you conjugate these different cycle structures with elements of A_5. An outline of the process is in the exercises. □

Example 15.4.11 Let G be the set of two by two invertible matrices of real numbers. That is,

$$G = \left\{ \begin{pmatrix} a & b \\ c & d \end{pmatrix} \mid a,b,c,d \in \mathbb{R}, ad - bc \neq 0 \right\}$$

We saw in Chapter 11 that G is a group with matrix multiplication.

This group has many subgroups, but consider just two: $H_1 = \left\{ \begin{pmatrix} a & 0 \\ 0 & a \end{pmatrix} \mid a \neq 0 \right\}$ and $H_2 = \left\{ \begin{pmatrix} a & 0 \\ 0 & d \end{pmatrix} \mid ad \neq 0 \right\}$. It is fairly simple to apply one of the conditions we have observed for normallity that H_1 a normal subgroup of G, while H_2 is not normal in G. □

15.4.2 Homomorphisms

Think of the word isomorphism. Chances are, one of the first images that comes to mind is an equation something like

$$\theta(x * y) = \theta(x) \diamond \theta(y)$$

An isomorphism must be a bijection, but the equation above is the algebraic property of an isomorphism. Here we will examine functions that satisfy equations of this type.

Definition 15.4.12 Homomorphism. Let $[G; *]$ and $[G'; \diamond]$ be groups. $\theta : G \to G'$ is a homomorphism if $\theta(x * y) = \theta(x) \diamond \theta(y)$ for all $x, y \in G$. ◊

Many homomorphisms are useful since they point out similarities between the two groups (or, on the universal level, two algebraic systems) involved.

Example 15.4.13 Decreasing modularity. Define $\alpha : \mathbb{Z}_6 \to \mathbb{Z}_3$ by $\alpha(n) = n \mod 3$. Therefore, $\alpha(0) = 0, \alpha(1) = 1, \alpha(2) = 2, \alpha(3) = 1 + 1 + 1 = 0$, $\alpha(4) = 1$, and $\alpha(5) = 2$. If $n, m \in \mathbb{Z}_6$. We could actually show that α is a homomorphism by checking all $6^2 = 36$ different cases for the formula

$$\alpha(n +_6 m) = \alpha(n) +_3 \alpha(m) \tag{15.4.1}$$

but we will use a line of reasoning that generalizes. We have already encountered Sun Tzu's Theorem, which implies that the function $\beta : \mathbb{Z}_6 \to \mathbb{Z}_3 \times \mathbb{Z}_2$ defined by $\beta(n) = (n \mod 3, n \mod 2)$. We need only observe that equating the first coordinates of both sides of the equation

$$\beta(n +_6 m) = \beta(n) + \beta(m) \tag{15.4.2}$$

gives us precisely the homomorphism property. □

Theorem 15.4.14 Group Homomorphism Properties. *If $\theta : G \to G'$ is a homomorphism, then:*

(a) $\theta(e) = \theta(\text{the identity of } G) = \text{the identity of } G' = e'$.

(b) $\theta(a^{-1}) = \theta(a)^{-1}$ for all $a \in G$.

(c) If $H \leq G$, then $\theta(H) = \{\theta(h) | h \in H\} \leq G'$.

Proof.

(a) Let a be any element of G. Then $\theta(a) \in G'$.

$$\begin{aligned} \theta(a) \diamond e' &= \theta(a) && \text{by the definition of } e' \\ &= \theta(a * e) && \text{by the definition of } e \\ &= \theta(a) \diamond \theta(e) && \text{by the fact that } \theta \text{ is a homomorphism} \end{aligned}$$

By cancellation, $e' = \theta(e)$.

(b) Again, let $a \in G$. $e' = \theta(e) = \theta(a * a^{-1}) = \theta(a) \diamond \theta(a^{-1})$. Hence, by the uniqueness of inverses, $\theta(a)^{-1} = \theta(a^{-1})$.

(c) Let $b_1, b_2 \in \theta(H)$. Then there exists $a_1, a_2 \in H$ such that $\theta(a_1) = b_1$, $\theta(a_2) = b_2$. Recall that a compact necessary and sufficient condition for $H \leq G$ is that $x * y^{-1} \in H$ for all $x, y \in H$. Now we apply the same condition in G':

$$\begin{aligned} b_1 \diamond b_2^{-1} &= \theta(a_1) \diamond \theta(a_2)^{-1} \\ &= \theta(a_1) \diamond \theta(a_2^{-1}) \\ &= \theta(a_1 * a_2^{-1}) \in \theta(H) \end{aligned}$$

since $a_1 * a_2^{-1} \in H$, and so we can conclude that $\theta(H) \leq G'$. ∎

Corollary 15.4.15 *Since a homomorphism need not be a surjection and part (c) of Theorem 15.4.14, p. 173 is true for the case of $H = G$, the range of θ, $\theta(G)$, is a subgroup of G'*

Example 15.4.16 If we define $\pi : \mathbb{Z} \to \mathbb{Z}/4\mathbb{Z}$ by $\pi(n) = n + 4\mathbb{Z}$, then π is a homomorphism. The image of the subgroup $4\mathbb{Z}$ is the single coset $0 + 4\mathbb{Z}$, the identity of the factor group. Homomorphisms of this type are called natural homomorphisms. The following theorems will verify that π is a homomorphism and also show the connection between homomorphisms and normal subgroups. The reader can find more detail and proofs in most abstract algebra texts. □

Theorem 15.4.17 *If $H \triangleleft G$, then the function $\pi : G \to G/H$ defined by $\pi(a) = aH$ is a homomorphism.*

Proof. We leave the proof of this theorem to the reader. ∎

Definition 15.4.18 Natural Homomorphism. If $H \triangleleft G$, then the function $\pi : G \to G/H$ defined by $\pi(a) = aH$ is called the natural homomorphism. ◊

Based on Theorem 15.4.17, p. 174, every normal subgroup gives us a homomorphism. Next, we see that the converse is true.

Definition 15.4.19 Kernel of a homomorphism. Let $\theta : G \to G'$ be a homomorphism, and let e and e' be the identities of G and G', respectively. The kernel of θ is the set $\ker \theta = \{a \in G \mid \theta(a) = e'\}$ ◊

Theorem 15.4.20 *Let $\theta : G \to G'$ be a homomorphism from G into G'. The kernel of θ is a normal subgroup of G.*

Proof. Let $K = \ker \theta$. We can see that K is a subgroup of G by letting $a, b \in K$ and verify that $a * b^{-1} \in K$ by computing $\theta(a * b^{-1}) = \theta(a) * \theta(b)^{-1} = e' * e'^{-1} = e'$. To prove normality, we let g be any element of G and $k \in K$. We compute

15.4. NORMAL SUBGROUPS AND GROUP HOMOMORPHISMS 175

$\theta(g * k * g^{-1})$ to verify that $g * k * g^{-1} \in K$.

$$\begin{aligned}\theta(g * k * g^{-1}) &= \theta(g) * \theta(k) * \theta(g^{-1}) \\ &= \theta(g) * \theta(k) * \theta(g)^{-1} \\ &= \theta(g) * e' * \theta(g)^{-1} \\ &= \theta(g) * \theta(g)^{-1} \\ &= e'\end{aligned}$$

■

Based on this most recent theorem, every homomorphism gives us a normal subgroup.

Theorem 15.4.21 Fundamental Theorem of Group Homomorphisms. Let $\theta : G \to G'$ be a homomorphism. Then $\theta(G)$ is isomorphic to $G/\ker \theta$.

Example 15.4.22 Define $\theta : \mathbb{Z} \to \mathbb{Z}_{10}$ by $\theta(n) = n \mod 10$. The three previous theorems imply the following:

- $\pi : \mathbb{Z} \to \mathbb{Z}/10\mathbb{Z}$ defined by $\pi(n) = n + 10\mathbb{Z}$ is a homomorphism.

- $\{n \in \mathbb{Z} | \theta(n) = 0\} = \{10n \mid n \in \mathbb{Z}\} = 10\mathbb{Z} \triangleleft \mathbb{Z}$.

- $\mathbb{Z}/10\mathbb{Z}$ is isomorphic to $\mathbb{Z}_{10}$.

□

Example 15.4.23 Let G be the same group of two by two invertible real matrices as in Example 15.4.11, p. 173. Define $\Phi : G \to G$ by $\Phi(A) = \dfrac{A}{\sqrt{|\det A|}}$. We will let the reader verify that Φ is a homomorphism. The theorems above imply the following.

- $\ker \Phi = \{A \in G | \Phi(A) = I\} = \left\{ \begin{pmatrix} a & 0 \\ 0 & a \end{pmatrix} \mid a \in \mathbb{R}, a \neq 0 \right\} \triangleleft G$. This verifies our statement in Example 15.4.11, p. 173. As in that example, let $\ker \Phi = H_1$.

- G/H_1 is isomorphic to $\{A \in G \mid \det A = 1\}$.

- $\pi : G \to G/H_1$ defined, naturally, by $\pi(A) = AH_1$ is a homomorphism.

□

For the remainder of this section, we will be examining certain kinds of homomorphisms that will play a part in our major application to homomorphisms, coding theory.

Example 15.4.24 Consider $\Phi : \mathbb{Z}_2^2 \to \mathbb{Z}_2^3$ defined by $\Phi(a,b) = (a,b,a+_2 b)$. If $(a_1, b_1), (a_2, b_2) \in \mathbb{Z}_2^2$,

$$\begin{aligned}\Phi((a_1,b_1)+(a_2,b_2)) &= \Phi(a_1 +_2 a_2, b_1 +_2 b_2) \\ &= (a_1 +_2 a_2, b_1 +_2 b_2, a_1 +_2 a_2 +_2 b_1 +_2 b_2) \\ &= (a_1, b_1, a_1 +_2 b_1) + (a_2, b_2, a_2 +_2 b_2) \\ &= \Phi(a_1,b_1) + \Phi(a_2,b_2)\end{aligned}$$

Since $\Phi(a,b)=(0,0,0)$ implies that $a = 0$ and $b = 0$, the kernel of Φ is $\{(0,0)\}$. By previous theorems, $\Phi(\mathbb{Z}_2^2) = \{(0,0,0),(1,0,1),(0,1,1),(1,1,0)\}$ is isomorphic to $\mathbb{Z}_2^2$. □

We can generalize the previous example as follows: If $n, m \geq 1$ and A is an

$m \times n$ matrix of 0's and 1's (elements of $\mathbb{Z}_2$), then $\Phi : \mathbb{Z}_2^m \to \mathbb{Z}_2^n$ defined by

$$\Phi(a_1, a_2, ..., a_m) = (a_1, a_2, ..., a_m) A$$

is a homomorphism. This is true because matrix multiplication is distributive over addition. The only new idea here is that computation is done in $\mathbb{Z}_2$. If $a = (a_1, a_2, ..., a_m)$ and $b = (b_1, b_2, ..., b_m)$, $(a + b)A = aA + bA$ is true by basic matrix laws. Therefore, $\Phi(a + b) = \Phi(a) + \Phi(b)$.

15.4.3 Exercises

1. Which of the following functions are homomorphisms? What are the kernels of those functions that are homomorphisms?
   (a) $\theta_1 : \mathbb{R}^* \to \mathbb{R}^+$ defined by $\theta_1(a) = |a|$.
   (b) $\theta_2 : \mathbb{Z}_5 \to \mathbb{Z}_2$ where $\theta_2(n) = \begin{cases} 0 & \text{if } n \text{ is even} \\ 1 & \text{if } n \text{ is odd} \end{cases}$.
   (c) $\theta_3 : \mathbb{R} \times \mathbb{R} \to \mathbb{R}$, where $\theta_3(a, b) = a + b$.
   (d) $\theta_4 : S_4 \to S_4$ defined by $\theta_4(f) = f \circ f = f^2$.

2. Which of the following functions are homomorphisms? What are the kernels of those functions that are homomorphisms?
   (a) $\alpha_1 : M_{2 \times 2}(\mathbb{R}) \to \mathbb{R}$, defined by $\alpha_1(A) = A_{11} A_{22} + A_{12} A_{21}$.
   (b) $\alpha_2 : (\mathbb{R}^*)^2 \to \mathbb{R}^*$ defined by $\alpha_2(a, b) = ab$.
   (c) $\alpha_3 : \{A \in M_{2 \times 2}(\mathbb{R}) | \det A \neq 0\} \to \mathbb{R}^*$, where $\alpha_3(A) = \det A$.
   (d) $\alpha_4 : S_4 \to S_4$ defined by $\alpha_4(f) = f^{-1}$.

3. Show that D_4 has one proper normal subgroup, but that $\langle (1,4)(2,3) \rangle$ is not normal.

4. Prove that the function Φ in Example 15.4.23, p. 175 is a homomorphism.

5. Define the two functions $\alpha : \mathbb{Z}_2^3 \to \mathbb{Z}_2^4$ and $\beta : \mathbb{Z}_2^4 \to \mathbb{Z}_2$ by $\alpha(a_1, a_2, a_3) = (a_1, a_2, a_3, a_1 +_2 a_2 +_2 a_3)$, and $\beta(b_1, b_2, b_3, b_4) = b_1 + b_2 + b_3 + b_4$ Describe the function $\beta \circ \alpha$. Is it a homomorphism?

6. Express Φ in Example 15.4.23, p. 175 in matrix form.

7. Prove that if G is an abelian group, then $q(x) = x^2$ defines a homomorphism from G into G. Is q ever an isomorphism?

8. Prove that if $\theta : G \to G'$ is a homomorphism, and $H \triangleleft G$, then $\theta(H) \triangleleft \theta(G)$. Is it also true that $\theta(H) \triangleleft G'$?

9. Prove that if $\theta : G \to G'$ is a homomorphism, and $H' \leq \theta(G)$, then $\theta^{-1}(H') = \{a \in G | \theta(a) \in H'\} \leq G$.

10. Following up on Example 15.4.10, p. 172, prove that A_5 is a simple group; i. e., it has no proper normal subgroups.
    (a) Make a list of the different cycle structures that occur in A_5 and how many elements have those structures.
    (b) Within each set of permutations with different cycle structures, identify which subsets are closed with respect to the conjugation operation. With this you will have a partition of A_5 into conjugate classes where for each class, C, $f, g \in C$ if and only if $\exists \phi \in A_5$ such that $\phi^{-1} \circ f \circ \phi = g$.

(c) Use the fact that a normal subgroup of A_5 needs to be a union of conjugate classes and verify that no such union exists.

15.5 Coding Theory, Linear Codes

A Transmission Problem. In this section, we will introduce the basic ideas involved in coding theory and consider solutions of a coding problem by means of linear codes.

Imagine a situation in which information is being transmitted between two points. The information takes the form of high and low pulses (for example, radio waves or electric currents), which we will label 1 and 0, respectively. As these pulses are sent and received, they are grouped together in blocks of fixed length. The length determines how much information can be contained in one block. If the length is r, there are 2^r different values that a block can have. If the information being sent takes the form of text, each block might be a character. In that case, the length of a block may be seven, so that $2^7 = 128$ block values can represent letters (both upper and lower case), digits, punctuation, and so on. During the transmission of data, noise can alter the signal so that what is received differs from what is sent. Figure 15.5.1, p. 177 illustrates the problem that can be encountered if information is transmitted between two points.

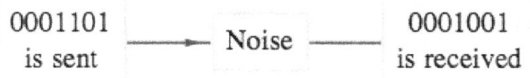

Figure 15.5.1 A noisy transmission

Noise is a fact of life for anyone who tries to transmit information. Fortunately, in most situations we could expect a high percentage of the pulses that are sent to be received properly. However, when large numbers of pulses are transmitted, there are usually some errors due to noise. For the remainder of the discussion, we will make assumptions about the nature of the noise and the message that we want to send. Henceforth, we will refer to the pulses as bits.

We will assume that our information is being sent along a **binary symmetric channel**. By this, we mean that any single bit that is transmitted will be received improperly with a certain fixed probability, p, independent of the bit value. The magnitude of p is usually quite small. To illustrate the process, we will assume that $p = 0.001$, which, in the real world, would be considered somewhat large. Since $1 - p = 0.999$, we can expect 99.9% of all bits to be properly received.

In addition to assuming $p = 0.001$ throughout, we will also suppose that our message consists of 3,000 bits of information. Two factors will be considered in evaluating a method of transmission. The first is the probability that the message is received with no errors. The second is the number of bits that will be transmitted in order to send the message. This quantity is called the rate of transmission:
$$\text{Rate} = \frac{\text{Message length}}{\text{Number of bits transmitted}}$$
As you might expect, as we devise methods to improve the probability of success, the rate will decrease.

Suppose that we ignore the noise and transmit the message without any coding. The probability of success is $0.999^{3000} = 0.0497124$. Therefore we only successfully receive the message in a totally correct form less than 5% of the time. The rate of $\frac{3000}{3000} = 1$ certainly doesn't offset this poor probability.

Our strategy for improving our chances of success will be to send an encoded message. The encoding will be done in such a way that small errors can be identified and corrected. This idea is illustrated in Figure 15.5.2, p. 178.

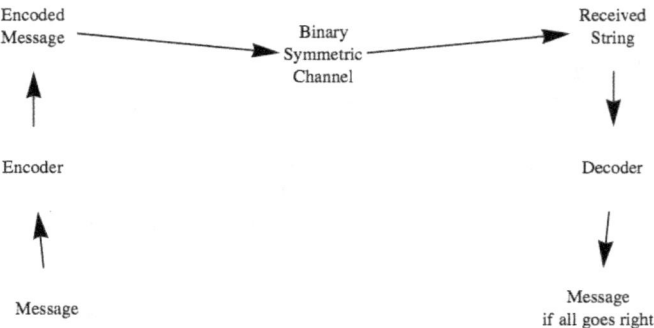

Figure 15.5.2 The Coding Process

In all of our examples, the functions that will correspond to our encoding devices will involve multiplication of messages by matrices using mod 2 arithmetic. First we will introduce some geometric ideas to make the process more intuitive.

15.5.1 Introduction

Although we'll be using algebra to help improve communications, the basic solution can be imagined from a geometric point of view. For any positive integer n, we define a distance function on the elements of the group $\mathbb{Z}_2^n$. This distance is called the **Hamming Distance**.

Definition 15.5.3 Hamming Distance. Given two elements of $\mathbb{Z}_2^n$, a and b, the Hamming Distance, $d_H(a,b)$ between them is the number of positions in which they differ. ◊

For example, $d_H((1,1,0,0),(1,1,0,1)) = 1$ since these two elements of $\mathbb{Z}_2^4$ differ in just the last position; and $d_H((1,1,0,0),(1,1,0,0)) = 0$. Notice that we can compute the distance between two bit strings by adding them coordinatewise in the Cartesian product and counting the number 1's that appear in the sum. For example $(1,1,0,0) + (1,0,0,1) = (0,1,0,1)$. The sum has two 1's, so the distance between $(1,1,0,0)$ and $(1,0,0,1)$ is 2. In addition, the location of the 1's in the sum tell us where the two bit strings differ.

When we look at groups like $\mathbb{Z}_2^4$ from a point of view, we refer to these sets as **metric spaces** or simply **spaces**. In the case of $\mathbb{Z}_2^4$, there are just $2^4 = 16$ points in the space and the maximum distance between the points is 4. More generally $\mathbb{Z}_2^n$ has 2^n points and the maximum distance between points in that space is n. Looking at the group $\mathbb{Z}_2^n$ from this geometric point of view is essentially the same as the n-cube 9.0.3, p. 17 we considered in discussing Hamiltonian graphs. In this section we will use n-tuples such as $(1,1,0,1)$ interchangeably with strings of bits such as 1101.

For any distance r in a space, the ball of radius r centered at a point a, denoted $B_r(a)$, is the set of all points whose distance from a is r or less. For example, in the space $\mathbb{Z}_2^4$,

$$B_1((1,1,0,0)) = \{(1,1,0,0),(0,1,0,0),(1,0,0,0),(1,1,1,0),(1,1,0,1)\}.$$

The ultimate goal of our encoding will be to take a set of possible messages, the **message space**, and distribute them in a larger space, the **code space**, in such a way that the encoded message, called a **code word** is at least a certain distance away from any other code word. The minimum distance between the code words will determine whether we can correct errors or just detect them. Now let's turn to some examples.

15.5.2 Error Detection

Suppose that each block of three bits $a = (a_1, a_2, a_3)$ is encoded with the function $e : \mathbb{Z}_2^3 \to \mathbb{Z}_2^4$, where

$$e(a) = (a_1, a_2, a_3, a_1 +_2 a_2 +_2 a_3)$$

The fourth bit of $e(a)$ is called the parity-check bit. When the encoded block is received, the four bits will probably all be correct (they are correct approximately 99.6% of the time under our assumed parameters), but the added bit that is sent will make it possible to detect single bit errors in the block. Note that when $e(a)$ is transmitted, the sum of its components is $a_1 +_2 a_2 +_2 a_3 +_2 (a_1 +_2 a_2 +_2 a_3) = 0$, since $a_i + a_i = 0$ in $\mathbb{Z}_2$.

If any single bit is garbled by noise, the sum of the received bits will be 1. A **parity error** occurs if the sum of the received bits is 1. Since more than one error is unlikely when p is small, a high percentage of all errors can be detected.

At the receiving end, the decoding function acts on the four-bit block $b = (b_1, b_2, b_3, b_4)$ with the function $d : \mathbb{Z}_2^4 \to \mathbb{Z}_2^4$, where

$$d(b) = (b_1, b_2, b_3, b_1 +_2 b_2 +_2 b_3 +_2 b_4)$$

Notice that the fourth bit of $d(b)$ is an indicator of whether there is a parity error - 0 if no error, and 1 if an error. If no parity error occurs, the first three bits are recorded as part of the message. If a parity error occurs, we will assume that a retransmission of that block can be requested. This request can take the form of automatically having the parity-check bit of $d(b)$ sent back to the source. If 1 is received, the previous block is retransmitted; if 0 is received, the next block is sent. This assumption of two-way communication is significant, but it is desirable to make this coding system useful. For our calculations, it is reasonable to expect that the probability of a transmission error in the opposite direction is also 0.001. Without going into the details, we will report that the probability of success in sending 3000 bits is approximately 0.990 and the rate is approximately 3/5. The rate includes the transmission of the parity-check bit to the source and is only approximate because the resent blocks will decrease the rate below 3/5 somewhat.

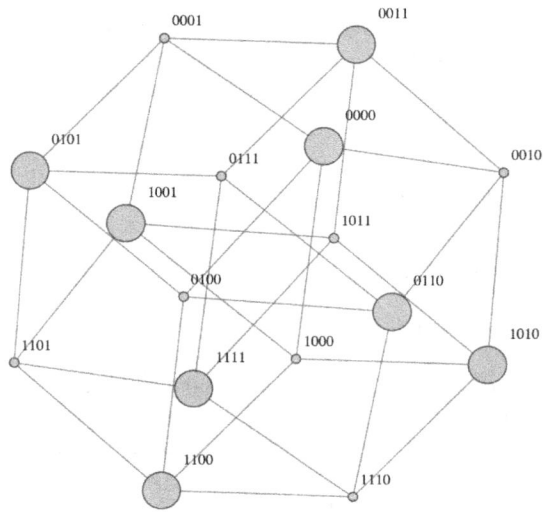

Figure 15.5.4 The 4-cube with code words displayed as larger vertices

Let's consider the geometry of this code. If we examine the 4-cube in Figure 15.5.4, p. 180, the code words are the strings of four bits with an even number of ones. These vertices are the larger ones. Notice that the ball of radius 1 centered around any of the code words consists of that code word and the smaller vertices that are connected to the code word with an edge of the 4-cube. Since there are no other code-words in the ball, a single bit error produces a non-code word and so an error can be detected.

15.5.3 Error Correction

Next, we will consider coding functions that allow us to correct errors at the receiving end so that only one-way communication is needed. Before we begin, recall that every element of $\mathbb{Z}_2^n$, $n \geq 1$, is its own inverse; that is, $-b = b$. Therefore, $a - b = a + b$.

Example 15.5.5 The Triple Repetition Code. Suppose we take each individual bit in our message and encode it by repeating it three times. In other words, if a is a single bit, $e(a) = (a, a, a)$. The code words for this code are $(0, 0, 0)$ and $(1, 1, 1)$. Let's look at the geometry behind this code. The message space has just two points, but the code space is $\mathbb{Z}_2^3$, which has 8 points, the vertices of the 3-cube, which appears in Figure 15.5.6, p. 180.

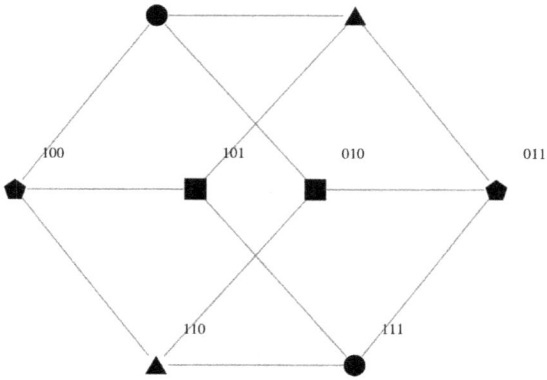

Figure 15.5.6 The 3-cube with code words displayed as circular vertices

15.5. CODING THEORY, LINEAR CODES

In the figure for this code, the code words are circular vertices. If we identify the balls of radius 1 centered around the two code words, you might notice that the two balls do not intersect. Each has a different vertex with triangular, square and pentagonal shapes. From a geometric point of view, this is why we can correct a single bit error. If any string of three bits in the code space is received it is in one of the two balls and the code word in that ball had to have been the one that was transmitted.

Regarding the actual correction process, the shapes have a meaning, as outlined in the following list.

- Circle: No correction needed
- Pentagon: Correct the first bit
- Square: Correct the second bit
- Triangle: Correct the third bit

Of course, once the correction is made, only the first bit is extracted from the code word since all bits will be equal. The simplicity of the final result masks an important property of all error correcting codes we consider. All of the possible points in the code space can be partitioned in such a way that each block in the partition corresponds with a specific correction that we can make to recover the correct code word.

If you have read about cosets, you will see that the partition we refer to is the set of left cosets of the set of code words. □

Triple repetition is effective, but not very efficient since its rate is quite low, $1/3$. Next we consider a slightly more efficient error correcting code based on matrix multiplication. Any such code that is computed with a matrix multiplication is called a **linear code**. We should point out that both the parity check code and the triple repetition code are linear codes. For the parity check code, the encoding function can be thought of as acting on a 1×3 row vector $a = (a_1, a_2, a_3)$ by multiplying times a 3×4 matrix:

$$e(a) = (a_1, a_2, a_3) \begin{pmatrix} 1 & 0 & 0 & 1 \\ 0 & 1 & 0 & 1 \\ 0 & 0 & 1 & 1 \end{pmatrix} = (a_1, a_2, a_3, a_1 +_2 c_2 +_2 a_3)$$

For triple repetition, the encoding function can be thought of as acting on a 1×1 matrix a by multiplying times a 1×3 matrix:

$$e(a) = (a) \begin{pmatrix} 1 & 1 & 1 \end{pmatrix} = \begin{pmatrix} a & a & a \end{pmatrix}$$

Example 15.5.7 A Somewhat More Efficient Linear Code. The encoding that we will consider here takes a block $a = (a_1, a_2, a_3)$ and produces a code word of length 6. As in the triple repetition code, each code word will differ from each other code word by at least three bits. As a result, any single error will not push a code word close enough to another code word to cause confusion. Now for the details.

Let
$$G = \begin{pmatrix} 1 & 0 & 0 & 1 & 1 & 0 \\ 0 & 1 & 0 & 1 & 0 & 1 \\ 0 & 0 & 1 & 0 & 1 & 1 \end{pmatrix}.$$

We call G the **generator matrix** for the code, and let $a = (a_1, a_2, a_3)$ be our message. Define $e : \mathbb{Z}_2^3 \to \mathbb{Z}_2^6$ by

$$e(a) = aG = (a_1, a_2, a_3, a_4, a_5, a_6)$$

where
$$a_4 = a_1 +_2 a_2$$
$$a_5 = a_1 +_2 a_3$$
$$a_6 = a_2 +_2 a_3$$

Notice that since matrix multiplication is distributive over addition, we have
$$e(a+b) = (a+b)G = aG + bG = e(a) + e(b)$$
for all $a, b \in \mathbb{Z}_2^3$. This equality, may look familiar from the definition of an isomorphism, but in this case the function e is not onto. If you've read about homomorphisms, this is indeed an example of one.

One way to see that any two distinct code words have a distance from one another of at least 3 is to consider the images of any two distinct messages. If a and b are distinct elements of $\mathbb{Z}_2^3$, then $c = a + b$ has at least one coordinate equal to 1. Now consider the difference between $e(a)$ and $e(b)$:
$$e(a) + e(b) = e(a+b)$$
$$= e(c)$$

Whether c has 1, 2, or 3 ones, $e(c)$ must have at least three ones. This can be seen by considering the three cases separately. For example, if c has a single one, two of the parity bits are also 1. Therefore, $e(a)$ and $e(b)$ differ in at least three bits. By the same logic as with triple repetition, a single bit error in any code word produces an element of the code space that is contained in on of the balls of radius 1 centered about a code word.

Now consider the problem of decoding received transmissions. Imagine that a code word, $e(a)$, is transmitted, and $b = (b_1, b_2, b_3, b_4, b_5, b_6)$ is received. At the receiving end, we know the formula for $e(a)$, and if no error has occurred in transmission,

$$\begin{aligned} b_1 &= a_1 \\ b_2 &= a_2 \\ b_3 &= a_3 \\ b_4 &= a_1 +_2 a_2 \\ b_5 &= a_1 +_2 a_3 \\ b_6 &= a_2 +_2 a_3 \end{aligned} \Rightarrow \begin{aligned} b_1 +_2 b_2 +_2 b_4 &= 0 \\ b_1 +_2 b_3 +_2 b_5 &= 0 \\ b_2 +_2 b_3 +_2 b_6 &= 0 \end{aligned}$$

The three equations on the right are called parity-check equations. If any of them are not true, an error has occurred. This error checking can be described in matrix form.

Let
$$H = \begin{pmatrix} 1 & 1 & 0 \\ 1 & 0 & 1 \\ 0 & 1 & 1 \\ 1 & 0 & 0 \\ 0 & 1 & 0 \\ 0 & 0 & 1 \end{pmatrix}$$

The matrix H is called the parity-check matrix for this code. Now define $p : \mathbb{Z}_2^6 \to \mathbb{Z}_2^3$ by $p(b) = bH$. We call $p(b)$ the **syndrome** of the received block. For example, $p(0,1,0,1,0,1) = (0,0,0)$ and $p(1,1,1,1,0,0) = (1,0,0)$

Note that p has a similar property as e, that $p(b_1 + b_2) = p(b_1) + p(b_2)$. If the syndrome of a block is $(0,0,0)$, we can be almost certain that the message block is (b_1, b_2, b_3).

15.5. CODING THEORY, LINEAR CODES

Next we turn to the method of correcting errors. Despite the fact that there are only eight code words, one for each three-bit block value, the set of possible received blocks is $\mathbb{Z}_2^6$, with 64 elements. Suppose that b is not a code word, but that it differs from a code word by exactly one bit. In other words, it is the result of a single error in transmission. Suppose that w is the code word that b is closest to and that they differ in the first bit. Then $b + w = (1,0,0,0,0,0)$ and

$$\begin{aligned}p(b) &= p(b) + p(w) \quad \text{since } p(w) = (0,0,0)\\ &= bH + wH\\ &= (b+w)H \quad \text{by the distributive property}\\ &= p(b+w)\\ &= p(1,0,0,0,0,0)\\ &= (1,1,0)\end{aligned}$$

This is the first row of H!

Note that we haven't specified b or w, only that they differ in the first bit. Therefore, if b is received, there was probably an error in the first bit and $p(b) = (1,1,0)$, the transmitted code word was probably $b + (1,0,0,0,0,0)$ and the message block was $(b_1 +_2 1, b_2, b_3)$. The same analysis can be done if b and w differ in any of the other five bits.

In general, if the syndrome of a received string of bits is the kth row of the parity check matrix, the error has occurred in the kth bit. $\square$

Probability Epilog. For the two error correction examples we've looked at, we can compare their probabilities of successfully receiving all 3000 bits correctly over a binary symmetric channel with $p = 0.001$.

For the triple repetition code, the probability is

$$\left(0.999^3 + 3 \cdot 0.999^2 \cdot 0.001\right)^{3000} = 0.991,$$

and the rate of this code is $\frac{1}{3}$ which means we need to transmit 9000 bits.

For the second code, the probability of success is

$$\left(0.999^6 + 6 \cdot 0.999^5 \cdot 0.001\right)^{1000} = 0.985,$$

and rate for this code is $\frac{1}{2}$, which means we need to transmit 6000 bits.

Clearly, there is a trade-off between accuracy and speed.

Example 15.5.8 Another Linear Code. Consider the linear code with generator matrix

$$G = \begin{pmatrix} 1 & 1 & 0 & 1 & 0 \\ 0 & 1 & 1 & 0 & 0 \\ 0 & 0 & 0 & 1 & 1 \end{pmatrix}.$$

Since G is 3×5, this code encodes three bits into five bits. The natural question to ask is what detection or correction does it afford? We can answer this question by constructing the parity check matrix. We observe that if $\mathbf{a} = (a_1, a_2, a_3)$ the encoding function is

$$e(\mathbf{a}) = \mathbf{a}G = (a_1, a_1 + a_2, a_2, a_1 + a_3, a_3)$$

where addition is mod 2 addition. If we receive five bits $(c_1, c_2, c_3, c_4, c_5)$ and no error has occurred, the following two equations would be true.

$$c_1 + c_2 + c_3 = 0 \qquad (15.5.1)$$
$$c_1 + c_4 + c_5 = 0 \qquad (15.5.2)$$

Notice that in general, the number of parity check equations is equal to the number of extra bits that are added by the encoding function. These equations are equivalent to the single matrix equation $(c_1, c_2, c_3, c_4, c_5)H = \mathbf{0}$, where

$$H = \begin{pmatrix} 1 & 1 \\ 1 & 0 \\ 1 & 0 \\ 0 & 1 \\ 0 & 1 \end{pmatrix}$$

At a glance, we can see that this code will not correct most single bit errors. Suppose an error $\mathbf{e} = (e_1, e_2, e_3, e_4, e_5)$ is added in the transmission of the five bits. Specifically, suppose that 1 is added (mod 2) in position j, where $1 \leq j \leq 5$ and the other coordinates of $\mathbf{e}$ are 0. Then when we compute the syndrome of our received transmission, we see that

$$\mathbf{c}H = (\mathbf{a}G + \mathbf{e})H = (\mathbf{a}G)H + \mathbf{e}H = \mathbf{e}H.$$

But $\mathbf{e}H$ is the jth row of H. If the syndrome is $(1, 1)$ we know that the error occurred in position 1 and we can correct it. However, if the error is in any other position we can't pinpoint its location. If the syndrome is $(1, 0)$, then the error could have occurred in either position 2 or position 3. This code does detect all single bit errors but only corrects one fifth of them. □

15.5.4 Exercises

1. If the error-detecting code is being used, how would you act on the following received blocks?

    (a) $(1, 0, 1, 1)$

    (b) $(1, 1, 1, 1)$

    (c) $(0, 0, 0, 0)$

2. Determine the parity check matrix for the triple repetition code.

3. If the error-correcting code from this section is being used, how would you decode the following blocks? Expect an error that cannot be fixed with one of these.

    (a) $(1, 0, 0, 0, 1, 1)$

    (b) $(1, 0, 1, 0, 1, 1)$

    (c) $(0, 1, 1, 1, 1, 0)$

    (d) $(0, 0, 0, 1, 1, 0)$

    (e) $(1, 0, 0, 0, 0, 1)$

    (f) $(1, 0, 0, 1, 0, 0)$

4. Suppose that the code words of a coding function have the property that any two of them have a Hamming distance of at least 5. How many bit errors could be corrected with such a code?

5. Consider the linear code defined by the generator matrix

$$G = \begin{pmatrix} 1 & 0 & 1 & 0 \\ 0 & 1 & 1 & 1 \end{pmatrix}$$

(a) What size blocks does this code encode and what is the length of the code words?

(b) What are the code words for this code?

(c) With this code, can you detect single bit errors? Can you correct all, some, or no single bit errors?

6. **Rectangular codes.** To build a rectangular code, you partition your message into blocks of length m and then factor m into $k_1 \cdot k_2$ and arrange the bits in a $k_1 \times k_2$ rectangular array as in the figure below. Then you add parity bits along the right side and bottom of the rows and columns. The code word is then read row by row.

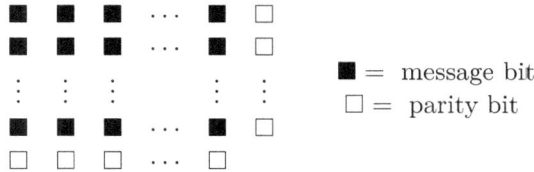

■ = message bit
□ = parity bit

For example, if m is 4, then our only choice is a 2 by 2 array. The message 1101 would be encoded as

$$\begin{array}{cc|c} 1 & 1 & 0 \\ 0 & 1 & 1 \\ \hline 1 & 0 & \end{array}$$

and the code word is the string 11001110.

(a) Suppose that you were sent four bit messages using this code and you received the following strings. What were the messages, assuming no more than one error in the transmission of coded data?

  (i) 11011000   (ii) 01110010   (iii) 10001111

(b) If you encoded n^2 bits in this manner, what would be the rate of the code?

(c) Rectangular codes are linear codes. For the 2 by 2 rectangular code, what are the generator and parity check matrices?

7. Suppose that the code in Example 15.5.8, p. 183 is expanded to add the column

$$\begin{pmatrix} 0 \\ 1 \\ 1 \end{pmatrix}$$

to the generator matrix G, can all single bit errors be corrected? Explain your answer.

8. Suppose that a linear code has parity check matrix

$$H = \begin{pmatrix} 1 & 1 & 1 \\ 1 & 1 & 0 \\ 1 & 0 & 1 \\ 0 & 1 & 1 \\ 1 & 0 & 0 \\ 0 & 1 & 0 \\ 0 & 0 & 1 \end{pmatrix}$$

Determine the generator matrix, G, and in so doing, identify the number of bits in each message block and the number of parity bits.

Hint. There is a parity check equation for each parity bit.

9. A code with minimum distance d is called *perfect* if every string of bits is within Hamming distance $r = \frac{d-1}{2}$ of some code word. For such a code, the spheres of radius r around the code words partition the set of all strings. This is analogous to packing objects into a box with no wasted space. Using just the number of bit strings of length n and the number of strings in a sphere of radius 1, for what values of n is it possible to find a perfect code of distance 3? You don't have to actually find the codes.

Chapter 16

An Introduction to Rings and Fields

> **field extension**
>
> **Field extensions** are simple. Let's say
> That field L is a subfield of K,
> Then it goes without mention,
> Field K's an extension
> Of L — like a shell, in a way.
>
> *zqms, The Omnificent English Dictionary In Limerick Form*

In our early elementary school days we began the study of mathematics by learning addition and multiplication on the set of positive integers. We then extended this to operations on the set of all integers. Subtraction and division are defined in terms of addition and multiplication. Later we investigated the set of real numbers under the operations of addition and multiplication. Hence, it is quite natural to investigate those structures on which we can define these two fundamental operations, or operations similar to them. The structures similar to the set of integers are called rings, and those similar to the set of real numbers are called fields.

In coding theory, highly structured codes are needed for speed and accuracy. The theory of finite fields is essential in the development of many structured codes. We will discuss basic facts about finite fields and introduce the reader to polynomial algebra.

16.1 Rings, Basic Definitions and Concepts

16.1.1 Basic Definitions

We would like to investigate algebraic systems whose structure imitates that of the integers.

Definition 16.1.1 Ring. A ring is a set R together with two binary operations, addition and multiplication, denoted by the symbols $+$ and $\cdot$ such that the following axioms are satisfied:

(1) $[R; +]$ is an abelian group.

(2) Multiplication is associative on R.

(3) Multiplication is distributive over addition; that is, for all $a, b, c \in R$, the left distributive law, $a \cdot (b + c) = a \cdot b + a \cdot c$, and the right distributive law, $(b + c) \cdot a = b \cdot a + c \cdot a$.

$\diamond$

Note 16.1.2

(1) A ring is denoted $[R; +, \cdot]$ or as just plain R if the operations are understood.

(2) The symbols $+$ and $\cdot$ stand for arbitrary operations, not just "regular" addition and multiplication. These symbols are referred to by the usual names. For simplicity, we may write ab instead of $a \cdot b$ if it is not ambiguous.

(3) For the abelian group $[R; +]$, we use additive notation. In particular, the group identity is designated by 0 rather than by e and is customarily called the "zero" of the ring. The group inverse is also written in additive notation: $-a$ rather than a^{-1}.

We now look at some examples of rings. Certainly all the additive abelian groups of Chapter 11 are likely candidates for rings.

Example 16.1.3 The ring of integers. $[\mathbb{Z}; +, \cdot]$ is a ring, where $+$ and $\cdot$ stand for regular addition and multiplication on $\mathbb{Z}$. From Chapter 11, we already know that $[\mathbb{Z}; +]$ is an abelian group, so we need only check parts 2 and 3 of the definition of a ring. From elementary algebra, we know that the associative law under multiplication and the distributive laws are true for $\mathbb{Z}$. This is our main example of an infinite ring. $\square$

Example 16.1.4 The ring of integers modulo n. $[\mathbb{Z}_n; +_n, \times_n]$ is a ring. The properties of modular arithmetic on $\mathbb{Z}_n$ were described in Section 11.4, and they give us the information we need to convince ourselves that $[\mathbb{Z}_n; +_n, \times_n]$ is a ring. This example is our main example of finite rings of different orders. $\square$

Definition 16.1.5 Commutative Ring. A ring in which multiplication is a commutative operation is called a commutative ring. $\diamond$

It is common practice to use the word "abelian" when referring to the commutative law under addition and the word "commutative" when referring to the commutative law under the operation of multiplication.

Definition 16.1.6 Unity of a Ring. A ring $[R; +, \cdot]$ that has a multiplicative identity is called a ring with unity. The multiplicative identity itself is called the unity of the ring. More formally, if there exists an element $1 \in R$, such that for all $x \in R$, $x \cdot 1 = 1 \cdot x = x$, then R is called a **ring with unity**. $\diamond$

Example 16.1.7 The rings in our first two examples were commutative rings with unity, the unity in both cases being the number 1. The ring $[M_{2 \times 2}(\mathbb{R}); +, \cdot]$ is a noncommutative ring with unity, the unity being the two by two identity matrix.

An example of a ring that is not a ring with unity is the ring of even integers, $[2\mathbb{Z}; +, \cdot]$. $\square$

16.1.2 Direct Products of Rings

Products of rings are analogous to products of groups or products of Boolean algebras.

Let $[R_i; +_i, \cdot_i]$, $i = 1, 2, \ldots, n$ be rings. Let $P = \underset{i=1}{\overset{n}{\times}} R_i$ and $a = (a_1, a_2, \ldots, a_n)$, $b = (b_1, b_2, \ldots, b_n) \in P$.

16.1. RINGS, BASIC DEFINITIONS AND CONCEPTS

From Chapter 11 we know that P is an abelian group under the operation of componentwise addition:

$$a + b = (a_1 +_1 b_1, a_2 +_2 b_2, ..., a_n +_n b_n)$$

We also define multiplication on P componentwise:

$$a \cdot b = (a_1 \cdot_1 b_1, a_2 \cdot_2 b_2, ..., a_n \cdot_n b_n)$$

To show that P is a ring under the above operations, we need only show that the (multiplicative) associative law and the distributive laws hold. This is indeed the case, and we leave it as an exercise. If each of the R_i is commutative, then P is commutative, and if each contains a unity, then P is a ring with unity, which is the n-tuple consisting of the unities of each of the R_i's.

Example 16.1.8 Since $[\mathbb{Z}_4; +_4, \times_4]$ and $[\mathbb{Z}_3; +_3, \times_3]$ are rings, then $\mathbb{Z}_4 \times \mathbb{Z}_3$ is a ring, where, for example,

$$(2, 1) + (2, 2) = (2 +_4 2, 1 +_3 2) = (0, 0)$$
$$\text{and}$$
$$(3, 2) \cdot (2, 2) = (3 \times_4 2, 2 \times_3 2) = (2, 1)$$

To determine the unity in the ring $\mathbb{Z}_4 \times \mathbb{Z}_3$, we look for the element (m, n) such that for all elements $(a, b) \in \mathbb{Z}_4 \times \mathbb{Z}_3$, $(a, b) = (a, b) \cdot (m, n) = (m, n) \cdot (a, b)$, or, equivalently,

$$(a \times_4 m, b \times_3 n) = (m \times_4 a, n \times_3 b) = (a, b)$$

So we want m such that $a \times_4 m = m \times_4 a = a$ in the ring $\mathbb{Z}_4$. The only element m in $\mathbb{Z}_4$ that satisfies this equation is $m = 1$. Similarly, we obtain value of 1 for n. So the unity of $\mathbb{Z}_4 \times \mathbb{Z}_3$, which is unique by Exercise 15 of this section, is $(1, 1)$. We leave to the reader to verify that this ring is commutative. □

16.1.3 Multiplicative Inverses in Rings

We now consider the extremely important concept of multiplicative inverses. Certainly many basic equations in elementary algebra (e.g., $2x = 3$) are solved with this concept.

Example 16.1.9 The equation $2x = 3$ has a solution in the ring $[\mathbb{Q}; +, \cdot]$ but does not have a solution in $[\mathbb{Z}; +, \cdot]$ since, to solve this equation, we multiply both sides of the equation $2x = 3$ by the multiplicative inverse of 2. This number, 2^{-1} exists in $\mathbb{Q}$ but does not exist in $\mathbb{Z}$. We formalize this important idea in a definition which by now should be quite familiar to you. □

Definition 16.1.10 Multiplicative Inverses. Let $[R; +, \cdot]$ be a ring with unity, 1. If $u \in R$ and there exists an element $v \in R$ such that $u \cdot v = v \cdot u = 1$, then u is said to have a multiplicative inverse, v. A ring element that possesses a multiplicative inverse is a unit of the ring. The set of all units of a ring R is denoted by $U(R)$. ◇

By Theorem 11.3.3, p. 30, the multiplicative inverse of a ring element is unique, if it exists. For this reason, we can use the notation u^{-1} for the multiplicative inverse of u, if it exists.

Example 16.1.11 In the rings $[\mathbb{R}; +, \cdot]$ and $[\mathbb{Q}; +, \cdot]$ every nonzero element has a multiplicative inverse. The only elements in $[\mathbb{Z}; +, \cdot]$ that have multiplicative inverses are -1 and 1. That is, $U(\mathbb{R}) = \mathbb{R}^*$, $U(\mathbb{Q}) = \mathbb{Q}^*$, and $U(\mathbb{Z}) = \{-1, 1\}$.
□

Example 16.1.12 Let us find the multiplicative inverses, when they exist, of each element of the ring $[\mathbb{Z}_6; +_6, \times_6]$. If $u = 3$, we want an element v such that $u \times_6 v = 1$. We do not have to check whether $v \times_6 u = 1$ since $\mathbb{Z}_6$ is commutative. If we try each of the six elements, 0, 1, 2, 3, 4, and 5, of $\mathbb{Z}_6$, we find that none of them satisfies the above equation, so 3 does not have a multiplicative inverse in $\mathbb{Z}_6$. However, since $5 \times_6 5 = 1$, 5 does have a multiplicative inverse in $\mathbb{Z}_6$, namely itself: $5^{-1} = 5$. The following table summarizes all results for $\mathbb{Z}_6$.

u	u^{-1}
0	does not exist
1	1
2	does not exist
3	does not exist
4	does not exist
5	5

It shouldn't be a surprise that the zero of a ring is never going to have a multiplicative inverse. □

16.1.4 More universal concepts, isomorphisms and subrings

Isomorphism is a universal concept that is important in every algebraic structure. Two rings are isomorphic as rings if and only if they have the same cardinality and if they behave exactly the same under corresponding operations. They are essentially the same ring. For this to be true, they must behave the same as groups (under +) and they must behave the same under the operation of multiplication.

Definition 16.1.13 Ring Isomorphism. Let $[R; +, \cdot]$ and $[R'; +', \cdot']$ be rings. Then R is isomorphic to R' if and only if there exists a function, $f : R \to R'$, called a ring isomorphism, such that

(1) f is a bijection

(2) $f(a + b) = f(a) +' f(b)$ for all $a, b \in R$

(3) $f(a \cdot b) = f(a) \cdot' f(b)$ for all $a, b \in R$.

◇

Conditions 1 and 2 tell us that f is a group isomorphism.

This leads us to the problem of how to show that two rings are not isomorphic. This is a universal concept. It is true for any algebraic structure and was discussed in Chapter 11. To show that two rings are not isomorphic, we must demonstrate that they behave differently under one of the operations. We illustrate through several examples.

Example 16.1.14 Consider the rings $[\mathbb{Z}; +, \cdot]$ and $[2\mathbb{Z}; +, \cdot]$. In Chapter 11 we showed that as groups, the two sets $\mathbb{Z}$ and $2\mathbb{Z}$ with addition were isomorphic. The group isomorphism that proved this was the function $f : \mathbb{Z} \to 2\mathbb{Z}$, defined by $f(n) = 2n$. Is f a ring isomorphism? We need only check whether $f(m \cdot n) = f(m) \cdot f(n)$ for all $m, n \in \mathbb{Z}$. In fact, this condition is not satisfied:

$$f(m \cdot n) = 2 \cdot m \cdot n \quad \text{and} \quad f(m) \cdot f(n) = 2m \cdot 2n = 4 \cdot m \cdot n$$

Therefore, f is not a ring isomorphism. This does not necessarily mean that the two rings $\mathbb{Z}$ and $2\mathbb{Z}$ are not isomorphic, but simply that f doesn't satisfy the conditions. We could imagine that some other function does. We could try to find another function that is a ring isomorphism, or we could try to show that

16.1. RINGS, BASIC DEFINITIONS AND CONCEPTS

$\mathbb{Z}$ and $2\mathbb{Z}$ are not isomorphic as rings. To do the latter, we must find something different about the ring structure of $\mathbb{Z}$ and $2\mathbb{Z}$.

We already know that they behave identically under addition, so if they are different as rings, it must have something to do with how they behave under the operation of multiplication. Let's begin to develop a checklist of how the two rings could differ:

(1) Do they have the same cardinality? Yes, they are both countable.

(2) Are they both commutative? Yes.

(3) Are they both rings with unity? No.

$\mathbb{Z}$ is a ring with unity, namely the number 1. $2\mathbb{Z}$ is not a ring with unity, $1 \notin 2\mathbb{Z}$. Hence, they are not isomorphic as rings. □

Example 16.1.15 Next consider whether $[2\mathbb{Z}; +, \cdot]$ and $[3\mathbb{Z}; +, \cdot]$ are isomorphic. Because of the previous example, we might guess that they are not. However, checklist items 1 through 3 above do not help us. Why? We add another checklist item:

4. Find an equation that makes sense in both rings, which is solvable in one and not the other.

The equation $x + x = x \cdot x$, or $2x = x^2$, makes sense in both rings. However, this equation has a nonzero solution, $x = 2$, in $2\mathbb{Z}$, but does not have a nonzero solution in $3\mathbb{Z}$. Thus we have an equation solvable in one ring that cannot be solved in the other, so they cannot be isomorphic. □

Another universal concept that applies to the theory of rings is that of a subsystem. A subring of a ring $[R; +, \cdot]$ is any nonempty subset S of R that is a ring under the operations of R. First, for S to be a subring of the ring R, S must be a subgroup of the group $[R; +]$. Also, S must be closed under $\cdot$, satisfy the associative law under $\cdot$, and satisfy the distributive laws. But since R is a ring, the associative and distributive laws are true for every element in R, and, in particular, for all elements in S, since $S \subseteq R$. We have just proven the following theorem:

Theorem 16.1.16 *A nonempty subset S of a ring $[R; +, \cdot]$ is a subring of R if and only if:*

(1) $[S; +]$ is a subgroup of the group $[R; +]$

(2) S is closed under multiplication: if $a, b \in S$, then $a \cdot b \in S$.

Example 16.1.17 The set of even integers, $2\mathbb{Z}$, is a subring of the ring $[\mathbb{Z}; +, \cdot]$ since $[2\mathbb{Z}; +]$ is a subgroup of the group $[\mathbb{Z}; +]$ and since it is also closed with respect to multiplication:

$$2m, 2n \in 2\mathbb{Z} \Rightarrow (2m) \cdot (2n) = 2(2 \cdot m \cdot n) \in 2\mathbb{Z}$$

□

Several of the basic facts that we are familiar with are true for any ring. The following theorem lists a few of the elementary properties of rings.

Theorem 16.1.18 Some Basic Properties. *Let $[R; +, \cdot]$ be a ring, with $a, b \in R$. Then*

(1) $a \cdot 0 = 0 \cdot a = 0$

(2) $a \cdot (-b) = (-a) \cdot b = -(a \cdot b)$

(3) $(-a) \cdot (-b) = a \cdot b$

Proof.

(1) $a \cdot 0 = a \cdot (0+0) = a \cdot 0 + a \cdot 0$, the last equality valid by the left distributive axiom. Hence if we add $-(a \cdot 0)$ to both sides of the equality above, we obtain $a \cdot 0 = 0$. Similarly, we can prove that $0 \cdot a = 0$.

(2) Before we begin the proof of this part, recall that the inverse of each element of the group $[R; +]$ is unique. Hence the inverse of the element $a \cdot b$ is unique and it is denoted $-(a \cdot b)$. Therefore, to prove that $a \cdot (-b) = -(a \cdot b)$, we need only show that $a \cdot (-b)$ inverts $a \cdot b$.

$$\begin{aligned} a \cdot (-b) + a \cdot b &= a \cdot (-b + b) && \text{by the left distributive axiom} \\ &= a \cdot 0 && \text{since } -b \text{ inverts } b \\ &= 0 && \text{by part 1 of this theorem} \end{aligned}$$

Similarly, it can be shown that $(-a) \cdot b = -(a \cdot b)$.

(3) We leave the proof of part 3 to the reader as an exercise. ∎

Example 16.1.19 We will compute $2 \cdot (-2)$ in the ring $[\mathbb{Z}_6; +_6, \times_6]$. $2 \times_6 (-2) = -(2 \times_6 2) = -4 = 2$, since the additive inverse of 4 (mod 6) is 2. Of course, we could have done the calculation directly as $2 \times_6 (-2) = 2 \times_6 4 = 2$. □

16.1.5 Integral Domains and Zero Divisors

As the example above illustrates, Theorem 16.1.18, p. 191 is a modest beginning in the study of which algebraic manipulations are possible when working with rings. A fact in elementary algebra that is used frequently in problem solving is the cancellation law. We know that the cancellation laws are true under addition for any ring, based on group theory. Are the cancellation laws true under multiplication, where the group axioms can't be counted on? More specifically, let $[R; +, \cdot]$ be a ring and let $a, b, c \in R$ with $a \neq 0$. When can we cancel the a's in the equation $a \cdot b = a \cdot c$? We can do so if a^{-1} exists, but we cannot assume that a has a multiplicative inverse. The answer to this question is found with the following definition and the theorem that follows.

Definition 16.1.20 Zero Divisor. Let $[R; +, \cdot]$ be a ring. If a and b are two nonzero elements of R such that $a \cdot b = 0$, then a and b are called zero divisors. ◇

Example 16.1.21

(a) In the ring $[\mathbb{Z}_8; +_8, \times_8]$, the numbers 4 and 2 are zero divisors since $4 \times_8 2 = 0$. In addition, 6 is a zero divisor because $6 \times_8 4 = 0$.

(b) In the ring $[M_{2 \times 2}(\mathbb{R}); +, \cdot]$ the matrices $A = \begin{pmatrix} 0 & 0 \\ 0 & 1 \end{pmatrix}$ and $B = \begin{pmatrix} 0 & 1 \\ 0 & 0 \end{pmatrix}$ are zero divisors since $AB = 0$.

(c) $[\mathbb{Z}; +, \cdot]$ has no zero divisors.

□

Now, here is why zero divisors are related to cancellation.

Theorem 16.1.22 Multiplicative Cancellation. *The multiplicative cancellation laws hold in a ring $[R; +, \cdot]$ if and only if R has no zero divisors.*

Proof. We prove the theorem using the left cancellation axiom, namely that if $a \neq 0$ and $a \cdot b = a \cdot c$, then $b = c$ for all $a, b, c \in R$. The proof using the right cancellation axiom is its mirror image.

($\Rightarrow$) Assume the left cancellation law holds in R and assume that a and b are two elements in R such that $a \cdot b = 0$. We must show that either $a = 0$ or $b = 0$. To do this, assume that $a \neq 0$ and show that b must be 0.

$$a \cdot b = 0 \Rightarrow a \cdot b = a \cdot 0$$
$$\Rightarrow b = 0 \quad \text{by the left cancellation law}.$$

($\Leftarrow$) Conversely, assume that R has no zero divisors and we will prove that the left cancellation law must hold. To do this, assume that $a, b, c \in R$, $a \neq 0$, such that $a \cdot b = a \cdot c$ and show that $b = c$.

$$a \cdot b = a \cdot c \Rightarrow a \cdot b - a \cdot c = 0$$
$$\Rightarrow a \cdot (b - c) = 0$$
$$\Rightarrow b - c = 0 \quad \text{since there are no zero divisors}$$
$$\Rightarrow b = c$$

∎

Hence, the only time that the cancellation laws hold in a ring is when there are no zero divisors. The commutative rings with unity in which the two conditions are true are given a special name.

Definition 16.1.23 Integral Domain. A commutative ring with unity containing no zero divisors is called an integral domain. ◊

In this chapter, Integral domains will be denoted genericaly by the letter D. We state the following two useful facts without proof.

Theorem 16.1.24 *If $m \in \mathbb{Z}_n$, $m \neq 0$, then m is a zero divisor if and only if m and n are not relatively prime; i.e., $\gcd(m, n) > 1$.*

Corollary 16.1.25 *If p is a prime, then $\mathbb{Z}_p$ has no zero divisors.*

Example 16.1.26 $[\mathbb{Z}; +, \cdot]$, $[\mathbb{Z}_p; +_p, \times_p]$ with p a prime, $[\mathbb{Q}; +, \cdot]$, $[\mathbb{R}; +, \cdot]$, and $[\mathbb{C}; +, \cdot]$ are all integral domains. The key example of an infinite integral domain is $[\mathbb{Z}; +, \cdot]$. In fact, it is from $\mathbb{Z}$ that the term integral domain is derived. Our main example of a finite integral domain is $[\mathbb{Z}_p; +_p, \times_p]$, when p is prime. □

We close this section with the verification of an observation that was made in Chapter 11, namely that the product of two algebraic systems may not be an algebraic system of the same type.

Example 16.1.27 Both $[\mathbb{Z}_2; +_2, \times_2]$ and $[\mathbb{Z}_3; +_3, \times_3]$ are integral domains. Consider the direct product $\mathbb{Z}_2 \times \mathbb{Z}_3$. It's true that $\mathbb{Z}_2 \times \mathbb{Z}_3$ is a commutative ring with unity (see Exercise 13). However, $(1, 0) \cdot (0, 2) = (0, 0)$, so $\mathbb{Z}_2 \times \mathbb{Z}_3$ has zero divisors and is therefore not an integral domain. □

16.1.6 Exercises

1. Review the definition of rings to show that the following are rings. The operations involved are the usual operations defined on the sets. Which of these rings are commutative? Which are rings with unity? For the rings with unity, determine the unity and all units.

    (a) $[\mathbb{Z}; +, \cdot]$

    (b) $[\mathbb{C}; +, \cdot]$

(c) $[\mathbb{Q}; +, \cdot]$

(d) $[M_{2\times 2}(\mathbb{R}); +, \cdot]$

(e) $[\mathbb{Z}_2; +_2, \times_2]$

2. Follow the instructions for Exercise 1 and the following rings:
   (a) $[\mathbb{Z}_6; +_6, \times_6]$ (d) $[\mathbb{Z}_8; +_8, \times_8]$
   (b) $[\mathbb{Z}_5; +_5, \times_5]$ (e) $[\mathbb{Z} \times \mathbb{Z}; +, \cdot]$
   (c) $[\mathbb{Z}_2{}^3; +, \cdot]$ (f) $[\mathbb{R}^2; +, \cdot]$

3. Show that the following pairs of rings are not isomorphic:
   (a) $[\mathbb{Z}; +, \cdot]$ and $[M_{2\times 2}(\mathbb{Z}); +, \cdot]$
   (b) $[3\mathbb{Z}; +, \cdot]$ and $[4\mathbb{Z}; +, \cdot]$.

4. Show that the following pairs of rings are not isomorphic:
   (a) $[\mathbb{R}; +, \cdot]$ and $[\mathbb{Q}; +, \cdot]$.
   (b) $[\mathbb{Z}_2 \times \mathbb{Z}_2; +, \cdot]$ and $[\mathbb{Z}_4; +, \cdot]$.

5.
   (a) Show that $3\mathbb{Z}$ is a subring of the ring $[\mathbb{Z}; +, \cdot]$
   (b) Find all subrings of $\mathbb{Z}_8$.
   (c) Find all subrings of $\mathbb{Z}_2 \times \mathbb{Z}_2$.

6. Verify the validity of Theorem 16.1.22, p. 192 by finding examples of elements a, b, and c, $a \neq 0$ in the following rings, where $a \cdot b = a \cdot c$ and yet $b \neq c$:
   (a) $\mathbb{Z}_8$
   (b) $M_{2\times 2}(\mathbb{R})$
   (c) $\mathbb{Z}_2{}^2$

7.
   (a) Determine all solutions of the equation $x^2 - 5x + 6 = 0$ in $\mathbb{Z}$. Can there be any more than two solutions to this equation (or any quadratic equation) in $\mathbb{Z}$?
   (b) Find all solutions of the equation in part a in $\mathbb{Z}_{12}$. Why are there more than two solutions?

8. Solve the equation $x^2 + 4x + 4 = 0$ in the following rings. Interpret 4 as $1 + 1 + 1 + 1$, where 1 is the unity of the ring.
   (a) in $\mathbb{Z}_8$
   (b) in $M_{2\times 2}(\mathbb{R})$
   (c) in $\mathbb{Z}$
   (d) in $\mathbb{Z}_3$

9. The relation "is isomorphic to" on rings is an equivalence relation. Explain the meaning of this statement.

10.
    (a) Let $R_1, R_2, \ldots, R_n$ be rings. Prove the multiplicative, associative,

and distributive laws for the ring

$$R = \underset{i=1}{\overset{n}{\times}} R_i$$

- (b) If each of the R_i is commutative, is R commutative?
- (c) Under what conditions will R be a ring with unity?
- (d) What will the units of R be when it has a unity?

11.
- (a) Prove that the ring $\mathbb{Z}_2 \times \mathbb{Z}_3$ is commutative and has unity.
- (b) Determine all zero divisors for the ring $\mathbb{Z}_2 \times \mathbb{Z}_3$.
- (c) Give another example illustrating the fact that the product of two integral domains may not be an integral domain. Is there an example where the product is an integral domain?

12. **Boolean Rings.** Let U be a nonempty set.
- (a) Verify that $[\mathcal{P}(U); \oplus, \cap]$ is a commutative ring with unity.
- (b) What are the units of this ring?

13.
- (a) For any ring $[R; +, \cdot]$, expand $(a+b)(c+d)$ for $a, b, c, d \in R$.
- (b) If R is commutative, prove that $(a+b)^2 = a^2 + 2ab + b^2$ for all $a, b \in R$.

14.
- (a) Let R be a commutative ring with unity. Prove by induction that for $n \geq 1$, $(a+b)^n = \sum_{k=0}^{n} \binom{n}{k} a^k b^{n-k}$
- (b) Simplify $(a+b)^5$ in $\mathbb{Z}_5$.
- (c) Simplify $(a+b)^{10}$ in $\mathbb{Z}_{10}$.

15. Prove part 3 of Theorem 16.1.18, p. 191.
16. Let U be a finite set. Prove that the Boolean ring $[\mathcal{P}(U); \oplus, \cap]$ is isomorphic to the ring $[\mathbb{Z}_2{}^n; +, \cdot]$. where $n = |U|$.

16.2 Fields

Although the algebraic structures of rings and integral domains are widely used and play an important part in the applications of mathematics, we still cannot solve the simple equation $ax = b$, $a \neq 0$ in all rings or all integral domains, for that matter. Yet this is one of the first equations we learn to solve in elementary algebra and its solubility is basic to innumerable questions. If we wish to solve a wide range of problems in a system we need at least all of the laws true for rings and the cancellation laws together with the ability to solve the equation $ax = b$, $a \neq 0$. We summarize the above in a definition and list theorems that will place this concept in the context of the previous section.

Definition 16.2.1 Field. A field is a commutative ring with unity such that each nonzero element has a multiplicative inverse. ◇

In this chapter, we denote a field generically by the letter F. The letters k, K and L are also conventionally used for fields.

Example 16.2.2 Some common fields. The most common infinite fields are $[\mathbb{Q}; +, \cdot]$, $[\mathbb{R}; +, \cdot]$, and $[\mathbb{C}; +, \cdot]$. □

Remark 16.2.3 Since every field is a ring, all facts and concepts that are true for rings are true for any field.

Theorem 16.2.4 Field ⇒ Integral Domain. *Every field is an integral domain.*
Proof. The proof is fairly easy and a good exercise, so we provide a hint. Starting with the assumption that $a \cdot b = 0$ if we assume that $a \neq 0$ then the existence of a^{-1} makes it possible to infer that $b = 0$. ∎

Of course the converse of Theorem 16.2.4, p. 196 is not true. Consider $[\mathbb{Z}; +, \cdot]$. However, the next theorem proves the converse in finite fields.

Theorem 16.2.5 Finite Integral Domain ⇒ Field. *Every finite integral domain is a field.*
Proof. We leave the details to the reader, but observe that if D is a finite integral domain, we can list all elements as $a_1, a_2, \ldots, a_n$, where $a_1 = 1$. Now, to show that any a_i has a multiplicative inverse, consider the n products $a_i \cdot a_1, a_i \cdot a_2, \ldots, a_i \cdot a_n$. What can you say about these products? ∎

If p is a prime, $p \mid (a \cdot b) \Rightarrow p \mid a$ or $p \mid b$. An immediate implication of this fact is the following corollary.

Corollary 16.2.6 *If p is a prime, then $\mathbb{Z}_p$ is a field.*

Example 16.2.7 A field of order 4. Corollary 16.2.6, p. 196 gives us a large number of finite fields, but we must be cautious. This does not tell us that all finite fields are of the form $\mathbb{Z}_p$, p a prime. To see this, let's try to construct a field of order 4.

First the field must contain the additive and multiplicative identities, 0 and 1, so, without loss of generality, we can assume that the field we are looking for is of the form $F = \{0, 1, a, b\}$. Since there are only two nonisomorphic groups of order 4, we have only two choices for the group table for $[F; +]$. If the additive group is isomorphic to $\mathbb{Z}_4$ then two of the nonzero elements of F would not be their own additive inverse (as are 1 and 3 in $\mathbb{Z}_4$). Let's assume $\beta \in F$ is one of those elements and $\beta + \beta = \gamma \neq 0$. An isomorphism between the additive groups F and $\mathbb{Z}_4$ would require that γ in F correspond with 2 in $\mathbb{Z}_4$. We could continue our argument and infer that $\gamma \cdot \gamma = 0$, producing a zero divisor, which we need to avoid if F is to be a field. We leave the remainder of the argument to the reader. We can thus complete the addition table so that $[F; +]$ is isomorphic to $\mathbb{Z}_2^2$:

+	0	1	a	b
0	0	1	a	b
1	1	0	b	a
a	a	b	0	1
b	b	a	1	0

Next, since 1 is the unity of F, the partial multiplication table must look like:

·	0	1	a	b
0	0	0	0	0
1	0	1	a	b
a	0	a	–	–
b	0	b	–	–

Hence, to complete the table, we have only four entries to find, and, since F must be commutative, this reduces our task to filling in three entries. Next,

16.2. FIELDS

each nonzero element of F must have a unique multiplicative inverse. The inverse of a must be either a itself or b. If $a^{-1} = a$, then $b^{-1} = b$. (Why?) But $a^{-1} = a \Rightarrow a \cdot a = 1$. And if $a \cdot a = 1$, then $a \cdot b$ is equal to a or b. In either case, by the cancellation law, we obtain $a = 1$ or $b = 1$, which is impossible. Therefore we are forced to conclude that $a^{-1} = b$ and $b^{-1} = a$. To determine the final two products of the table, simply note that, $a \cdot a \neq a$ because the equation $x^2 = x$ has only two solutions, 0 and 1 in any field. We also know that $a \cdot a$ cannot be 1 because a doesn't invert itself and cannot be 0 because a can't be a zero divisor. This leaves us with one possible conclusion. that $a \cdot a = b$ and similarly $b \cdot b = a$. Hence, our multiplication table for F is:

·	0	1	a	b
0	0	0	0	0
1	0	1	a	b
a	0	a	b	1
b	0	b	1	a

We leave it to the reader to verify that $[F; +, \cdot]$, as described above, is a field. Hence, we have produced a field of order 4. This construction would be difficult to repeat for larger fields. In section 16.4 we will introduce a different approach to constructing fields that will be far more efficient. □

Even though not all finite fields are isomorphic to $\mathbb{Z}_p$ for some prime p, it can be shown that every field F must have either:

- a subfield isomorphic to $\mathbb{Z}_p$ for some prime p, or

- a subfield isomorphic to $\mathbb{Q}$.

One can think of all fields as being constructed from either $\mathbb{Z}_p$ or $\mathbb{Q}$.

Example 16.2.8 $[\mathbb{R}; +, \cdot]$ is a field, and it contains a subfield isomorphic to $[\mathbb{Q}; +, \cdot]$, namely $\mathbb{Q}$ itself. □

Example 16.2.9 The field F that we constructed in Example 16.2.7, p. 196 has a subfield isomorphic to $\mathbb{Z}_p$ for some prime p. From the tables, we note that the subset $\{0, 1\}$ of $\{0, 1, a, b\}$ under the given operations of F behaves exactly like $[\mathbb{Z}_2; +_2, \times_2]$. Hence, F has a subfield isomorphic to $\mathbb{Z}_2$. □

We close this section with a brief discussion of isomorphic fields. Again, since a field is a ring, the definition of isomorphism of fields is the same as that of rings. It can be shown that if f is a field isomorphism, then $f(a^{-1}) = f(a)^{-1}$; that is, inverses are mapped onto inverses under any field isomorphism. A major question to try to solve is: How many different non-isomorphic finite fields are there of any given order? If p is a prime, it seems clear from our discussions that all fields of order p are isomorphic to $\mathbb{Z}_p$. But how many nonisomorphic fields are there, if any, of order 4, 6, 8, 9, etc? The answer is given in the following theorem, whose proof is beyond the scope of this text.

Theorem 16.2.10

(1) Any finite field F has order p^n for a prime p and a positive integer n.

(2) For any prime p and any positive integer n there is a field of order p^n.

(3) Any two fields of order p^n are isomorphic.

Galois. The field of order p^n is frequently referred to as the Galois field of order p^n and it is denoted by $GF(p^n)$. Evariste Galois (1811-32) was a pioneer in the field of abstract algebra.

Figure 16.2.11 French stamp honoring Evariste Galois

This theorem tells us that there is a field of order $2^2 = 4$, and there is only one such field up to isomorphism. That is, all such fields of order 4 are isomorphic to F, which we constructed in the example above.

Exercises

1. Write out the addition, multiplication, and "inverse" tables for each of the following fields'.

    (a) $[\mathbb{Z}_2; +_2, \times_2]$

    (b) $[\mathbb{Z}_3; +_3, \times_3]$

    (c) $[\mathbb{Z}_5; +_5, \times_5]$

2. Show that the set of units of the fields in Exercise 1 form a group under the operation of the multiplication of the given field. Recall that a unit is an element which has a multiplicative inverse.

3. Complete the proof of Theorem 16.2.5, p. 196 that every finite integral domain is a field.

4. Write out the operation tables for $\mathbb{Z}_2^2$. Is $\mathbb{Z}_2^2$ a ring? An integral domain? A field? Explain.

5. Determine all values x from the given field that satisfy the given equation:

    (a) $x + 1 = -1$ in $\mathbb{Z}_2$, $\mathbb{Z}_3$ and $\mathbb{Z}_5$

    (b) $2x + 1 = 2$ in $\mathbb{Z}_3$ and $\mathbb{Z}_5$

    (c) $3x + 1 = 2$ in $\mathbb{Z}_5$

6.
    (a) Prove that if p and q are prime, then $\mathbb{Z}_p \times \mathbb{Z}_q$ is never a field.

    (b) Can $\mathbb{Z}_{p^n}$ be a field for any prime p and any positive integer $n \geq 2$?

7. Determine all solutions to the following equations over $\mathbb{Z}_2$. That is, find all elements of $\mathbb{Z}_2$ that satisfy the equations.

    (a) $x^2 + x = 0$

    (b) $x^2 + 1 = 0$

    (c) $x^3 + x^2 + x + 1 = 0$

    (d) $x^3 + x + 1 = 0$

8. Determine the number of different fields, if any, of all orders 2 through 15. Wherever possible, describe these fields via a known field.

9. Let $\mathbb{Q}(\sqrt{2}) = \{a + b\sqrt{2} \mid a, b \in \mathbb{Q}\}$.

    (a) Prove that $[\mathbb{Q}(\sqrt{2}); +, \cdot]$ is a field.

    (b) Show that $\mathbb{Q}$ is a subfield of $\mathbb{Q}(\sqrt{2})$. For this reason, $\mathbb{Q}(\sqrt{2})$ is called an extension field of $\mathbb{Q}$.

(c) Show that all the roots of the equation $x^2 - 4x + \frac{7}{2} = 0$ lie in the extension field $\mathbb{Q}\left(\sqrt{2}\right)$.

(d) Do the roots of the equation $x^2 - 4x + 3 = 0$ lie in this field? Explain.

16.3 Polynomial Rings

In the previous sections we examined the solutions of a few equations over different rings and fields. To solve the equation $x^2 - 2 = 0$ over the field of the real numbers means to find all solutions of this equation that are in this particular field $\mathbb{R}$. This statement can be replaced as follows: Determine all $a \in \mathbb{R}$ such that the polynomial $f(x) = x^2 - 2$ is equal to zero when evaluated at $x = a$. In this section, we will concentrate on the theory of polynomials. We will develop concepts using the general setting of polynomials over rings since results proven over rings are true for fields (and integral domains). The reader should keep in mind that in most cases we are just formalizing concepts that he or she learned in high school algebra over the field of reals.

Definition 16.3.1 Polynomial over a Ring. Let $[R; +, \cdot]$ be a ring. A polynomial, $f(x)$, over R is an expression of the form

$$f(x) = \sum_{i=0}^{n} a_i x^i = a_0 + a_1 x + a_2 x^2 + \cdots + a_n x^n$$

where $n \geq 0$, and $a_0, a_1, a_2, \ldots, a_n \in R$. If $a_n \neq 0$, then the degree of $f(x)$ is n. If $f(x) = 0$, then the degree of $f(x)$ is undefined, but for convenience we say that $\deg 0 = -\infty$. If the degree of $f(x)$ is n, we write $\deg f(x) = n$. The set of all polynomials in the indeterminate x with coefficients in R is denoted by $R[x]$. ◊

Note 16.3.2

- The symbol x is an object called an **indeterminate**, which is not an element of the ring R.

- Note that $R \subseteq R[x]$. The elements of R are called constant polynomials, with the nonzero elements of R being the polynomials of degree 0.

- R is called the ground, or base, ring for $R[x]$.

- In the definition above, we have written the terms in increasing degree starting with the constant. The ordering of terms can be reversed without changing the polynomial. For example, $1 + 2x - 3x^4$ and $-3x^4 + 2x + 1$ are the same polynomial.

- A term of the form x^k in a polynomial is understood to be $1x^k$.

- It is understood that if $\deg f(x) = n$, then coefficients of powers of x higher than n are equal to the zero of the base ring.

Definition 16.3.3 Polynomial Addition. Let $f(x) = a_0 + a_1 x + a_2 x^2 + \cdots + a_m x^m$ and $g(x) = b_0 + b_1 x + b_2 x^2 + \cdots + b_n x^n$ be elements in $R[x]$ so that $a_i \in R$ and $b_i \in R$ for all i. Let k be the maximum of m and n. Then $f(x) + g(x) = c_0 + c_1 x + c_2 x^2 + \cdots + c_k x^k$, where $c_i = a_i + b_i$ for $i = 0, 1, 2, \ldots, k$. ◊

Definition 16.3.4 Polynomial Multiplication. Let $f(x) = a_0 + a_1 x + a_2 x^2 + \cdots + a_m x^m$ and $g(x) = b_0 + b_1 x + b_2 x^2 + \cdots + b_n x^n$. Then

$$f(x) \cdot g(x) = d_0 + d_1 x + d_2 x^2 + \cdots + d_p x^p \quad \text{where } p = m + n \text{ and}$$
$$d_s = \sum_{i=0}^{s} a_i b_{s-i} = a_0 b_s + a_1 b_{s-1} + a_2 b_{s-2} + \cdots + a_{s-1} b_1 + a_s b_0$$
$$\text{for } 0 \leq s \leq p$$

$\Diamond$

The important fact to keep in mind is that addition and multiplication in $R[x]$ depends on addition and multiplication in R. The powers of x merely serve the purpose of "place holders." All computations involving coefficients are done over the given ring. Powers of the indeterminate are computed formally applying the rule of adding exponents when multiplying powers.

Example 16.3.5 $f(x) = 3$, $g(x) = 2 - 4x + 7x^2$, and $h(x) = 2 + x^4$ are all polynomials in $\mathbb{Z}[x]$. Their degrees are 0, 2, and 4, respectively. $\square$

Addition and multiplication of polynomials are performed as in high school algebra. However, we must do our computations in the ground ring of the polynomials.

Example 16.3.6 In $\mathbb{Z}_3[x]$, if $f(x) = 1 + x$ and $g(x) = 2 + x$, then

$$\begin{aligned} f(x) + g(x) &= (1 + x) + (2 + x) \\ &= (1 +_3 2) + (1 +_3 1)\, x \\ &= 0 + 2x \\ &= 2x \end{aligned}$$

and

$$\begin{aligned} f(x)g(x) &= (1 + x) \cdot (2 + x) \\ &= 1 \times_3 2 + (1 \times_3 1 +_3 1 \times_3 2)x + (1 \times_3 1)x^2 \\ &= 2 + 0x + x^2 \\ &= 2 + x^2 \end{aligned}$$

However, for the same polynomials as above, $f(x)$ and $g(x)$ in the more familiar setting of $\mathbb{Z}[x]$, we have

$$f(x) + g(x) = (1 + x) + (2 + x) = (1 + 2) + (1 + 1)x = 3 + 2x$$

and

$$\begin{aligned} f(x)g(x) &= (1 + x) \cdot (2 + x) \\ &= 1 \cdot 2 + (1 \cdot 1 + 1 \cdot 2)x + (1 \cdot 1)x^2 \\ &= 2 + 3x + x^2 \end{aligned}$$

$\square$

Example 16.3.7 Let $f(x) = 2 + x^2$ and $g(x) = -1 + 4x + 3x^2$. We will compute $f(x) \cdot g(x)$ in $\mathbb{Z}[x]$. Of course this product can be obtained by the usual methods of high school algebra. We will, for illustrative purposes, use the above definition. Using the notation of the above definition, $a_0 = 2$, $a_1 = 0$, $a_2 = 1$, $b_0 = -1$, $b_1 = 4$, and $b_2 = 3$. We want to compute the coefficients d_0, d_1, d_2, d_3, and d_4. We will compute d_3, the coefficient of the x^3 term of the product, and leave the remainder to the reader (see Exercise 2 of this section).

16.3. POLYNOMIAL RINGS

Since the degrees of both factors is 2, $a_i = b_i = 0$ for $i \geq 3$. The coefficient of x^3 is

$$d_3 = a_0 b_3 + a_1 b_2 + a_2 b_1 + a_3 b_0 = 2 \cdot 0 + 0 \cdot 3 + 1 \cdot 4 + 0 \cdot (-1) = 4$$

□

The proofs of the following theorem are not difficult but rather long, so we omit them.

Theorem 16.3.8 Properties of Polynomial Rings. *Let $[R; +, \cdot]$ be a ring. Then:*

(1) *$R[x]$ is a ring under the operations of polynomial addition and multiplication.*

(2) *If R is a commutative ring, then $R[x]$ is a commutative ring.*

(3) *If R is a ring with unity, 1, then $R[x]$ is a ring with unity (the unity in $R[x]$ is $1 + 0x + 0x^2 + \cdots$).*

(4) *If R is an integral domain, then $R[x]$ is an integral domain.*

(5) *If F is a field, then $F[x]$ is not a field. However, $F[x]$ is an integral domain.*

Next we turn to division of polynomials, which is not an operation since the result is a pair of polynomials, not a single one. From high school algebra we all learned the standard procedure for dividing a polynomial $f(x)$ by a second polynomial $g(x)$. This process of polynomial long division is referred to as the division property for polynomials. Under this scheme we continue to divide until the result is a quotient $q(x)$ and a remainder $r(x)$ whose degree is strictly less than that of the divisor $g(x)$. This property is valid over any field. Before giving a formal description, we consider some examples.

Example 16.3.9 Polynomial Division. Let $f(x) = 1 + x + x^3$ and $g(x) = 1 + x$ be two polynomials in $\mathbb{Z}_2[x]$. Let us divide $f(x)$ by $g(x)$. Keep in mind that we are in $\mathbb{Z}_2[x]$ and that, in particular, $-1 = 1$ in $\mathbb{Z}_2$. This is a case where reordering the terms in decreasing degree is preferred.

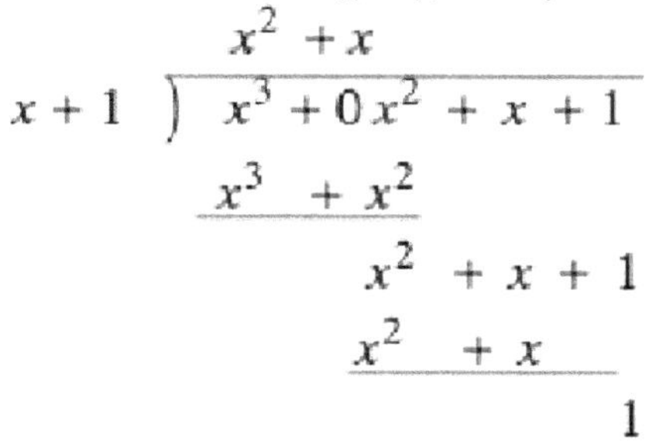

$$\begin{array}{r} x^2 + x \\ x+1 \overline{\smash{\big)}\, x^3 + 0x^2 + x + 1} \\ \underline{x^3 + x^2 } \\ x^2 + x + 1 \\ \underline{x^2 + x } \\ 1 \end{array}$$

Figure 16.3.10

Therefore,
$$\frac{x^3 + x + 1}{x + 1} = x^2 + x + \frac{1}{x+1}$$

or equivalently,
$$x^3 + x + 2 = (x^2 + x) \cdot (x + 1) + 1$$

That is, $f(x) = g(x) \cdot q(x) + r(x)$ where $q(x) = x^2 + x$ and $r(x) = 1$. Notice that $\deg(r(x)) = 0$, which is strictly less than $\deg(g(x)) = 1$. □

Example 16.3.11 Let $f(x) = 1 + x^4$ and $g(x) = 1 + x$ be polynomials in $\mathbb{Z}_2[x]$. Let us divide $f(x)$ by $g(x)$:

$$
\begin{array}{r}
x^3 + x^2 + x + 1 \\
x+1 \overline{\smash{)}\ x^4 + 0x^3 + 0x^2 + 0x + 1} \\
\underline{x^4 + x^3 } \\
x^3 + 1 \\
\underline{x^3 + x^2 } \\
x^2 + 1 \\
\underline{x^2 + x } \\
x + 1 \\
\underline{x + 1} \\
0
\end{array}
$$

Figure 16.3.12

Thus $x^4 + 1 = (x^3 + x^2 + x + 1)(x + 1)$. Since we have 0 as a remainder, $x + 1$ must be a factor of $x^4 + 1$. Also, since $x + 1$ is a factor of $x^4 + 1$, 1 is a zero (or root) of $x^4 + 1$. Of course we could have determined that 1 is a root of $f(x)$ simply by computing $f(1) = 1^4 +_2 1 = 1 +_2 1 = 0$. □

Before we summarize the main results suggested by the previous examples, we should probably consider what could have happened if we had attempted to perform divisions of polynomials in the ring $\mathbb{Z}[x]$ rather than in the polynomials over the field $\mathbb{Z}_2$. For example, $f(x) = x^2 - 1$ and $g(x) = 2x - 1$ are both elements of the ring $\mathbb{Z}[x]$, yet $x^2 - 1 = (\frac{1}{2}x + \frac{1}{2})(2x - 1) - \frac{3}{4}$ The quotient and remainder are not a polynomials over $\mathbb{Z}$ but polynomials over the field of rational numbers. For this reason it would be wise to describe all results over a field F rather than over an arbitrary ring R so that we don't have to expand our possible set of coefficients.

Theorem 16.3.13 Division Property for Polynomials. *Let $[F; +, \cdot]$ be a field and let $f(x)$ and $g(x)$ be two elements of $F[x]$ with $g(x) \neq 0$. Then there exist unique polynomials $q(x)$ and $r(x)$ in $F[x]$ such that $f(x) = g(x)q(x) + r(x)$, where $\deg r(x) < \deg g(x)$.*
Proof. This theorem can be proven by induction on $\deg f(x)$. ■

Theorem 16.3.14 The Factor Theorem. *Let $[F; +, \cdot]$ be a field. An element $a \in F$ is a zero of $f(x) \in F[x]$ if and only if $x - a$ is a factor of $f(x)$ in $F[x]$.*
Proof.

($\Rightarrow$) Assume that $a \in F$ is a zero of $f(x) \in F[x]$. We wish to show that $x - a$ is a factor of $f(x)$. To do so, apply the division property to $f(x)$ and $g(x) = x - a$. Hence, there exist unique polynomials $q(x)$ and $r(x)$ from $F[x]$ such that $f(x) = (x - a) \cdot q(x) + r(x)$ and the $\deg r(x) < \deg(x - a) = 1$, so $r(x) = c \in F$, that is, $r(x)$ is a constant. Also, the fact that a is a zero of $f(x)$ means that $f(a) = 0$. So $f(x) = (x - a) \cdot q(x) + c$ becomes $0 = f(a) = (a - a)q(a) + c$. Hence $c = 0$, so $f(x) = (x - a) \cdot q(x)$, and $x - a$ is a factor of $f(x)$. The reader should note that a critical point of the proof of this half of the theorem was the part of the division property that stated that $\deg r(x) < \deg g(x)$.

($\Leftarrow$) We leave this half to the reader as an exercise. ■

Theorem 16.3.15 *A nonzero polynomial $f(x) \in F[x]$ of degree n can have at most n zeros.*
Proof. Let $a \in F$ be a zero of $f(x)$. Then $f(x) = (x - a) \cdot q_1(x)$, $q_1(x) \in F[x]$, by the Factor Theorem. If $b \in F$ is a zero of $q_1(x)$, then again by Factor Theorem,

16.3. POLYNOMIAL RINGS

$f(x) = (x - a)(x - b)q_2(x)$, $q_2(x) \in F[x]$. Continue this process, which must terminate in at most n steps since the degree of $q_k(x)$ would be $n - k$. ∎

From The Factor Theorem, p. 202, we can get yet another insight into the problems associated with solving polynomial equations; that is, finding the zeros of a polynomial. The initial important idea here is that the zero a is from the ground field F. Second, a is a zero only if $(x - a)$ is a factor of $f(x)$ in $F[x]$; that is, only when $f(x)$ can be factored (or reduced) to the product of $(x - a)$ times some other polynomial in $F[x]$.

Example 16.3.16 Consider the polynomial $f(x) = x^2 - 2$ taken as being in $\mathbb{Q}[x]$. From high school algebra we know that $f(x)$ has two zeros (or roots), namely $\pm\sqrt{2}$, and $x^2 - 2$ can be factored as $(x - \sqrt{2})(x + \sqrt{2})$. However, we are working in $\mathbb{Q}[x]$, these two factors are not in the set of polynomials over the rational numbers, $\mathbb{Q}$ since $\sqrt{2} \notin \mathbb{Q}$. Therefore, $x^2 - 2$ does not have a zero in $\mathbb{Q}$ since it cannot be factored over $\mathbb{Q}$. When this happens, we say that the polynomial is irreducible over $\mathbb{Q}$. □

The problem of factoring polynomials is tied hand-in-hand with that of the reducibility of polynomials. We give a precise definition of this concept.

Definition 16.3.17 Reducibility over a Field. Let $[F; +, \cdot]$ be a field and let $f(x) \in F[x]$ be a nonconstant polynomial. $f(x)$ is **reducible** over F if and only if it can be factored as a product of two nonconstant polynomials in $F[x]$. A polynomial is **irreducible** over F if it is not reducible over F. ◇

Example 16.3.18 The polynomial $f(x) = x^4 + 1$ is reducible over $\mathbb{Z}_2$ since $x^4 + 1 = (x + 1)(x^3 + x^2 + x - 1)$. □

Example 16.3.19 Is the polynomial $f(x) = x^3 + x + 1$ reducible over $\mathbb{Z}_2$? Since a factorization of a cubic polynomial can only be as a product of linear and quadratic factors, or as a product of three linear factors, $f(x)$ is reducible if and only if it has at least one linear factor. From the Factor Theorem, $x - a$ is a factor of $x^3 + x + 1$ over $\mathbb{Z}_2$ if and only if $a \in \mathbb{Z}_2$ is a zero of $x^3 + x + 1$. So $x^3 + x + 1$ is reducible over $\mathbb{Z}_2$ if and only if it has a zero in $\mathbb{Z}_2$. Since $\mathbb{Z}_2$ has only two elements, 0 and 1, this is easy enough to check. $f(0) = 0^3 +_2 0 +_2 1 = 1$ and $f(1) = 1^3 +_2 1 +_2 1 = 1$, so neither 0 nor 1 is a zero of $f(x)$ over $\mathbb{Z}_2$. Hence, $x^3 + x + 1$ is irreducible over $\mathbb{Z}_2$. □

From high school algebra we know that $x^3 + x + 1$ has three zeros from some field. Can we find this field? To be more precise, can we construct the field that contains $\mathbb{Z}_2$ and all zeros of $x^3 + x + 1$? We will consider this task in the next section.

We close this section with a final analogy. Prime numbers play an important role in mathematics. The concept of irreducible polynomials (over a field) is analogous to that of a prime number. Just think of the definition of a prime number. A useful fact concerning primes is: If p is a prime and if $p \mid ab$, then $p \mid a$ or $p \mid b$. We leave it to the reader to think about the veracity of the following: If $p(x)$ is an irreducible polynomial over F, $a(x), b(x) \in F[x]$ and $p(x) \mid a(x)b(x)$, then $p(x) \mid a(x)$ or $p(x) \mid b(x)$.

Exercises

1. Let $f(x) = 1 + x$ and $g(x) = 1 + x + x^2$. Compute the following sums and products in the indicated rings.

    (a) $f(x) + g(x)$ and $f(x) \cdot g(x)$ in $\mathbb{Z}[x]$

    (b) $f(x) + g(x)$ and $f(x) \cdot g(x)$ in $\mathbb{Z}_2[x]$

(c) $(f(x) \cdot g(x)) \cdot f(x)$ in $\mathbb{Q}[x]$

(d) $(f(x) \cdot g(x)) \cdot f(x)$ in $\mathbb{Z}_2[x]$

(e) $f(x) \cdot f(x) + f(x) \cdot g(x)$ in $\mathbb{Z}_2[x]$

2. Complete the calculations started in Example 16.3.7, p. 200.

3. Prove that:

    (a) The ring $\mathbb{R}$ is a subring of the ring $\mathbb{R}[x]$.

    (b) The ring $\mathbb{Z}[x]$ is a subring of the $\mathbb{Q}[x]$.

    (c) The ring $\mathbb{Q}[x]$ is a subring of the ring $\mathbb{R}[x]$.

4.

    (a) Find all zeros of $x^4 + 1$ in $\mathbb{Z}_3$.

    (b) Find all zeros of $x^5 + 1$ in $\mathbb{Z}_5$.

5. Determine which of the following are reducible over $\mathbb{Z}_2$. Explain.

    (a) $f(x) = x^3 + 1$

    (b) $g(x) = x^3 + x^2 + x$.

    (c) $h(x) = x^3 + x^2 + 1$.

    (d) $k(x) = x^4 + x^2 + 1$. (Be careful.)

6. Prove the second half of The Factor Theorem, p. 202.

7. Give an example of the contention made in the last paragraph of this section.

8. Determine all zeros of $x^4 + 3x^3 + 2x + 4$ in $\mathbb{Z}_5[x]$.

9. Show that $x^2 - 3$ is irreducible over $\mathbb{Q}$ but reducible over the field of real numbers.

10. The definition of a vector space given in Chapter 13 holds over any field F, not just over the field of real numbers, where the elements of F are called scalars.

    (a) Show that $F[x]$ is a vector space over F.

    (b) Find a basis for $F[x]$ over F.

    (c) What is the dimension of $F[x]$ over F?

11. Prove Theorem 16.3.13, p. 202, the Division Property for Polynomials

12.

    (a) Show that the field $\mathbb{R}$ of real numbers is a vector space over $\mathbb{R}$. Find a basis for this vector space. What is dim $\mathbb{R}$ over $\mathbb{R}$?

    (b) Repeat part a for an arbitrary field F.

    (c) Show that $\mathbb{R}$ is a vector space over $\mathbb{Q}$.

16.4 Field Extensions

From high school algebra we realize that to solve a polynomial equation means to find its roots (or, equivalently, to find the zeros of the polynomials). From Example 16.3.16, p. 203 and Example 16.3.19, p. 203 we know that the zeros may not lie in the given ground field. Hence, to solve a polynomial really

16.4. FIELD EXTENSIONS

involves two steps: first, find the zeros, and second, find the field in which the zeros lie. For economy's sake we would like this field to be the smallest field that contains all the zeros of the given polynomial. To illustrate this concept, let us reconsider the examples from the previous section..

Example 16.4.1 Extending the Rational Numbers. Let $f(x) = x^2 - 2 \in \mathbb{Q}[x]$. It is important to remember that we are considering $x^2 - 2$ over $\mathbb{Q}$, no other field. We would like to find all zeros of $f(x)$ and the smallest field, call it S for now, that contains them. The zeros are $x = \pm\sqrt{2}$, neither of which is an element of $\mathbb{Q}$. The set S we are looking for must satisfy the conditions:

(1) S must be a field.

(2) S must contain $\mathbb{Q}$ as a subfield,

(3) S must contain all zeros of $f(x) = x^2 - 2$

By the last condition $\sqrt{2}$ must be an element of S, and, if S is to be a field, the sum, product, difference, and quotient of elements in S must be in S. So operations involving this number, such as $\sqrt{2}$, $\left(\sqrt{2}\right)^2$, $\left(\sqrt{2}\right)^3$, $\sqrt{2} + \sqrt{2}$, $\sqrt{2} - \sqrt{2}$, and $\frac{1}{\sqrt{2}}$ must all be elements of S. Further, since S contains $\mathbb{Q}$ as a subset, any element of $\mathbb{Q}$ combined with $\sqrt{2}$ under any field operation must be an element of S. Hence, every element of the form $a + b\sqrt{2}$, where a and b can be any elements in $\mathbb{Q}$, is an element of S. We leave to the reader to show that $S = \{a + b\sqrt{2} \mid a, b \in \mathbb{Q}\}$ is a field (see Exercise 1 of this section). We note that the second zero of $x^2 - 2$, namely $-\sqrt{2}$, is an element of this set. To see this, simply take $a = 0$ and $b = -1$. The field S is frequently denoted as $\mathbb{Q}\left(\sqrt{2}\right)$, and it is referred to as an extension field of $\mathbb{Q}$. Note that the polynomial $x^2 - 2 = \left(x - \sqrt{2}\right)\left(x + \sqrt{2}\right)$ factors into linear factors, or **splits**, in $\mathbb{Q}\left(\sqrt{2}\right)[x]$; that is, all coefficients of both factors are elements of the field $\mathbb{Q}\left(\sqrt{2}\right)$. □

Example 16.4.2 Extending $\mathbb{Z}_2$. Consider the polynomial $g(x) = x^2 + x + 1 \in \mathbb{Z}_2[x]$. Let's repeat the steps from the previous example to factor $g(x)$. First, $g(0) = 1$ and $g(1) = 1$, so none of the elements of $\mathbb{Z}_2$ are zeros of $g(x)$. Hence, the zeros of $g(x)$ must lie in an extension field of $\mathbb{Z}_2$. By Theorem 16.3.15, p. 202, $g(x) = x^2 + x + 1$ can have at most two zeros. Let a be a zero of $g(x)$. Then the extension field S of $\mathbb{Z}_2$ must contain, besides a, $a \cdot a = a^2$, a^3, $a + a$, $a + 1$, and so on. But, since $g(a) = 0$, we have $a^2 + a + 1 = 0$, or equivalently, $a^2 = -(a+1) = a+1$ (remember, we are working in an extension of $\mathbb{Z}_2$). We can use this recurrence relation to reduce powers of a. So far our extension field, S, of $\mathbb{Z}_2$ must contain the set $\{0, 1, a, a+1\}$, and we claim that this the complete extension. For S to be a field, all possible sums, products, and differences of elements in S must be in S. Let's try a few: $a + a = a(1 +_2 1) = a \cdot 0 = 0 \in S$ Since $a + a = 0$, $-a = a$, which is in S. Adding three a's together doesn't give us anything new: $a + a + a = a \in S$ In fact, na is in S for all possible positive integers n. Next,

$$\begin{aligned} a^3 &= a^2 \cdot a \\ &= (a+1) \cdot a \\ &= a^2 + a \\ &= (a+1) + a \\ &= 1 \end{aligned}$$

Therefore, $a^{-1} = a + 1 = a^2$ and $(a+1)^{-1} = a$.

It is not difficult to see that a^n is in S for all positive n. Does S contain all zeros of $x^2 + x + 1$? Remember, $g(x)$ can have at most two distinct zeros and we called one of them a, so if there is a second, it must be $a+1$. To see if $a+1$ is indeed a zero of $g(x)$, simply compute $f(a+1)$:

$$\begin{aligned} f(a+1) &= (a+1)^2 + (a+1) + 1 \\ &= a^2 + 1 + a + 1 + 1 \\ &= a^2 + a + 1 \\ &= 0 \end{aligned}$$

Therefore, $a+1$ is also a zero of x^2+x+1. Hence, $S = \{0, 1, a, a+1\}$ is the smallest field that contains $\mathbb{Z}_2 = \{0, 1\}$ as a subfield and contains all zeros of x^2+x+1. This extension field is denoted by $\mathbb{Z}_2(a)$. Note that x^2+x+1 splits in $\mathbb{Z}_2(a)$; that is, it factors into linear factors in $\mathbb{Z}_2(a)$. We also observe that $\mathbb{Z}_2(a)$ is a field containing exactly four elements. By Theorem 16.2.10, p. 197, we expected that $\mathbb{Z}_2(a)$ would be of order p^2 for some prime p and positive integer n. Also recall that all fields of order p^n are isomorphic. Hence, we have described all fields of order $2^2 = 4$ by finding the extension field of a polynomial that is irreducible over $\mathbb{Z}_2$. □

The reader might feel somewhat uncomfortable with the results obtained in Example 16.4.2, p. 205. In particular, what is a? Can we describe it through a known quantity? All we know about a is that it is a zero of $g(x)$ and that $a^2 = a+1$. We could also say that $a(a+1) = 1$, but we really expected more. However, should we expect more? In Example 16.4.1, p. 205, $\sqrt{2}$ is a number we are more comfortable with, but all we really know about it is that $\alpha = \sqrt{2}$ is the number such that $\alpha^2 = 2$. Similarly, the zero that the reader will obtain in Exercise 2 of this section is the imaginary number i. Here again, this is simply a symbol, and all we know about it is that $i^2 = -1$. Hence, the result obtained in Example 16.4.2, p. 205 is not really that strange.

The reader should be aware that we have just scratched the surface in the development of topics in polynomial rings. One area of significant applications is in coding theory.

Example 16.4.3 An Error Correcting Polynomial Code. An important observation regarding the previous example is that the nonzero elements of $GF(4)$ can be represented two ways. First as a linear combination of 1 and a. There are four such linear combinations, one of which is zero. Second, as powers of a. There are three distinct powers and the each match up with a nonzero linear combination:

$$\begin{aligned} a^0 &= 1 \cdot 1 + 0 \cdot a \\ a^1 &= 0 \cdot 1 + 1 \cdot a \\ a^2 &= 1 \cdot 1 + 1 \cdot a \end{aligned}$$

Next, we briefly describe the field $GF(8)$ and how an error correcting code can be build on a the same observation about that field.

First, we start with the irreducible polynomial $p(x) = x^3 + x + 1$ over $\mathbb{Z}_2$. There is another such cubic polynomial, but its choice produces essentially the same result. Just as we did in the previous example, we assume we have a zero of $p(x)$ and call it β. Since we have assumed that $p(\beta) = \beta^3 + \beta + 1 = 0$, we get the recurrence relation for powers $\beta^3 = \beta + 1$ that lets us reduce the seven powers β^k, $0 \leq k \leq 6$, to linear combinations of 1, β, and β^2. Higher powers will reduce to these seven, which make up the elements of a field with $2^3 = 8$ elements when we add zero to the set. We leave as an exercise for you to set up a table relating powers of β with the linear combinations.

With this information we are now in a position to take blocks of four bits and encode them with three parity bits to create an error correcting code. If the bits

16.4. FIELD EXTENSIONS

are $b_3 b_4 b_5 b_6$, then we reduce the expression $B_m = b_3 \cdot \beta^3 + b_4 \cdot \beta^4 + b_5 \cdot \beta^5 + b_6 \cdot \beta^6$ using the recurrence relation to an expression $B_p = b_0 \cdot 1 + b_1 \cdot \beta + b_2 \cdot \beta^2$. Since we are equating equals within $GF(8)$, we have $B_p = B_m$, or $B_p + B_m = 0$. The encoded message is $b_0 b_1 b_2 b_3 b_4 b_5 b_6$, which is a representation of 0 in $GF(8)$. If the transmitted sequence of bits is received as $c_0 c_1 c_2 c_3 c_4 c_5 c_6$ we reduce $C = c_0 \cdot 1 + c_1 \cdot \beta + c_2 \cdot \beta^2 + c_3 \cdot \beta^3 + c_4 \cdot \beta^4 + c_5 \cdot \beta^5 + c_6 \cdot \beta^6$ using the recurrence. If there was no transmission error, the result is zero. If the reduced result is zero it is most likely that the original message was $c_3 c_4 c_5 c_6$. If bit k is switched in the transmission, then

$$C = B_p + B_m + \beta^k = \beta^k$$

Therefore if we reduce C with the recurrence, we get the linear combination of 1, β, and β^2 that is equal to β^k and so we can identify the location of the error and correct it. □

Exercises

1.
   (a) Use the definition of a field to show that $\mathbb{Q}(\sqrt{2})$ is a field.
   (b) Use the definition of vector space to show that $\mathbb{Q}\left(\sqrt{2}\right)$ is a vector space over $\mathbb{Q}$.
   (c) Prove that $\{1, \sqrt{2}\}$ is a basis for the vector space $\mathbb{Q}\left(\sqrt{2}\right)$ over $\mathbb{Q}$, and, therefore, the dimension of $\mathbb{Q}(\sqrt{2})$ over $\mathbb{Q}$ is 2.

2.
   (a) Determine the splitting field of $f(x) = x^2 + 1$ over $\mathbb{R}$. This means consider the polynomial $f(x) = x^2 + 1 \in \mathbb{R}[x]$ and find the smallest field that contains $\mathbb{R}$ and all the zeros of $f(x)$. Denote this field by $\mathbb{R}(i)$.
   (b) $\mathbb{R}(i)$ is more commonly referred to by a different name. What is it?
   (c) Show that $\{1, i\}$ is a basis for the vector space $\mathbb{R}(i)$ over $\mathbb{R}$. What is the dimension of this vector space (over $\mathbb{R}$)?

3. Determine the splitting field of $x^4 - 5x^2 + 6$ over $\mathbb{Q}$.

4.
   (a) Factor $x^2 + x + 1$ into linear factors in $\mathbb{Z}_2(a)$.
   (b) Write out the field tables for the field $\mathbb{Z}_2(a)$ and compare the results to the tables of Example 16.2.7, p. 196.
   (c) Cite a theorem and use it to show why the results of part b were to be expected.

5.
   (a) Show that $x^3 + x + 1$ is irreducible over $\mathbb{Z}_2$.
   (b) Determine the splitting field of $x^3 + x + 1$ over $\mathbb{Z}_2$.
   (c) By Theorem 16.2.10, p. 197, you have described all fields of order 2^3.

6.
   (a) List all polynomials of degree 1, 2, 3, and 4 over $\mathbb{Z}_2 = GF(2)$.
   (b) From your list in part a, identify all irreducible polynomials of degree

1, 2, 3, and 4.

(c) Determine the splitting fields of each of the polynomials in part b.

(d) What is the order of each of the splitting fields obtained in part c? Explain your results using Theorem 16.2.10, p. 197.

7. Is the polynomial code described in this section a linear code?

16.5 Power Series

16.5.1 Definition

Earlier in this chapter, we found that a polynomial of degree n over a ring R is an expression of the form

$$f(x) = \sum_{i=0}^{n} a_i x^i = a_0 + a_1 x + a_2 x^2 + \cdots + a_n x^n$$

where $n \geq 0$, each of the a_i are elements of R and $a_n \neq 0$. In Section 8.5 we defined a generating function of a sequence s with terms $s_0, s_1, s_2, \ldots$ as the infinite sum

$$G(s, z) = \sum_{i=0}^{\infty} s_i z^i = s_0 + s_1 z + s_2 z^2 + \cdots$$

The main difference between these two expressions, disregarding notation, is that the latter is an infinite expression and the former is a finite expression. In this section we will extend the algebra of polynomials to the algebra of infinite expressions like $G(s, z)$, which are called power series.

Definition 16.5.1 Power Series. Let $[R; +, \cdot]$ be a ring. A power series over R is an expression of the form

$$f(x) = \sum_{i=0}^{\infty} a_i x^i = a_0 + a_1 x + a_2 x^2 + \cdots$$

where $a_1, a_2, a_3, \ldots \in R$. The set of all such expressions is denoted by $R[[x]]$.
◇

Our first observation in our comparison of $R[x]$ and $R[[x]]$ is that every polynomial is a power series and so $R[x] \subseteq R[[x]]$. This is true because a polynomial $a_0 + a_1 x + a_2 x^2 + \cdots + a_n x^n$ of degree n in $R[x]$, can be thought of as an infinite expression where $a_i = 0$ for $i > n$. In addition, we will see that $R[[x]]$ is a ring with subring $R[x]$.

$R[[x]]$ is given a ring structure by defining addition and multiplication on power series as we did in $R[x]$, with the modification that, since we are dealing with infinite expressions, the sums and products will remain infinite expressions that we can determine term by term, as was done in with polynomials.

Definition 16.5.2 Power Series Addition. Given power series

$$f(x) = \sum_{i=0}^{\infty} a_i x^i = a_0 + a_1 x + a_2 x^2 + \cdots \text{ and}$$
$$g(x) = \sum_{i=0}^{\infty} b_i x^i = b_0 + b_1 x + b_2 x^2 + \cdots$$

their sum is

$$f(x) + g(x) = \sum_{i=0}^{\infty} (a_i + b_i) x^i$$
$$= (a_0 + b_0) + (a_1 + b_1) x + (a_2 + b_2) x^2 + (a_3 + b_3) x^3 + \cdots$$

16.5. POWER SERIES

Definition 16.5.3 Power Series Multiplication. Given power series

$$f(x) = \sum_{i=0}^{\infty} a_i x^i = a_0 + a_1 x + a_2 x^2 + \cdots \text{ and}$$
$$g(x) = \sum_{i=0}^{\infty} b_i x^i = b_0 + b_1 x + b_2 x^2 + \cdots$$

their product is

$$f(x) \cdot g(x) = \sum_{i=0}^{\infty} d_i x^i \quad \text{where } d_i = \sum_{j=0}^{i} a_j b_{i-j}$$
$$= (a_0 \cdot b_0) + (a_0 \cdot b_1 + a_1 \cdot b_0)x + (a_0 \cdot b_2 + a_1 \cdot b_1 + c_2 \cdot b_0)x^2 + \cdots$$

Example 16.5.4 Some Power Series Calcuations. Let

$$f(x) = \sum_{i=0}^{\infty} i x^i = 0 + 1x + 2x^2 + 3x^3 + \cdots \text{ and}$$
$$g(x) = \sum_{i=0}^{\infty} 2^i x^i = 1 + 2x + 4x^2 + 8x^3 + \cdots$$

be elements in $\mathbb{Z}[[x]]$. Let us compute $f(x) + g(x)$ and $f(x) \cdot g(x)$. First the sum:

$$f(x) + g(x) = \sum_{i=0}^{\infty} i x^i + \sum_{i=0}^{\infty} 2^i x^i$$
$$= \sum_{i=0}^{\infty} \left(i + 2^i\right) x^i$$
$$= 1 + 3x + 6x^2 + 11x^3 + \cdots$$

The product is a bit more involved:

$$f(x) \cdot g(x) = \left(\sum_{i=0}^{\infty} i x^i\right) \cdot \left(\sum_{i=0}^{\infty} 2^i x^i\right)$$
$$= \left(0 + 1x + 2x^2 + 3x^3 + \cdots\right) \cdot \left(1 + 2x + 4x^2 + 8x^3 + \cdots\right)$$
$$= 0 \cdot 1 + (0 \cdot 2 + 1 \cdot 1)x + (0 \cdot 4 + 1 \cdot 2 + 2 \cdot 1)x^2 + \cdots$$
$$= x + 4x^2 + 11x^3 + \cdots$$
$$= \sum_{i=0}^{\infty} d_i x^i \quad \text{where } d_i = \sum_{j=0}^{i} j 2^{i-j}$$

We can compute any value of d_i, with the amount of time/work required increasing as i increases.

```
def d(i):
    s=0
    for j in range(1,i+1):
        s+=j*2^(i-j)
    return s
d(20)
```

2097130

A closed-form expression for d_i would be desirable. Using techniques from Chapter 8, the formula is $d_i = 2^{i+1} - i - 2$, which we leave it to the reader to derive. Hence, $f(x) \cdot g(x) = \sum_{i=0}^{\infty} (2^{i+1} - i - 2)x^i$ □

16.5.2 Properties, Units

We have seen that addition and multiplication in $R[[x]]$ is virtually identical to that in $R[x]$. The following theorem parallels Theorem 16.3.8, p. 201, establishing the ring properties of $R[[x]]$.

Theorem 16.5.5 Properties of Power Series. *Let $[R; +, \cdot]$ be a ring. Then:*

(1) $R[[x]]$ is a ring under the operations of power series addition and multiplication, which depend on the operations in R.

(2) If R is a commutative ring, then $R[[x]]$ is a commutative ring.

(3) If R is a ring with unity, 1, then $R[[x]]$ is a ring with unity (the unity in $R[x]$ is $1 + 0x + 0x^2 + \cdots$).

(4) If R is an integral domain, then $R[[x]]$ is an integral domain.

(5) If F is a field, then $F[[x]]$ is not a field. However, $F[[x]]$ is an integral domain.

We are most interested in the situation when the set of coefficients is a field. The theorem above indicates that when F is a field, $F[[x]]$ is an integral domain. A reason that $F[[x]]$ is not a field is the same as one that we can cite for $F[x]$, namely that x does not have multiplicative inverse in $F[[x]]$.

With all of these similarities, one might wonder if the rings of polynomials and power series over a field are isomorphic. It turns out that they are not. The difference between $F[x]$ and $F[[x]]$ becomes apparent when one studies which elements are units in each. First we prove that the only units in $F[x]$ are the nonzero constants; that is, the nonzero elements of F.

Theorem 16.5.6 Polynomial Units. *Let $[F; +, \cdot]$ be a field. Polynomial $f(x)$ is a unit in $F[x]$ if and only if it is a nonzero constant polynomial.*
Proof.

($\Rightarrow$) Let $f(x)$ be a unit in $F[x]$. Then $f(x)$ has a multiplicative inverse, call it $g(x)$, such that $f(x) \cdot g(x) = 1$. Hence, the $\deg(f(x) \cdot g(x)) = \deg(1) = 0$. But $\deg(f(x) \cdot g(x)) = \deg f(x) + \deg g(x)$. So $\deg f(x) + \deg g(x) = 0$, and since the degree of a polynomial is always nonnegative, this can only happen when the $\deg f(x) = \deg g(x) = 0$. Hence, $f(x)$ is a constant, an element of F, which is a unit if and only if it is nonzero.

($\Leftarrow$) If $f(x)$ is a nonzero element of F, then it is a unit since F is a field. Thus it has an inverse, which is also in $F[x]$ and so $f(x)$ is a unit of $F[x]$. ∎

Before we proceed to categorize the units in $F[[x]]$, we remind the reader that two power series $a_0 + a_1 x + a_2 x^2 + \cdots$ and $b_0 + b_1 x + b_2 x^2 + \cdots$ are equal if and only if corresponding coefficients are equal, $a_i = b_i$ for all $i \geq 0$.

Theorem 16.5.7 Power Series Units. *Let $[F; +, \cdot]$ be a field. Then $f(x) = \sum_{i=0}^{\infty} a_i x^i$ is a unit of $F[[x]]$ if and only if $a_0 \neq 0$.*
Proof.

($\Rightarrow$) If $f(x)$ is a unit of $F[[x]]$, then there exists $g(x) = \sum_{i=0}^{\infty} b_i x^i$ in $F[[x]]$ such that

$$\begin{aligned} f(x) \cdot g(x) &= (a_0 + a_1 x + a_2 x^2 + \cdots) \cdot (b_0 + b_1 x + b_2 x^2 + \cdots) \\ &= 1 \\ &= 1 + 0x + 0x^2 + \cdots \end{aligned}.$$

Since corresponding coefficients in the equation above must be equal, $a_0 \cdot b_0 = 1$, which implies that $a_0 \neq 0$.

($\Leftarrow$) Assume that $a_0 \neq 0$. To prove that $f(x)$ is a unit of $F[[x]]$ we need to find $g(x) = \sum_{i=0}^{\infty} b_i x^i$ in $F[[x]]$ such that $f(x) \cdot g(x) = \sum_{i=0}^{\infty} d_i x^i = 1$. If we use the formula for the coefficients $f(x) \cdot g(x)$ and equate coefficients, we get

$$d_0 = a_0 \cdot b_0 = 1 \quad \Rightarrow \quad b_0 = a_0^{-1}$$
$$d_1 = a_0 \cdot b_1 + a_1 \cdot b_0 = 0 \quad \Rightarrow \quad b_1 = -a_0^{-1} \cdot (a_1 \cdot b_0)$$
$$d_2 = a_0 b_2 + a_1 b_1 + a_2 b_0 = 0 \quad \Rightarrow \quad b_2 = -a_0^{-1} \cdot (a_1 \cdot b_1 + a_2 \cdot b_0)$$
$$\vdots$$
$$d_s = a_0 \cdot b_s + a_1 \cdot b_{s-1} + \cdots + a_s \cdot b_0 = 0 \quad \Rightarrow \quad b_s = -a_0^{-1} \cdot (a_1 \cdot b_{s-1} + a_2 \cdot b_{s-2} + \cdots + a_s \cdot b_0)$$

Therefore the powers series $\sum_{i=0}^{\infty} b_i x^i$ is an expression whose coefficients lie in F and that satisfies the statement $f(x) \cdot g(x) = 1$. Hence, $g(x)$ is the multiplicative inverse of $f(x)$ and $f(x)$ is a unit. ∎

Example 16.5.8 Let $f(x) = 1 + 2x + 3x^2 + 4x^3 + \cdots = \sum_{i=0}^{\infty}(i+1)x^i$ be an element of $\mathbb{Q}[[x]]$. Then, by Theorem 16.5.7, p. 210, since $a_0 = 1 \neq 0$, $f(x)$ is a unit and has an inverse, call it $g(x)$. To compute $g(x)$, we follow the procedure outlined in the above theorem. Using the formulas for the b_i's, we obtain

$$b_0 = 1$$
$$b_1 = -1(2 \cdot 1) = -2$$
$$b_2 = -1(2 \cdot (-2) + 3 \cdot 1) = 1$$
$$b_3 = -1(2 \cdot 1 + 3 \cdot (-2) + 4 \cdot 1) = 0$$
$$b_4 = -1(2 \cdot 0 + 3 \cdot 1 + 4 \cdot (-2) + 5 \cdot 1) = 0$$
$$b_5 = -1(2 \cdot 0 + 3 \cdot 0 + 4 \cdot (1) + 5 \cdot (-2) + 6 \cdot 1) = 0$$
$$\vdots$$

For $s \geq 3$, we have

$$b_s = -1(2 \cdot 0 + 3 \cdot 0 + \cdots (s-2) \cdot 0 + (s-1) \cdot 1 + s \cdot (-2) + (s+1) \cdot 1) = 0$$

Hence, $g(x) = 1 - 2x + x^2$ is the multiplicative inverse of $f(x)$. □

Certainly $F[[x]]$ contains a wider variety of units than $F[x]$. Yet $F[[x]]$ is not a field, since $x \in F[[x]]$ is not a unit. So concerning the algebraic structure of $F[[x]]$, we know that it is an integral domain that contains $F[x]$. If we allow our power series to take on negative exponents; that is, consider expressions of the form $f(x) = \sum_{i=-\infty}^{\infty} a_i x^i$ where all but a finite number of terms with a negative index equal zero. These expressions are called extended power series. The set of all such expressions is a field, call it E. This set does contain, for example, the inverse of x, which is x^{-1}. It can be shown that each nonzero element of E is a unit.

16.5.3 Exercises

1. Let $f(x) = \sum_{i=0}^{\infty} a_i x^i$ and $g(x) = \sum_{i=0}^{\infty} b_i x^i$ be elements of $R[[x]]$. Let $f(x) \cdot g(x) = \sum_{i=0}^{\infty} d_i x^i = 1$. Apply basic algebra to $(a_0 + a_1 x + a_2 x^2 + \cdots) \cdot (b_0 + b_1 x + b_2 x^2 + \cdots)$ to derive the formula $d_s = \sum_{i=0}^{s} a_i b_{s-i}$ for the coefficients of $f(x) \cdot g(x)$. Hence, to show that $f(x) \cdot g(x) = \sum_{s=0}^{\infty} \left(\sum_{i=0}^{s} a_i b_{s-i}\right) x^s$

2.
    (a) Prove that for any integral domain D, the following can be proven: $f(x) = \sum_{i=0}^{\infty} a_i x^i$ is a unit of $D[[x]]$ if and only if a_0 is a unit in D.

    (b) Compare the statement in part a to that in Theorem 16.5.7, p. 210.

(c) Give an example of the statement in part a in $\mathbb{Z}[[x]]$.

3. Use the formula for the product to verify that the expression $g(x)$ of Example 16.5.8, p. 211 is indeed the inverse of $f(x)$.

4.
   (a) Determine the inverse of $f(x) = 1 + x + x^2 + \cdots = \sum_{i=0}^{\infty} x^i$ in $\mathbb{Q}[[x]]$.

   (b) Repeat part a with $f(x)$ taken in $\mathbb{Z}_2[[x]]$.

   (c) Use the method outlined in Chapter 8 to show that the power series $f(x) = \sum_{i=0}^{\infty} x^i$ is the rational generating function $\frac{1}{1-x}$. What is the inverse of this function? Compare your results with those in part a.

5.
   (a) Determine the inverse of $h(x) = \sum_{i=0}^{\infty} 2^i x^i$ in $\mathbb{Q}[[x]]$.

   (b) Use the procedures in Chapter 8 to find a rational generating function for $h(x)$ in part a. Find the multiplicative inverse of this function.

6. Let $a(x) = 1 + 3x + 9x^2 + 27x^3 + \cdots = \sum_{i=0}^{\infty} 3^i x^i$ and $b(x) = 1 + x + x^2 + x^3 + \cdots = \sum_{i=0}^{\infty} x^i$ both in $\mathbb{R}[[x]]$.

   (a) What are the first four terms (counting the constant term as the 0th term) of $a(x) + b(x)$?

   (b) Find a closed form expression for $a(x)$.

   (c) What are the first four terms of $a(x)b(x)$?

7. Write as an extended power series:

   (a) $\left(x^4 - x^5\right)^{-1}$

   (b) $\left(x^2 - 2x^3 + x^4\right)^{-1}$

8. Derive the closed form expression $d_i = 2^{i+1} - i - 2$ for the coefficients of the product $f(x) \cdot g(x)$ in Example 16.5.4, p. 209.

Appendix A

Algorithms

algorithm

Using step-by-step math operations,
It performs with exact calculations.
An algorithm's job
Is to work out a "prob"
With repeated precise computations.

Jesse Frankovich, The Omnificent English Dictionary In Limerick Form

Computer programs, bicycle assembly instructions, knitting instructions, and recipes all have several things in common. They all tell us how to do something; and the usual format is as a list of steps or instructions. In addition, they are usually prefaced with a description of the raw materials that are needed (the input) to produce the end result (the output). We use the term algorithm to describe such lists of instructions. We assume that the reader may be unfamiliar with algorithms, so the first section of this appendix will introduce some of the components of the algorithms that appear in this book. Since we would like our algorithms to become computer programs in many cases, the notation will resemble a computer language such as Python or Sage; but our notation will be slightly less formal. In some cases we will also translate the pseudocode to Sage. Our goal will be to give mathematically correct descriptions of how to accomplish certain tasks. To this end, the second section of this appendix is an introduction to the Invariant Relation Theorem, which is a mechanism for algorithm verification that is related to Mathematical Induction

A.1 An Introduction to Algorithms

Most of the algorithms in this book will contain a combination of three kinds of steps: the assignment step, the conditional step, and the loop.

A.1.1 Assignments

In order to assign a value to a variable, we use an assignment step, which takes the form:

$$\text{Variable} = \text{Expression to be computed}$$

The equals sign in most languages is used for assignment but some languages may use variations such as := or a left pointing arrow. Logical equality, which produces a boolean result and would be used in conditional or looping steps, is most commonly expressed with a double-equals, ==.

An example of an assignment is k = n - 1 which tells us to subtract 1 from the value of n and assign that value to variable k. During the execution of an algorithm, a variable may take on only one value at a time. Another example of an assignment is k = k - 1. This is an instruction to subtract one from the value of k and then reassign that value to k.

A.1.2 Conditional steps

Frequently there are steps that must be performed in an algorithm if and only if a certain condition is met. The conditional or "if ... then" step is then employed. For example, suppose that in step 2 of an algorithm we want to assure that the values of variables x and y satisfy the condition x <= y. The following step would accomplish this objective.

```
2. If x > y:
      2.1 t = x
      2.2 x = y
      2.3 y = t
```

Listing A.1.1

Steps 2.1 through 2.3 would be bypassed if the condition x > y were false before step 2.

One slight variation is the "if ... then ... else" step, which allows us to prescribe a step to be taken if the condition is false. For example, if you wanted to exercise today, you might look out the window and execute the following algorithm.

```
1. If it is cold or raining:
              exercise indoors
       else:
              go outside and run
2. Rest
```

Listing A.1.2

A.1.3 Loops

The conditional step tells us to do something once if a logical condition is true. A loop tells us to repeat one or more steps, called the body of the loop, while the logical condition is true. Before every execution of the body, the condition is tested. The following flow diagram serves to illustrate the steps in a While loop.

A.1. AN INTRODUCTION TO ALGORITHMS

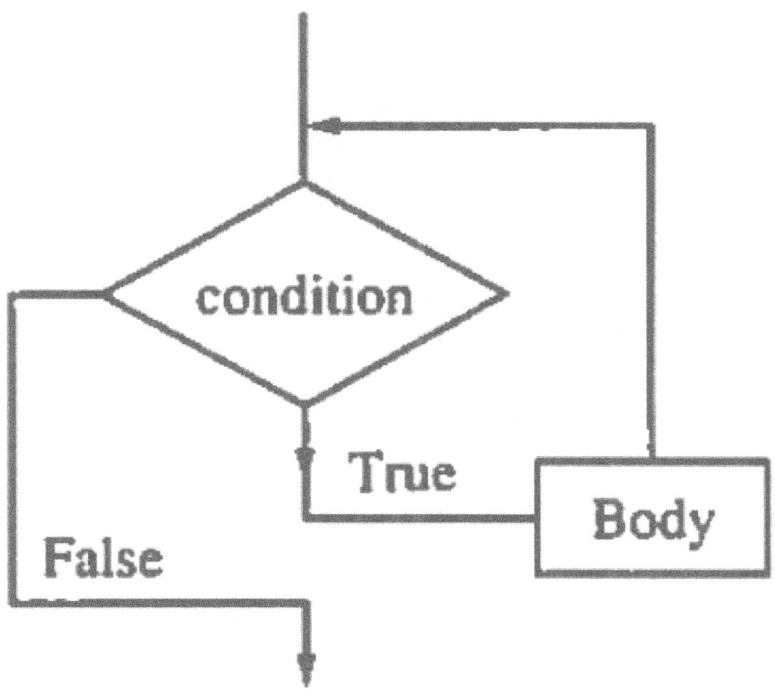

Figure A.1.3 Flow diagram for a while loop

Suppose you wanted to solve the equation $f(x) = 0$. The following initial assignment and loop could be employed.

```
1. c = your first guess
2. While f(c) != 0:
            c = another guess
```

Listing A.1.4

Caution: One must always guard against the possibility that the condition of a While loop will never become false. Such "infinite loops" are the bane of beginning programmers. The loop above could very well be such a situation, particularly if the equation has no solution, or if the variable takes on real values

In cases where consecutive integer values are to be assigned to a variable, a different loop construction, a *For loop*, is often employed. For example, suppose we wanted to assign variable k each of the integer values from m to n and for each of these values perform some undefined steps. We could accomplish this with a While loop:

```
1. k := m
2. While k <= n:
        2.1 execute some steps
        2.2 k = k + 1
```

Listing A.1.5

Alternatively, we can perform these steps with a For loop.

```
For k = m to n:
        execute some   steps
```

Listing A.1.6

For loops such as this one have the advantage of being shorter than the equivalent While loop. The While loop construction has the advantage of being able to handle more different situations than the For loop.

A.1.4 Exercises

1. What are the inputs and outputs of the algorithms listed in the first sentence of this section?
2. What is wrong with this algorithm?
    ```
    Input: a and b, integers
    Output: the value of c will be a - b
    (1) c = 0
    (2) While a > b:
                (2.1) a := a - 1
                (2.2) c := c + 1
    ```
   Listing A.1.7
3. Describe, in words, what the following algorithm does:
    ```
    Input: k, a positive integer
    Output: s = ?
    (1) s = 0
    (2) While k > 0:
            (2.1) s = s + k
            (2.2) k = k - 1
    ```
   Listing A.1.8
4. Write While loops to replace the For loops in the following partial algorithms:

    (a) (1) S = 0

    (2) for k = 1 to 5: S = S + k^2

    (b) The floor of a number is the greatest integer less than or equal to that number.

    (1) m = a positive integer greater than 1

    (2) B = floor(sqrt(m))

    (3) for i = 2 to B: if i divides evenly into m, jump to step 5

    (4) print "m is a prime" and exit

    (5) print "m is composite" and exit

5. Describe in words what the following algorithm does:
    ```
    Input: n, a positive integer
    Output: k?
    (1) f= 0
    (2) k=n
    (3) While k is even:
            (3.1) f = f+ 1
            (3.2) k = k div 2
    ```
   Listing A.1.9
6. Fix the algorithm in Exercise 2.

A.2 The Invariant Relation Theorem

A.2.1 Two Exponentiation Algorithms

Consider the following algorithm implemented in Sage to compute $a^m \bmod n$, given an arbitrary integer a, non-negative exponent m, and a modulus n, $n \geq 0$. The default sample evaluation computes $2^5 \bmod 7 = 32 \bmod 7 = 4$, but you can edit the final line for other inputs.

```
def slow_exp(a,m,n):
    b=1
    k=m
    while k>0:
        b=(b*a)%n   # % is integer remainder (mod) operation
        k-=1
    return b
slow_exp(2,5,7)
```

4

It should be fairly clear that this algorithm will successfully compute $a^m \pmod{n}$ since it mimics the basic definition of exponentiation. However, this algorithm is highly inefficient. The algorithm that is most commonly used for the task of exponentiation is the following one, also implemented in Sage.

```
def fast_exp(a,m,n):
    t=a
    b=1
    k=m
    while k>0:
        if k%2==1: b=(b*t)%n
        t=(t^2)%n
        k=k//2   # // is the integer quotient operation
    return b
fast_exp(2,5,7)
```

The only difficulty with the "fast algorithm" is that it might not be so obvious that it always works. When implemented, it can be verified by example, but an even more rigorous verification can be done using the Invariant Relation Theorem. Before stating the theorem, we define some terminology.

A.2.2 Proving the correctness of the fast algorithm

Definition A.2.1 Pre and Post Values. Given a variable x, the pre value of x, denoted $\grave{x}$, is the value before an iteration of a loop. The post value, denoted $\acute{x}$, is the value after the iteration. $\Diamond$

Example A.2.2 Pre and post values in the fast exponentiation algorithm. In the fast exponentiation algorithm, the relationships between the pre and post values of the three variables are as follows.

$$\acute{b} \equiv \grave{b} \grave{t}^{\grave{k} \bmod 2} \pmod{n}$$

$$\acute{t} \equiv \grave{t}^2 \pmod{n}$$

$$\acute{k} = \grave{k} // 2$$

□

Definition A.2.3 Invariant Relation. Given an algorithm's inputs and a set of variables that are used in the algorithm, an *invariant relation* is a set of one or more equations that are true prior to entering a loop and remain true in every iteration of the loop. ◇

Example A.2.4 Invariant Relation for Fast Exponentiation. We claim that the invariant relation in the fast algorithm is $bt^k = a^m (mod\, n)$. We will prove that this is indeed true below. □

Theorem A.2.5 The Invariant Relation Theorem. *Given a loop within an algorithm, if R is a relation with the properties*

(a) R is true before entering the loop

(b) the truth of R is maintained in any iteration of the loop

(c) the condition for exiting the loop will always be reached in a finite number of iterations.

then R will be true upon exiting the loop.

Proof. The condition that the loop ends in a finite number of iterations lets us apply mathematical induction with the induction variable being the number of iterations. We leave the details to the reader. ∎

We can verify the correctness of the fast exponentiation algorithm using the Invariant Relation Theorem. First we note that prior to entering the loop, $bt^k = 1a^m = a^m (mod\, n)$. Assuming the relation is true at the start of any iteration, that is $\grave{b}\grave{t}^{\grave{k}} = a^m (mod\, n)$, then

$$b\acute{t}^{\acute{k}} \equiv (\grave{b}\grave{t}^{\grave{k}\, mod\, 2})(\grave{t}^2)^{\grave{k}//2} (mod\, n)$$
$$\equiv \grave{b}\grave{t}^{2(\grave{k}//2)+\grave{k}\, mod\, 2} (mod\, n)$$
$$\equiv \grave{b}\grave{t}^{\grave{k}} (mod\, n)$$
$$\equiv a^m (mod\, n)$$

Finally, the value of k will decrease to zero in a finite number of steps because the number of binary digits of k decreases by one with each iteration. At the end of the loop,

$$b = bt^0 = bt^k \equiv a^m (mod\, n)$$

which verifies the correctness of the algorithm.

A.2.3 Exercises

1. How are the pre and post values in the slow exponentiation algorithm related? What is the invariant relation between the variables in the slow algorithm?

2. Verify the correctness of the following algorithm to compute the greatest common divisor of two integers that are not both zero.

```
def gcd(a,b):
    r0=a
    r1=b
    while r1 !=0:
        t= r0 % r1
        r0=r1
        r1=t
    return r0
gcd(1001,154)   #test
```

77

Hint. The invariant of this algorithm is $gcd(r0, r1) = gcd(a, b)$.

3. Verify the correctness of the Binary Conversion Algorithm, つ.1 in Chapter 1.

4. A dragon has 100 heads. A knight can cut off 15, 17, 20, or 5 heads, respectively, with one blow of his sword. In each of these cases 24, 2, 14, or 17 new heads grow on its shoulders, respectively. If all heads are blown off, the dragon dies. Can the dragon ever die? (problem attributed to Biswaroop Roy)

Appendix B

Hints and Solutions to Selected Exercises

For the most part, solutions are provided here for odd-numbered exercises.

7 · Functions

· Exercises

7.7. Answer. For each of the $|A|$ elements of A, there are $|B|$ possible images, so there are $|B| \cdot |B| \cdot \ldots \cdot |B| = |B|^{|A|}$ functions from A into B.

11 · Algebraic Structures
11.1 · Operations
11.1.4 · Exercises

11.1.4.1. Answer.

(a) Commutative, and associative. Notice that zero is the identity for addition, but it is not a positive integer.

(b) Commutative, associative, and has an identity (1)

(c) Commutative, associative, has an identity (1), and is idempotent

(d) Commutative, associative, and idempotent

(e) None. Notice that 2@(3@3) = 134217728, while (2@3)@3 = 512; and $a@1 = a$, while $1@a = 1$.

11.1.4.3. Answer.

$$a, b \in A \cap B \Rightarrow a, b \in A \text{ by the definition of intersection}$$
$$\Rightarrow a * b \in A \text{ by the closure of } A \text{ with respect to } *$$

Similarly, $a, b \in A \cap B \Rightarrow a * b \in B$. Therefore, $a * b \in A \cap B$. The set of positive integers is closed under addition, and so is the set of negative integers, but $1 + -1 = 0$. Therefore, their union, the nonzero integers, is not closed under addition.

11.1.4.5. Answer.

(a) $*$ is commutative since $|a - b| = |b - a|$ for all $a, b \in \mathbb{N}$

(b) $*$ is not associative. Take $a = 1$, $b = 2$, and $c = 3$, then $(a * b) * c = ||1 - 2| - 3| = 2$, and $a * (b * c) = |1 - |2 - 3|| = 0$.

(c) Zero is the identity for $*$ on $\mathbb{N}$, since $a * 0 = |a - 0| = a = |0 - a| = 0 * a$.

(d) Each element of $\mathbb{N}$ inverts itself since $a * a = |a - a| = 0$.

(e) $*$ is not idempotent, since, for $a \neq 0$, $a * a = 0 \neq a$.

11.2 · Algebraic Systems
11.2.4 · Exercises

11.2.4.1. Answer. The terms "generic" and "trade" for prescription drugs are analogous to "generic" and "concrete" algebraic systems. Generic aspirin, for example, has no name, whereas Bayer, Tylenol, Bufferin, and Anacin are all trade or specific types of aspirins. The same can be said of a generic group $[G; *]$ where G is a nonempty set and $*$ is a binary operation on G, When examples of typical domain elements can be given along with descriptions of how operations act on them, such as $\mathbb{Q}*$ or $M_{2 \times 2}(\mathbb{R})$, then the system is concrete (has a specific name, as with the aspirin). Generic is a way to describe a general algebraic system, whereas a concrete system has a name or symbols making it distinguishable from other systems.

11.2.4.3. Answer. The systems in parts b, d, e, and f are groups.

11.2.4.5. Answer.

(a) Elements are $I = \begin{pmatrix} 1 & 0 \\ 0 & 1 \end{pmatrix}$, and $T = \begin{pmatrix} 0 & 1 \\ 1 & 0 \end{pmatrix}$, the group is abelian.

Operation table is
·	I	T
I	I	T
T	T	I

(b)

	I	R_1	R_2	F_1	F_2	F_3
I	I	R_1	R_2	F_1	F_2	F_3
R_1	R_1	R_2	I	F_2	F_3	F_1
R_2	R_2	I	R_1	F_3	F_1	F_2
F_1	F_1	F	F_2	I	R_2	R_1
F_2	F_2	F_1	F_3	R_1	I	R_2
F_3	F_3	F_2	F_1	R_2	R_1	I

This group is non-abelian since, for example, $F_1 F_2 = R_2$ and $F_2 F_1 = R_2$.

(c) $4! = 24$, $n!$.

11.2.4.7. Answer. The identity is e. $a * b = c$, $a * c = b$, $b * c = a$, and $[V; *]$ is abelian. (This group is commonly called the Klein-4 group.)

11.2.4.8. Solution. Yes, this is a group. You might see some similarities with the group of three by three rook matrices.

11.3 · Some General Properties of Groups
11.3.3 · Exercises

11.3.3.1. Answer.

(a) f is injective:
$$f(x) = f(y) \Rightarrow a * x = a * y$$
$$\Rightarrow x = y \text{ by left cancellation}$$

f is surjective: For all $b \in G$, $f(x) = b$ has the solution $a^{-1} * b$.

(b) Functions of the form $f(x) = a + x$, where a is any integer, are bijections

11.3.3.3. Answer. Basis: $(n=2)$ $(a_1 * a_2)^{-1} = a_2^{-1} * a_1^{-1}$ by Theorem 11.3.7, p. 31.

Induction: Assume that for some $n \geq 2$,
$$(a_1 * a_2 * \cdots * a_n)^{-1} = a_n^{-1} * \cdots * a_2^{-1} * a_1^{-1}$$

We must show that
$$(a_1 * a_2 * \cdots * a_n * a_{n+1})^{-1} = a_{n+1}^{-1} * a_n^{-1} * \cdots * a_2^{-1} * a_1^{-1}$$

This can be accomplished as follows:

$$\begin{aligned}
(a_1 * a_2 * \cdots * a_n * a_{n+1})^{-1} &= ((a_1 * a_2 * \cdots * a_n) * a_{n+1})^{-1} \text{ by the associative law}\\
&= a_{n+1}^{-1} * (a_1 * a_2 * \cdots * a_n)^{-1} \text{ by the basis}\\
&= a_{n+1}^{-1} * (a_n^{-1} * \cdots * a_2^{-1} * a_1^{-1}) \text{ by the induction hypothesis}\\
&= a_{n+1}^{-1} * a_n^{-1} * \cdots * a_2^{-1} * a_1^{-1} \text{ by the associative law}
\end{aligned}$$

11.3.3.5. Answer. In this answer, we will refer to Lemma 11.3.13, p. 33 simply as "the lemma."

(a) Let $p(n)$ be $a^{-n} = (a^{-1})^n$, where a is any element of group $[G; *]$. First we will prove that $p(n)$ is true for all $n \geq 0$.

Basis: If $n = 0$, Using the definition of the zero exponent, $(a^0)^{-1} = e^{-1} = e$, while $(a^{-1})^0 = e$. Therefore, $p(0)$ is true.

Induction: Assume that for some $n \geq 0$, $p(n)$ is true.

$$\begin{aligned}
(a^{n+1})^{-1} &= (a^n * a)^{-1} \text{ by the definition of exponentiation}\\
&= a^{-1} * (a^n)^{-1} \text{ by the lemma}\\
&= a^{-1} * (a^{-1})^n \text{ by the induction hypothesis}\\
&= (a^{-1})^{n+1} \text{ by the lemma}
\end{aligned}$$

If n is negative, then $-n$ is positive and

$$\begin{aligned}
a^{-n} &= \left(\left((a^{-1})^{-1}\right)^{-n}\right)\\
&= (a^{-1})^{-(-n)} \text{ since the property is true for positive numbers}\\
&= (a^{-1})^n
\end{aligned}$$

(b) For $m > 1$, let $p(m)$ be $a^{n+m} = a^n * a^m$ for all $n \geq 1$. The basis for this proof follows directly from the basis for the definition of exponentiation.

Induction: Assume that for some $m > 1$, $p(m)$ is true. Then

$$\begin{aligned}
a^{n+(m+1)} &= a^{(n+m)+1} \text{ by the associativity of integer addition}\\
&= a^{n+m} * a^1 \text{ by the definition of exponentiation}\\
&= (a^n * a^m) * a^1 \text{ by the induction hypothesis}\\
&= a^n * (a^m * a^1) \text{ by associativity}\\
&= a^n * a^{m+1} \text{ by the definition of exponentiation}
\end{aligned}$$

To complete the proof, you need to consider the cases where m and/or n are negative.

(c) Let $p(m)$ be $(a^n)^m = a^{nm}$ for all integers n.

Basis: $(a^m)^0 = e$ and $a^{m \cdot 0} = a^0 = e$ therefore, $p(0)$ is true.

Induction; Assume that $p(m)$ is true for some $m > 0$,

$$\begin{aligned}(a^n)^{m+1} &= (a^n)^m * a^n \text{ by the definition of exponentiation} \\ &= a^{nm} * a^n \text{ by the induction hypothesis} \\ &= a^{nm+n} \text{ by part (b) of this proof} \\ &= a^{n(m+1)}\end{aligned}$$

Finally, if m is negative, we can verify that $(a^n)^m = a^{nm}$ using many of the same steps as the positive case.

11.4 · Greatest Common Divisors and the Integers Modulo n
11.4.2 · The Euclidean Algorithm

Investigation 11.4.1 Solution. If quotient in division is 1, then we get the slowest possible completion. If $a = b + r$, then working backwards, each remainder would be the sum of the two previous remainders. This described a sequence like the Fibonacci sequence and indeed, the greatest common divisor of two consecutive Fibonacci numbers will take the most steps to reach a final value of 1.

11.4.6 · Exercises

11.4.6.1. Answer.

(a) $2^2 \cdot 3 \cdot 5$

(b) $3^2 \cdot 5 \cdot 7$

(c) 19^4

(d) 12112

11.4.6.3. Answer.

(a) 2   (d) 0   (g) 1

(b) 5   (e) 2   (h) 3

(c) 0   (f) 2   (i) 0

11.4.6.5. Answer.

(a) 1

(b) 1

(c) $m(4) = r(4)$, where $m = 11q + r$, $0 \le r < 11$

11.4.6.7. Answer. Since the solutions, if they exist, must come from $\mathbb{Z}_2$, substitution is the easiest approach.

(a) 1 is the only solution, since $1^2 +_2 1 = 0$ and $0^2 +_2 1 = 1$

(b) No solutions, since $0^2 +_2 0 +_2 1 = 1$, and $1^2 +_2 1 +_2 1 = 1$

11.4.6.10. Hint. Prove by induction on m that you can divide any positive integer into m. That is, let $p(m)$ be "For all n greater than zero, there exist unique integers q and r such that ... ." In the induction step, divide n into $m - n$.

11.4.6.11. Solution. The given conditions can be converted to a system of

linear equations:
$$f(1) = 11 \Rightarrow a +_{17} b = 11$$
$$f(2) = 4 \Rightarrow 2 \times_{17} a +_{17} b = 4$$

If we subtract the first equation from the second, we get $a = 4 +_{17} (-11) = 4 +_{17} 6 = 10$. This implies that $b = 1$, and $f(i) = 10 \times +_{17}i + 1$. To get a formula for the inverse of f we solve $f(j) = i$ for j, using the fact that the multiplicative inverse of 10 (mod 17) is 12.

$$f(j) = i \Rightarrow 10 \times +_{17}j + 1 = i$$
$$\Rightarrow 10 \times +_{17}j = i +_{17} 16$$
$$\Rightarrow j = 12 \times_{17} (i +_{17} 16)$$

Therefore $f^{-1}(i) = 12 \times_{17} (i +_{17} 16) = 12 \times_{17} i +_{17} 5$.

11.4.6.12. Solution. This system is a monoid with identity 6 (surprise!). However it is not a group since 0 has no inverse.

11.4.6.13. Solution. By Bézout's lemma, p. 37, 450 is an element of $\mathbb{U}_{2021}$. It's inverse in the group is 759 because

$$450 \cdot 759 = 2021 \cdot 169 + 1 \quad \Rightarrow \quad 450 \times_{2021} 759 = 1.$$

11.5 · Subsystems
11.5.5 · Exercises

11.5.5.1. Answer. Only a and c are subgroups.

11.5.5.3. Answer. $\{I, R_1, R_2\}$, $\{I, F_1\}$, $\{I, F_2\}$, and $\{I, F_3\}$ are all the proper subgroups of R_3.

11.5.5.5. Answer.

(a) $\langle 1 \rangle = \langle 5 \rangle = \mathbb{Z}_6$, $\langle 2 \rangle = \langle 4 \rangle = \{2, 4, 0\}$, $\langle 3 \rangle = \{3, 0\}$, $\langle 0 \rangle = \{0\}$

(b) $\langle 1 \rangle = \langle 5 \rangle = \langle 7 \rangle = \langle 11 \rangle = \mathbb{Z}_{12}$, $\langle 2 \rangle = \langle 10 \rangle = \{2, 4, 6, 8, 10, 0\}$, $\langle 3 \rangle = \langle 9 \rangle = \{3, 6, 9, 0\}$, $\langle 4 \rangle = \langle 8 \rangle = \{4, 8, 0\}$, $\langle 6 \rangle = \{6, 0\}$, $\langle 0 \rangle = \{0\}$

(c) $\langle 1 \rangle = \langle 3 \rangle = \langle 5 \rangle = \langle 7 \rangle = \mathbb{Z}_8$, $\langle 2 \rangle = \langle 6 \rangle = \{2, 4, 6, 0\}$, $\langle 4 \rangle = \{4, 0\}$, $\langle 0 \rangle = \{0\}$

(d) Based on the ordering diagrams for parts a through c in Figure B.0.1, p. 225, we would expect to see an ordering diagram similar to the one for divides on $\{1, 2, 3, 4, 6, 8, 12, 24\}$ (the divisors of 24) if we were to examine the subgroups of $\mathbb{Z}_{24}$. This is indeed the case.

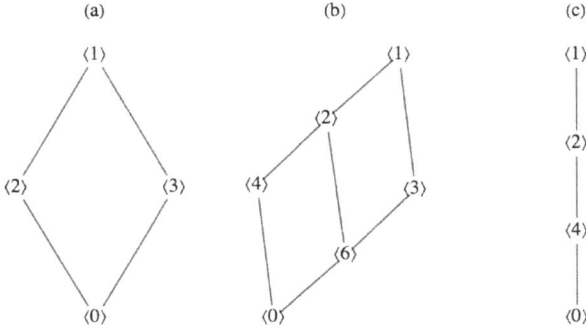

Figure B.0.1 Figure for exercise 5

11.5.5.7. Hint. Use an indirect argument.

Answer. Assume that H and K are subgroups of group G, and that, as in Figure 11.5.12, p. 48, there are elements $x \in H - K$ and $y \in K - H$. Consider the product $x * y$. Where could it be placed in the Venn diagram? If we can prove that it must lie in the outer region, $H^c \cap K^c = (H \cup K)^c$, then we have proven that $H \cup K$ is not closed under $*$ and cannot be a subgroup of G, Assume that $x * y \in H$. Since x is in H, x^{-1} is in H and so by closure $x^{-1} * (x * y) = y \in H$ which is a contradiction. Similarly, $x * y \notin K$.

One way to interpret this theorem is that no group is the union of two groups.

11.6 · Direct Products
11.6.3 · Exercises

11.6.3.1. Answer. Table of $\mathbb{Z}_2 \times \mathbb{Z}_3$:

+	(0,0)	(0,1)	(0,2)	(1,0)	(1,1)	(1,2)
(0,0)	(0,0)	(0,1)	(0,2)	(1,0)	(1,1)	(1,2)
(0,1)	(0,1)	(0,2)	(0,0)	(1,1)	(1,2)	(1,0)
(0,2)	(0,2)	(0,0)	(0,1)	(1,2)	(1,0)	(1,1)
(1,0)	(1,0)	(1,1)	(1,2)	(0,0)	(0,1)	(0,2)
(1,1)	(1,1)	(1,2)	(1,0)	(0,1)	(0,2)	(0,0)
(1,2)	(1,2)	(1,0)	(1,1)	(0,2)	(0,0)	(0,1)

The only two proper subgroups are $\{(0,0),(1,0)\}$ and $\{(0,0),(0,1),(0,2)\}$

11.6.3.3. Algebraic properties of the n-cube.

Answer.

(a) (i) $a + b$ could be $(1,0)$ or $(0,1)$. (ii) $a + b = (1,1)$.

(b) (i) $a + b$ could be $(1,0,0), (0,1,0)$, or $(0,0,1)$. (ii) $a + b = (1,1,1)$.

(c) (i) $a + b$ has exactly one 1. (ii) $a + b$ has all $1's$.

11.6.3.5. Answer.

(a) No, 0 is not an element of $\mathbb{Z} \times \mathbb{Z}$.

(b) Yes.

(c) No, (0, 0) is not an element of this set.

(d) No, the set is not closed: $(1,1) + (2,4) = (3,5)$ and $(3,5)$ is not in the set.

(e) Yes.

11.7 · Isomorphisms
11.7.4 · Exercises

11.7.4.1. Answer.

(a) Yes, $f(n,x) = (x,n)$ for $(n,x) \in \mathbb{Z} \times \mathbb{R}$ is an isomorphism.

(b) No, $\mathbb{Z}_2 \times \mathbb{Z}$ has a two element subgroup while $\mathbb{Z} \times \mathbb{Z}$ does not.

(c) No. $\mathbb{Q} \times \mathbb{Q}$ is countable and $\mathbb{R}$ is not. Therefore, no bijection can exist between them.

(d) Yes.

(e) No.

(f) Yes, one isomorphism is defined by $f(a_1, a_2, a_3, a_4) = \begin{pmatrix} a_1 & a_2 \\ a_3 & a_4 \end{pmatrix}$.

(g) Yes, one isomorphism is defined by $f(a_1, a_2) = (a_1, 10^{a_2})$.

(h) Yes.

(i) Yes $f(k) = k(1,1)$.

11.7.4.3. Answer. Consider three groups G_1, G_2, and G_3 with operations $*, \diamond$, and $\star$, respectively. We want to show that if G_1 is isomorphic to G_2, and if G_2 is isomorphic to G_3, then G_1 is isomorphic to G_3.

G_1 isomorphic to $G_2 \Rightarrow$ there exists an isomorphism $f : G_1 \to G_2$

G_2 isomorphic to $G_3 \Rightarrow$ there exists an isomorphism $g : G_2 \to G_3$

If we compose g with f, we get the function $g \circ f : G_1 \to G_3$, By Theorem 7.0.1, p. 13 and Theorem 7.0.2, p. 13, $g \circ f$ is a bijection, and if $a, b \in G_1$,

$$\begin{aligned}(g \circ f)(a * b) &= g(f(a * b)) \\ &= g(f(a) \diamond f(b)) \quad \text{since } f \text{ is an isomorphism} \\ &= g(f(a)) \star g(f(b)) \quad \text{since } g \text{ is an isomorphism} \\ &= (g \circ f)(a) \star (g \circ f)(b) \end{aligned}$$

Therefore, $g \circ f$ is an isomorphism from G_1 into G_3, proving that "is isomorphic to" is transitive.

11.7.4.5. Answer. By Theorem 11.7.14, p. 59(a), $T(0)$ must be 1. $T(2) = T(1 +_4 1) = T(1) \times_5 T(1) = 3 \times_5 3 = 4$. Since T is a bijection, $T(3) = 2$.

11.7.4.7. Answer. Let G be an infinite cyclic group generated by a. Then, using multiplicative notation, $G = \{a^n \mid n \in \mathbb{Z}\}$. The map $T : G \to \mathbb{Z}$ defined by $T(a^n) = n$ is an isomorphism. This is indeed a function, since $a^n = a^m$ implies $n = m$. Otherwise, a would have a finite order and would not generate G.

(a) T is one-to-one, since $T(a^n) = T(a^m)$ implies $n = m$, so $a^n = a^m$.

(b) T is onto, since for any $n \in \mathbb{Z}$, $T(a^n) = n$.

(c) $T(a^n * a^m) = T(a^{n+m}) = n + m = T(a^n) + T(a^m)$

11.7.4.11. Answer. $\mathbb{Z}_8$, $\mathbb{Z}_2 \times \mathbb{Z}_4$, and $\mathbb{Z}_2^3$. One other is the fourth dihedral group, introduced in Section 15.3.

11.7.4.13. Answer. Each 3 is the order of an element whose inverse is it's square; i. e., if a has order 3, $a^2 = a^{-1}$ is distinct from a and also has order 3 and contributes a second matching 3.

12 · More Matrix Algebra
12.1 · Systems of Linear Equations
12.1.7 · Exercises

12.1.7.1. Answer.

(a) $\{(4/3, 1/3)\}$

(b) $\{(\frac{1}{2}x_3 - 3, -4x_3 + 11, x_3) \mid x_3 \in \mathbb{R}\}$

(c) $\{(-5, 14/5, 8/5)\}$

(d) $\{(6.25 - 2.5x_3, -0.75 + 0.5x_3, x_3) \mid x_3 \in \mathbb{R}\}$

12.1.7.3. Answer.

(a) Basic variables: x_1, x_2 and x_4. Free variable: x_3. Solution set: $\{(1.2 + 5x_3, 2.6 - 4x_3, 4.5) \mid x_3 \in \mathbb{R}\}$

(b) Basic variables: x_1 and x_2. Free variable: x_3. The solution set is empty because the last row of the matrix converts to the inconsistent equation $0 = 1$.

(c) Basic variables: x_1 and x_2. Free variable: x_3. Solution set: $\{(-6x_3 + 5, 2x_3 + 1, x_3) \mid x_3 \in \mathbb{R}\}$

(d) Basic variables: x_1, x_2 and x_3. Free variable: x_4. Solution set: $\{(3x_4 + 1, -2x_4 + 2, x_4 + 1, x$

12.1.7.5. Answer.

(a) $\{(3,0)\}$

(b) $\{(3,0,4)\}$

12.1.7.7. Answer. Proof: Since b is the $n \times 1$ matrix of $0\{'\}$s, let's call it $\mathbf{0}$. Let S be the set of solutions to $AX = \mathbf{0}$. If X_1 and X_2 be in S. Then

$$A(X_1 + X_2) = AX_1 + AX_2 = \mathbf{0} + \mathbf{0} = \mathbf{0}$$

so $X_1 + X_2 \in S$; or in other words, S is closed under addition in $\mathbb{R}^n$. The identity of $\mathbb{R}^n$ is $\mathbf{0}$, which is in S. Finally, let X be in S. Then

$$A(-X) = -(AX) = -\mathbf{0} = \mathbf{0}$$

and so $-X$ is also in S.

12.2 · Matrix Inversion
12.2.3 · Exercises

12.2.3.3. Answer.

(a) $\begin{pmatrix} \frac{15}{11} & \frac{30}{11} \\ \frac{3}{11} & -\frac{5}{11} \end{pmatrix}$

(b) $\begin{pmatrix} -20 & \frac{21}{2} & \frac{9}{2} & -\frac{3}{2} \\ 2 & -1 & 0 & 0 \\ -4 & 2 & 1 & 0 \\ 7 & -\frac{7}{2} & -\frac{3}{2} & \frac{1}{2} \end{pmatrix}$

(c) The inverse does not exist. When the augmented matrix is row-reduced (see below), the last row of the first half cannot be manipulated to match the identity matrix.

(d) $\begin{pmatrix} 1 & 0 & 0 \\ -3 & 1 & 1 \\ -4 & 1 & 2 \end{pmatrix}$

(e) The inverse does not exist.

(f) $\begin{pmatrix} 9 & -36 & 30 \\ -36 & 192 & -180 \\ 30 & -180 & 180 \end{pmatrix}$

12.2.3.5. Answer. The solutions are in the solution section of Section 12.1, exercise 1, We illustrate with the outline of the solution to part (c). The matrix version of the system is

$$\begin{pmatrix} 1 & 1 & 2 \\ 1 & 2 & -1 \\ 1 & 3 & 1 \end{pmatrix} \begin{pmatrix} x_1 \\ x_2 \\ x_3 \end{pmatrix} = \begin{pmatrix} 1 \\ -1 \\ 5 \end{pmatrix}$$

We compute the inverse of the matrix of coefficients and get

$$A^{-1} = \begin{pmatrix} 1 & 1 & 2 \\ 1 & 2 & -1 \\ 1 & 3 & 1 \end{pmatrix}^{-1} = \frac{1}{5} \begin{pmatrix} 5 & 5 & -5 \\ -2 & -1 & 3 \\ 1 & -2 & 1 \end{pmatrix}$$

and

$$\begin{pmatrix} x_1 \\ x_2 \\ x_3 \end{pmatrix} = A^{-1} \begin{pmatrix} 1 \\ -1 \\ 5 \end{pmatrix} = \begin{pmatrix} -5 \\ \frac{14}{5} \\ \frac{8}{5} \end{pmatrix}$$

12.3 · An Introduction to Vector Spaces
12.3.3 · Exercises

12.3.3.3. Answer. The dimension of $M_{2\times 3}(\mathbb{R})$ is 6 and yes, $M_{m\times n}(\mathbb{R})$ is also a vector space of dimension $m \cdot n$. One basis for $M_{m\times n}(\mathbb{R})$ is $\{A_{ij} \mid 1 \leq i \leq m, 1 \leq j \leq n\}$ where A_{ij} is the $m \times n$ matrix with entries all equal to zero except for in row i, column j where the entry is 1.

12.3.3.7. Answer. If the matrices are named B, A_1, A_2, A_3, and A_4, then

$$B = \frac{8}{3}A_1 + \frac{5}{3}A_2 + \frac{-5}{3}A_3 + \frac{23}{3}A_4$$

12.3.3.9. Answer.

(a) If $x_1 = (1,0)$, $x_2 = (0,1)$, and $y = (b_1, b_2)$, then $y = b_1 x_1 + b_2 x_2$. If $x_1 = (3,2)$, $x_2 = (2,1)$, and $y = (b_1, b_2)$, then $y = (-b_1 + 2b_2) x_1 + (2b_1 - 3b_2) x_2$.

(b) If $y = (b_1, b_2)$ is any vector in $\mathbb{R}^2$, then $y = (-3b_1 + 4b_2) x_1 + (-b_1 + b_2) x_2 + (0) x_3$

(c) One solution is to add any vector(s) to x_1, x_2, and x_3 of part b.

(d) 2, n

(e) $\begin{pmatrix} x & y \\ z & w \end{pmatrix} = xA_1 + yA_2 + zA_3 + wA_4$

(f) $a_0 + a_1 x + a_2 x^2 + a_3 x^3 = a_0(1) + a_1(x) + a_2(x^2) + a_3(x^3)$.

12.3.3.11. Answer.

(a) The set is linearly independent: let a and b be scalars such that $a(4,1) + b(1,3) = (0,0)$, then $4a + b = 0$ and $a + 3b = 0$ which has $a = b = 0$ as its only solutions. The set generates all of $\mathbb{R}^2$: let (a,b) be an arbitrary vector in $\mathbb{R}^2$. We want to show that we can always find scalars β_1 and β_2 such that $\beta_1(4,1) + \beta_2(1,3) = (a,b)$. This is equivalent to finding scalars such that $4\beta_1 + \beta_2 = a$ and $\beta_1 + 3\beta_2 = b$. This system has a unique solution $\beta_1 = \frac{3a-b}{11}$, and $\beta_2 = \frac{4b-a}{11}$. Therefore, the set generates $\mathbb{R}^2$.

12.3.3.13. Answer. The answer to the last part is that the three vector spaces are all isomorphic to one another. Once you have completed part (a) of this exercise, the following translation rules will give you the answer to parts

(b) and (c),
$$(a,b,c,d) \leftrightarrow \begin{pmatrix} a & b \\ c & d \end{pmatrix} \leftrightarrow a + bx + cx^2 + dx^2$$

12.4 · The Diagonalization Process
12.4.4 · Exercises

12.4.4.1. Answer.

(a) Any nonzero multiple of $\begin{pmatrix} 1 \\ -1 \end{pmatrix}$ is an eigenvector associated with $\lambda = 1$.

(b) Any nonzero multiple of $\begin{pmatrix} 1 \\ 2 \end{pmatrix}$ is an eigenvector associated with $\lambda = 4$.

(c) Let $x_1 = \begin{pmatrix} a \\ -a \end{pmatrix}$ and $x_2 = \begin{pmatrix} b \\ 2b \end{pmatrix}$. You can verify that $c_1 x_1 + c_2 x_2 = \begin{pmatrix} 0 \\ 0 \end{pmatrix}$ if and only if $c_1 = c_2 = 0$. Therefore, $\{x_1, x_2\}$ is linearly independent.

12.4.4.3. Answer. Part c: You should obtain $\begin{pmatrix} 4 & 0 \\ 0 & 1 \end{pmatrix}$ or $\begin{pmatrix} 1 & 0 \\ 0 & 4 \end{pmatrix}$, depending on how you order the eigenvalues.

12.4.4.5. Answer.

(a) If $P = \begin{pmatrix} 2 & 1 \\ 3 & -1 \end{pmatrix}$, then $P^{-1}AP = \begin{pmatrix} 4 & 0 \\ 0 & -1 \end{pmatrix}$.

(b) If $P = \begin{pmatrix} 1 & 1 \\ 7 & 1 \end{pmatrix}$, then $P^{-1}AP = \begin{pmatrix} 5 & 0 \\ 0 & -1 \end{pmatrix}$.

(c) If $P = \begin{pmatrix} 1 & 0 \\ 0 & 1 \end{pmatrix}$, then $P^{-1}AP = \begin{pmatrix} 3 & 0 \\ 0 & 4 \end{pmatrix}$.

(d) If $P = \begin{pmatrix} 1 & -1 & 1 \\ -1 & 4 & 2 \\ -1 & 1 & 1 \end{pmatrix}$, then $P^{-1}AP = \begin{pmatrix} -2 & 0 & 0 \\ 0 & 1 & 0 \\ 0 & 0 & 0 \end{pmatrix}$.

(e) A is not diagonalizable. Five is a double root of the characteristic equation, but has an eigenspace with dimension only 1.

(f) If $P = \begin{pmatrix} 1 & 1 & 1 \\ -2 & 0 & 1 \\ 1 & -1 & 1 \end{pmatrix}$, then $P^{-1}AP = \begin{pmatrix} 3 & 0 & 0 \\ 0 & 1 & 0 \\ 0 & 0 & 0 \end{pmatrix}$.

12.4.4.7. Answer. This is a direct application of the definition of matrix multiplication. Let $A_{(i)}$ be the ith row of A, and let $P^{(j)}$ be the jth column of P. Then the jth column of the product AP is

$$\begin{pmatrix} A_{(1)} P^{(j)} \\ A_{(2)} P^{(j)} \\ \vdots \\ A_{(n)} P^{(j)} \end{pmatrix}$$

Hence, $(AP)^{(j)} = A\left(P^{(j)}\right)$ for $j = 1, 2, \ldots, n$. Thus, each column of AP depends on A and the jth column of P.

12.5 · Some Applications
12.5.5 · Exercises

12.5.5.4. Hint. The characteristic polynomial of the adjacency matrix is $\lambda^4 - 4\lambda^2$.

12.5.5.5. Answer.

(a) Since $A = A^1 = \begin{pmatrix} 1 & 1 & 0 \\ 1 & 0 & 1 \\ 0 & 1 & 1 \end{pmatrix}$, there are 0 paths of length 1 from: node c to node a, node b to node b, and node a to node c; and there is 1 path of length 1 for every other pair of nodes.

(b) The characteristic polynomial is $|A - cI| = \begin{vmatrix} 1-c & 1 & 0 \\ 1 & -c & 1 \\ 0 & 1 & 1-c \end{vmatrix} = -c^3 + 2c^2 + c - 2$

Solving the characteristic equation $-c^3 + 2c^2 + c - 2 = 0$ we find solutions 1, 2, and -1.

If $c = 1$, we find the associated eigenvector by finding a nonzero solution to $\begin{pmatrix} 0 & 1 & 0 \\ 1 & -1 & 1 \\ 0 & 1 & 0 \end{pmatrix} \begin{pmatrix} x_1 \\ x_2 \\ x_3 \end{pmatrix} = \begin{pmatrix} 0 \\ 0 \\ 0 \end{pmatrix}$ One of these, which will be the first column of P, is $\begin{pmatrix} 1 \\ 0 \\ -1 \end{pmatrix}$

If $c = 2$, the system $\begin{pmatrix} -1 & 1 & 0 \\ 1 & -2 & 1 \\ 0 & 1 & -1 \end{pmatrix} \begin{pmatrix} x_1 \\ x_2 \\ x_3 \end{pmatrix} = \begin{pmatrix} 0 \\ 0 \\ 0 \end{pmatrix}$ yields eigenvectors, including $\begin{pmatrix} 1 \\ 1 \\ 1 \end{pmatrix}$, which will be the second column of P.

If $c = -1$, then the system determining the eigenvectors is $\begin{pmatrix} 2 & 1 & 0 \\ 1 & 1 & 1 \\ 0 & 1 & 2 \end{pmatrix} \begin{pmatrix} x_1 \\ x_2 \\ x_3 \end{pmatrix} = \begin{pmatrix} 0 \\ 0 \\ 0 \end{pmatrix}$ and we can select $\begin{pmatrix} 1 \\ -2 \\ 1 \end{pmatrix}$, although any nonzero multiple of this vector could be the third column of P.

(c) Assembling the results of part (b) we have $P = \begin{pmatrix} 1 & 1 & 1 \\ 0 & 1 & -2 \\ -1 & 1 & 1 \end{pmatrix}$.

$$A^4 = P \begin{pmatrix} 1^4 & 0 & 0 \\ 0 & 2^4 & 0 \\ 0 & 0 & (-1)^4 \end{pmatrix} P^{-1} = P \begin{pmatrix} 1 & 0 & 0 \\ 0 & 16 & 0 \\ 0 & 0 & 1 \end{pmatrix} P^{-1}$$

$$= \begin{pmatrix} 1 & 16 & 1 \\ 0 & 16 & -2 \\ -1 & 16 & 1 \end{pmatrix} \begin{pmatrix} \frac{1}{2} & 0 & -\frac{1}{2} \\ \frac{1}{3} & \frac{1}{3} & \frac{1}{3} \\ \frac{1}{6} & -\frac{1}{3} & \frac{1}{6} \end{pmatrix}$$

$$= \begin{pmatrix} 6 & 5 & 5 \\ 5 & 6 & 5 \\ 5 & 5 & 6 \end{pmatrix}$$

Hence there are five different paths of length 4 between distinct vertices, and six different paths that start and end at the same vertex. The reader can verify these facts from Figure 12.5.4, p. 93

12.5.5.7. Answer.

(a) $e^A = \begin{pmatrix} e & e \\ 0 & 0 \end{pmatrix}$, $e^B = \begin{pmatrix} 0 & 0 \\ 0 & e^2 \end{pmatrix}$, and $e^{A+B} = \begin{pmatrix} e & e^2 - e \\ 0 & e^2 \end{pmatrix}$

(b) Let $\mathbf{0}$ be the zero matrix, $e^{\mathbf{0}} = I + \mathbf{0} + \frac{\mathbf{0}^2}{2} + \frac{\mathbf{0}^3}{6} + \ldots = I$.

(c) Assume that A and B commute. We will examine the first few terms in the product $e^A e^B$. The pattern that is established does continue in general. In what follows, it is important that $AB = BA$. For example, in the last step, $(A+B)^2$ expands to $A^2 + AB + BA + B^2$, not $A^2 + 2AB + B^2$, if we can't assume commutativity.

$$e^A e^B = \left(\sum_{k=0}^{\infty} \frac{A^k}{k!}\right)\left(\sum_{k=0}^{\infty} \frac{B^k}{k!}\right)$$
$$= \left(I + A + \frac{A^2}{2} + \frac{A^3}{6} + \cdots\right)\left(I + B + \frac{B^2}{2} + \frac{B^3}{6} + \cdots\right)$$
$$= I + A + B + \frac{A^2}{2} + AB + \frac{B^2}{2} + \frac{A^3}{6} + \frac{A^2 B}{2} + \frac{AB^2}{2} + \frac{B^3}{6} + \cdots$$
$$= I + (A+B) + \frac{1}{2}\left(A^2 + 2AB + B^2\right) + \frac{1}{6}\left(A^3 + 3A^2 B + 3AB^2 + B^3\right) + \cdots$$
$$= I + (A+B) + \frac{1}{2}(A+B)^2 + \frac{1}{6}(A+B)^3 + \cdots$$
$$= e^{A+B}$$

(d) Since A and $-A$ commute, we can apply part d;

$$e^A e^{-A} = e^{A+(-A)} = e^{\mathbf{0}} = I$$

12.6 · Linear Equations over the Integers Mod 2
12.6.2 · Exercises

12.6.2.1. Answer.

(a) $\{(0,0,0),(1,1,1)\}$

(b) $\{(1,1,1,0)\}$

12.6.2.2. Answer. As suggested here is the augmented matrix with both right sides, and its row reduction:

$$\begin{pmatrix} 1 & 1 & 0 & 1 & 1 & 1 \\ 1 & 0 & 1 & 1 & 0 & 0 \\ 0 & 1 & 1 & 0 & 1 & 0 \end{pmatrix} \longrightarrow \begin{pmatrix} \mathbf{1} & 0 & 1 & 1 & 0 & 0 \\ 0 & \mathbf{1} & 1 & 0 & 1 & 0 \\ 0 & 0 & 0 & 0 & 0 & 1 \end{pmatrix}$$

There are only two basic variables here because the left side of the last equation is the sum of the left sides of the first two equations.

(a) Ignoring the last column of both matrices, we see that the last equation of the first system reduces to $0 = 0$, which is always true, and the first two equations yield two free variables, x_3 and x_4. The general solution is the set of quadruples $\{(x_3 +_2 x_4, x_3 +_2 1, x_3, x_4) \mid x_3, x_4 \in \mathbb{Z}_2\}$. The cardinality of the solution set is 4.

(b) If we replace the fifth column with the sixth one, the last row indicates that $0 = 1$, which means that the solution set is empty.

12.6.2.3. Answer.

(a) Row reduction produces a solution with one free variable, x_3.

$$\begin{aligned}(x_1, x_2, x_3, x_4, x_5) &= (x_3, x_3, x_3, 0, 0) \\ &= x_3(1, 1, 1, 0, 0)\end{aligned}$$

The solution set has only two elements. It is $\{(0,0,0,0,0), (1,1,1,0,0)\}$. Since $\mathbb{Z}_2^5$ is a finite group, the solution set is a subgroup because it is closed with respect to coordinatewise mod 2 addition.

(b) The row-reduced augmented matrix of coefficients provides the solution

$$\begin{aligned}(x_1, x_2, x_3, x_4, x_5) &= (x_3, 1 + x_3, x_3, 1, 0) \\ &= (0, 1, 0, 1, 0) + x_3(1, 1, 1, 0, 0)\end{aligned}$$

Therefore, the solution to this system is a shift of the solution set to the homogeneous system by the vector $(0, 1, 0, 1, 0)$, which is $\{(0, 1, 0, 1, 0), (1, 0, 1, 1, 0)\}$

13 · Boolean Algebra
13.1 · Posets Revisited

· **Exercises**

13.1.1. Answer.

(a) 1, 5

(b) 5

(c) 30

(d) 30

(e) See the Sage cell below with the default input displaying a Hasse diagram for D_{12}.

```
Posets.DivisorLattice(12).show()
```

13.1.3. Answer.

- Solution for Hasse diagram (b):

○

∨	a_1	a_2	a_3	a_4	a_5
a_1	a_1	a_2	a_3	a_4	a_5
a_2	a_2	a_2	a_4	a_4	a_5
a_3	a_3	a_4	a_3	a_4	a_5
a_4	a_4	a_4	a_4	a_4	a_5
a_5	a_5	a_5	a_5	a_5	a_5

∧	a_1	a_2	a_3	a_4	a_5
a_1	a_1	a_1	a_1	a_1	a_1
a_2	a_1	a_2	a_1	a_2	a_2
a_3	a_1	a_1	a_3	a_3	a_3
a_4	a_1	a_2	a_3	a_4	a_4
a_5	a_1	a_2	a_3	a_4	a_5

a_1 is the least element and a_5 is the greatest element.

- Partial solution for Hasse diagram (f):

  ○ $\text{lub}(a_2, a_3)$ and $\text{lub}(a_4, a_5)$ do not exist.

  ○ No greatest element exists, but a_1 is the least element.

13.1.5. Answer. If 0 and $0'$ are distinct least elements, then

$$\left.\begin{array}{ll} 0 \leq 0' & \text{since } 0 \text{ is a least element} \\ 0' \leq 0 & \text{since } 0' \text{ is a least element} \end{array}\right\} \Rightarrow 0 = 0' \text{ by antisymmetry, a contradiction}$$

13.1.7. Answer.

(a) The sum of elements in $A \cap B = \{2, 3, 6\}$ is odd and disqualifies the set from being an element of the poset.

(b) The following correctly complete the statements in this part.

  (i) ... $A \subseteq R$ and $B \subseteq R$

  (ii) ... for all $A \in \mathcal{P}_0$, $R \subseteq A$

(c) Any set that contains the union of $A \cup B = \{1, 2, 3, 6, 7\}$ but also contains 3 or 5, but not both will be an upper bound. You can create several by including on not including 4 or 8.

(d) The least upper bound doesn't exist. Notice that the union of A and B isn't in $\mathcal{P}_0$. One of the two sets $\{1, 2, 3, 5, 6, 7\}$ and $\{1, 2, 3, 6, 7, 9\}$ is contained within every upper bound of A and B but neither is contained within the other.

13.2 · Lattices

· Exercises

13.2.5. Answer. One reasonable definition would be this: Let $[L; \vee, \wedge]$ be a lattice and let K be a nonempty subset of L. Then K is a sublattice of L if and only if K is closed under both $\vee$ and $\wedge$

13.3 · Boolean Algebras

· Exercises

13.3.1. Answer.

B	Complement of B
∅	A
{a}	{b, c}
{b}	{a, c}
{c}	{a, b}
{a, b}	{c}
{a, c}	{b}
{b, c}	{a}
A	∅

This lattice is a Boolean algebra since it is a distributive complemented lattice.

13.3.3. Answer. a and g.

13.3.5. Answer.

(a) $S^* : a \vee b = a$ if $a \geq b$

(b) The dual of $S : A \cap B = A$ if $A \subseteq B$ is $S^* : A \cup B = A$ if $A \supseteq B$

(c) Yes

(d) The dual of $S : p \wedge q \Leftrightarrow p$ if $p \Rightarrow q$ is $S^* : p \vee q \Leftrightarrow p$ if $q \Rightarrow p$

(e) Yes

13.3.7. Answer. $[B; \wedge, \vee, -]$ is isomorphic to $[B'; \wedge, \vee, \tilde{\ }]$ if and only if there exists a function $T : B \to B'$ such that

(a) T is a bijection;

(b) $T(a \wedge b) = T(a) \wedge T(b)$ for all $a, b \in B$

(c) $T(a \vee b) = T(a) \vee T(b)$ for all $a, b \in B$

(d) $T(\bar{a}) = \widetilde{T(a)}$ for all $a \in B$.

13.4 · Atoms of a Boolean Algebra

· **Exercises**

13.4.1. Answer.

(a) For $a = 3$ we must show that for each $x \in D_{30}$ one of the following is true: $x \wedge 3 = 3$ or $x \wedge 3 = 1$. We do this through the following table:

x	verification
1	$1 \wedge 3 = 1$
2	$2 \wedge 3 = 1$
3	$3 \wedge 3 = 3$
5	$5 \wedge 3 = 1$
6	$6 \wedge 3 = 3$
10	$20 \wedge 3 = 1$
15	$15 \wedge 3 = 3$
30	$30 \wedge 3 = 3$

For $a = 5$, a similar verification can be performed.

(b) $6 = 2 \vee 3$, $10 = 2 \vee 5$, $15 = 3 \vee 5$, and $30 = 2 \vee 3 \vee 5$.

13.4.3. Answer. If $B = D_{30}$ 30 then $A = \{2, 3, 5\}$ and D_{30} is isomorphic to

$\mathcal{P}(A)$, where
$$\begin{array}{ll} 1 \leftrightarrow \emptyset & 5 \leftrightarrow \{5\} \\ 2 \leftrightarrow \{2\} & 10 \leftrightarrow \{2,5\} \\ 3 \leftrightarrow \{3\} & 15 \leftrightarrow \{3,5\} \\ 6 \leftrightarrow \{2,3\} & 30 \leftrightarrow \{2,3,5\} \end{array}$$
and
$$\begin{array}{l}\text{Join} \leftrightarrow \text{Union} \\ \text{Meet} \leftrightarrow \text{Intersection} \\ \text{Complement} \leftrightarrow \text{Set Complement}\end{array}$$

13.4.5. Hint. Assume that $[B; \vee, \wedge, -]$ is a Boolean algebra of order 3 where $B = \{0, x, 1\}$ and show that this cannot happen by investigating the possibilities for its operation tables.

Answer. Assume that $x \neq 0$ or 1 is the third element of a Boolean algebra. Then there is only one possible set of tables for join and meet, all following from required properties of the Boolean algebra.

$\vee$	0	x	1
0	0	x	1
x	x	x	1
1	1	1	1

$\wedge$	0	x	1
0	0	0	0
x	0	x	x
1	0	x	1

Next, to find the complement of x we want y such that $x \wedge y = 0$ and $x \vee y = 1$. No element satisfies both conditions; hence the lattice is not complemented and cannot be a Boolean algebra. The lack of a complement can also be seen from the ordering diagram from which $\wedge$ and $\vee$ must be derived.

13.4.7. Answer. Let X be any countably infinite set, such as the integers. A subset of X is **cofinite** if it is finite or its complement is finite. The set of all cofinite subsets of X is:

(a) Countably infinite - this might not be obvious, but here is a hint. Assume $X = \{x_0, x_1, x_2, \ldots\}$. For each finite subset A of X, map that set to the integer $\sum_{i=0}^{\infty} \chi_A(x_i) 2^i$ You can do a similar thing to sets that have a finite complement, but map them to negative integers. Only one minor adjustment needs to be made to accommodate both the empty set and X.

(b) Closed under union

(c) Closed under intersection, and

(d) Closed under complementation.

Therefore, if $B = \{A \subseteq X : A \text{ is cofinite}\}$, then B is a countable Boolean algebra under the usual set operations.

13.4.8. Hint. "Copy" the corresponding proof for groups in Section 11.6.

13.5 · Finite Boolean Algebras as n-tuples of 0's and 1's

· Exercises

13.5.1. Answer.

(a)

$\vee$	(0,0)	(0,1)	(1,0)	(1,1)
(0,0)	(0,0)	(0,1)	(1,0)	(1,1)
(0,1)	(0,1)	(0,1)	(1,1)	(1,1)
(1,0)	(1,0)	(1,1)	(1,0)	(1,1)
(1,1)	(1,1)	(1,1)	(1,1)	(1,1)

$\wedge$	(0,0)	(0,1)	(1,0)	(1,1)
(0,0)	(0,0)	(0,0)	(0,0)	(0,0)
(0,1)	(0,0)	(0,1)	(0,0)	(0,1)
(1,0)	(0,0)	(0,0)	(1,0)	(1,0)
(1,1)	(0,0)	(0,1)	(1,0)	(1,1)

u	$\overline{u}$
(0,0)	(1,1)
(0,1)	(1,0)
(1,0)	(0,1)
(1,1)	(0,0)

(b) The graphs are isomorphic.

(c) (0, 1) and (1,0)

13.5.3. Answer.

(a) $(1,0,0,0)$, $(0,1,0,0)$, $(0,0,1,0)$, and $(0,0,0,1)$ are the atoms.

(b) The n-tuples of bits with exactly one 1.

13.6 · Boolean Expressions
· **Exercises**
13.6.1. Answer.

(a)
$f_1(x_1, x_2) = 0$
$f_2(x_1, x_2) = (\overline{x_1} \wedge \overline{x_2})$
$f_3(x_1, x_2) = (\overline{x_1} \wedge x_2)$
$f_4(x_1, x_2) = (x_1 \wedge \overline{x_2})$
$f_5(x_1, x_2) = (x_1 \wedge x_2)$
$f_6(x_1, x_2) = ((\overline{x_1} \wedge \overline{x_2}) \vee (\overline{x_1} \wedge x_2)) = \overline{x_1}$
$f_7(x_1, x_2) = ((\overline{x_1} \wedge \overline{x_2}) \vee (x_1 \wedge \overline{x_2})) = \overline{x_2}$
$f_8(x_1, x_2) = ((\overline{x_1} \wedge \overline{x_2}) \vee (x_1 \wedge x_2)) = ((x_1 \wedge x_2) \vee (\overline{x_1} \wedge \overline{x_2}))$
$f_9(x_1, x_2) = ((\overline{x_1} \wedge x_2) \vee (x_1 \wedge \overline{x_2})) = ((x_1 \wedge \overline{x_2}) \vee (\overline{x_1} \wedge x_2))$
$f_{10}(x_1, x_2) = ((\overline{x_1} \wedge x_2) \vee (x_1 \wedge x_2)) = x_2$
$f_{11}(x_1, x_2) = ((x_1 \wedge \overline{x_2}) \vee (x_1 \wedge x_2)) = x_1$
$f_{12}(x_1, x_2) = ((\overline{x_1} \wedge \overline{x_2}) \vee (\overline{x_1} \wedge x_2) \vee (x_1 \wedge \overline{x_2})) = (\overline{x_1} \vee \overline{x_2})$
$f_{13}(x_1, x_2) = ((\overline{x_1} \wedge \overline{x_2}) \vee (\overline{x_1} \wedge x_2) \vee (x_1 \wedge x_2)) = (\overline{x_1} \vee x_2)$
$f_{14}(x_1, x_2) = ((\overline{x_1} \wedge \overline{x_2}) \vee (x_1 \wedge \overline{x_2}) \vee (x_1 \wedge x_2)) = (x_1 \vee \overline{x_2})$
$f_{15}(x_1, x_2) = ((\overline{x_1} \wedge x_2) \vee (x_1 \wedge \overline{x_2}) \vee (x_1 \wedge x_2)) = (x_1 \vee x_2)$
$f_{16}(x_1, x_2) = ((\overline{x_1} \wedge \overline{x_2}) \vee (\overline{x_1} \wedge x_2) \vee (x_1 \wedge \overline{x_2}) \vee (x_1 \wedge x_2)) = 1$

(b) The truth table for the functions in part (a) are

x_1	x_2	f_1	f_2	f_3	f_4	f_5	f_6	f_7	f_8
0	0	0	1	0	0	0	1	1	1
0	1	0	0	1	0	0	1	0	0
1	0	0	0	0	1	0	0	1	0
1	1	0	0	0	0	1	0	0	1

x_1	x_2	f_9	f_{10}	f_{11}	f_{12}	f_{13}	f_{14}	f_{15}	f_{16}
0	0	0	0	0	1	1	1	0	1
0	1	1	1	0	1	1	0	1	1
1	0	1	0	1	1	0	1	1	1
1	1	0	1	1	0	1	1	1	1

(c) (i) $g_1(x_1, x_2) = f_{15}(x_1, x_2)$

(ii) $g_2(x_1, x_2) = f_{12}(x_1, x_2)$

(iii) $g_3(x_1, x_2) = f_{12}(x_1, x_2)$

(iv) $g_4(x_1, x_2) = f_{16}(x_1, x_2)$

13.6.3. Answer.

(a) The number of elements in the domain of f is $16 = 4^2 = |B|^2$

(b) With two variables, there are $4^3 = 256$ different Boolean functions. With three variables, there are $4^8 = 65536$ different Boolean functions.

(c) $f(x_1, x_2) = (1 \wedge \overline{x_1} \wedge \overline{x_2}) \vee (1 \wedge \overline{x_1} \wedge x_2) \vee (1 \wedge x_1 \wedge \overline{x_2}) \vee (0 \wedge x_1 \wedge x_2)$

(d) Consider $f : B^2 \to B$, defined by $f(0,0) = 0$, $f(0,1) = 1$, $f(1,0) = a$, $f(1,1) = a$, and $f(0,a) = b$, with the images of all other pairs in B^2 defined arbitrarily. This function is not a Boolean function. If we assume that it is Boolean function then f can be computed with a Boolean expression $M(x_1, x_2)$. This expression can be put into minterm normal form: $M(x_1, x_2) = (c_1 \wedge \overline{x_1} \wedge \overline{x_2}) \vee (c_2 \wedge \overline{x_1} \wedge x_2) \vee (c_3 \wedge x_1 \wedge \overline{x_2}) \vee (c_4 \wedge x_1 \wedge x_2)$

$$f(0,0) = 0 \Rightarrow M(0,0) = 0 \Rightarrow c_1 = 0$$
$$f(0,1) = 1 \Rightarrow M(0,0) = 1 \Rightarrow c_2 = 1$$
$$f(1,0) = a \Rightarrow M(0,0) = a \Rightarrow c_3 = a$$
$$f(1,1) = a \Rightarrow M(0,0) = a \Rightarrow c_4 = a$$

Therefore, $M(x_1, x_2) = (\overline{x_1} \wedge x_2) \vee (a \wedge x_1 \wedge \overline{x_2}) \vee (a \wedge x_1 \wedge x_2)$ and so, using this formula, $M(0, a) = (\overline{0} \wedge a) \vee (a \wedge 0 \wedge \overline{a}) \vee (a \wedge 0 \wedge a) = a$ This contradicts $f(0,a) = b$, and so f is not a Boolean function.

13.7 · A Brief Introduction to Switching Theory and Logic Design

· **Exercises**

13.7.1. Answer.

(1) Associative, commutative, and idempotent laws.

(2) Distributive law.

(3) Idempotent and complement laws.

(4) Null and identity laws

(5) Distributive law.

(6) Null and identity laws.

13.7.2. Answer.
$$(x_1 \cdot \overline{x_2}) + (x_1 \cdot x_2) + (\overline{x_1} \cdot x_2).$$

13.7.3. Answer. A simpler boolean expression for the function is $\overline{x_2} \cdot (x_1 + x_3)$.

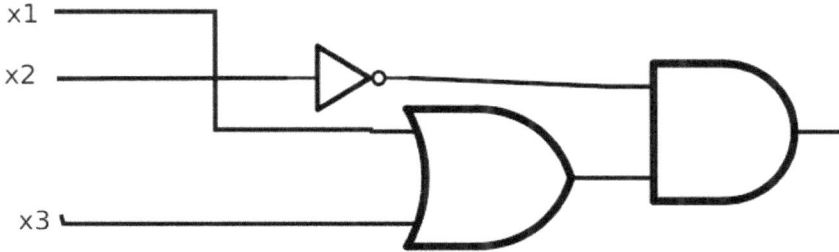

Figure B.0.2 An even simpler circuit

14 · Monoids and Automata
14.1 · Monoids

· **Exercises**

14.1.1. Answer.

(a) S_1 is not a submonoid since the identity of $[\mathbb{Z}_8; \times_8]$, which is 1, is not in S_1. S_2 is a submonoid since $1 \in S_2$ and S_2 is closed under multiplication; that is, for all $a, b \in S_2$, $a \times_8 b$ is in S_2.

(b) The identity of $\mathbb{N}^\mathbb{N}$ is the identity function $i : \mathbb{N} \to \mathbb{N}$ defined by $i(a) = a$, $\forall a \in \mathbb{N}$. If $a \in \mathbb{N}$, $i(a) = a \leq a$, thus the identity of $\mathbb{N}^\mathbb{N}$ is in S_1. However, the image of 1 under any function in S_2 is 2, and thus the identity of $\mathbb{N}^\mathbb{N}$ is not in S_2, so S_2 is not a submonoid. The composition of any two functions in S_1, f and g, will be a function in S_1:

$$(f \circ g)(n) = f(g(n)) \leq g(n) \text{ since } f \text{ is in } S_1$$
$$\leq n \text{ since } g \text{ is in } S_1 \Rightarrow f \circ g \in S_1$$

and the two conditions of a submonoid are satisfied and S_1 is a submonoid of $\mathbb{N}^\mathbb{N}$.

(c) The first set is a submonoid, but the second is not since the null set has a non-finite complement.

14.1.3. Answer. The set of $n \times n$ real matrices is a monoid under matrix multiplication. This follows from the laws of matrix algebra in Chapter 5. To prove that the set of stochastic matrices is a monoid over matrix multiplication, we need only show that the identity matrix is stochastic (this is obvious) and that the set of stochastic matrices is closed under matrix multiplication. Let A and B be $n \times n$ stochastic matrices.

$$(AB)_{ij} = \sum_{k=1}^{n} a_{ik} b_{kj}$$

The sum of the jth column is

$$\sum_{j=1}^{n}(AB)_{ij} = \sum_{k=1}^{n} a_{1k}b_{kj} + \sum_{k=1}^{n} a_{1k}b_{kj} + \cdots + \sum_{k=1}^{n} a_{nk}b_{kj}$$

$$= \sum_{k=1}^{n}(a_{1k}b_{kj} + a_{1k}b_{kj} + \cdots + a_{nk}b_{kj})$$

$$= \sum_{k=1}^{n} b_{kj}(a_{1k} + a_{1k} + \cdots + a_{nk})$$

$$= \sum_{k=1}^{n} b_{kj} \quad \text{since } A \text{ is stochastic}$$

$$= 1 \quad \text{since } B \text{ is stochastic}$$

14.1.5. Answer. Let $f, g, h \in M$, and $a \in B$.

$$((f*g)*h)(a) = (f*g)(a) \wedge h(a)$$
$$= (f(a) \wedge g(a)) \wedge h(a)$$
$$= f(a) \wedge (g(a) \wedge h(a))$$
$$= f(a) \wedge (g*h)(a)$$
$$= (f*(g*h))(a)$$

Therefore $(f*g)*h = f*(g*h)$ and $*$ is associative.

The identity for $*$ is the function $u \in M$ where $u(a) = 1 =$ the "one" of B. If $a \in B$, $(f*u)(a) = f(a) \wedge u(a) = f(a) \wedge 1 = f(a)$. Therefore $f*u = f$. Similarly, $u*f = f$.

There are $2^2 = 4$ functions in M for $B = B_2$. These four functions are named in the text. See Figure 14.1.4, p. 130. The table for $*$ is

$*$	z	i	t	u
z	z	z	z	z
i	z	i	z	i
t	z	z	t	t
u	z	i	t	u

14.2 · Free Monoids and Languages

· **Exercises**

14.2.1. Answer.

(a) For a character set of 350 symbols, the number of bits needed for each character is the smallest n such that 2^n is greater than or equal to 350. Since $2^9 = 512 > 350 > 2^8$, 9 bits are needed,

(b) $2^{12} = 4096 > 3500 > 2^{11}$; therefore, 12 bits are needed.

14.2.3. Answer. This grammar defines the set of all strings over B for which each string is a palindrome (same string if read forward or backward).

14.2.5. Answer.

(a) Terminal symbols: The null string, 0, and 1. Nonterminal symbols: S, E. Starting symbol: S. Production rules: $S \to 00S$, $S \to 01S$, $S \to 10S$, $S \to 11S$, $S \to E$, $E \to 0$, $E \to 1$ This is a regular grammar.

(b) Terminal symbols: The null string, 0, and 1. Nonterminal symbols: S, A, B, C Starting symbol: S Production rules: $S \to 0A$, $S \to 1A$, $S \to \lambda$,

$A \to 0B$, $A \to 1B$, $A \to \lambda$, $B \to 0C$, $B \to 1C$, $B \to A$, $C \to 0$, $C \to 1$, $C \to \lambda$ This is a regular grammar.

(c) See Exercise 3. This language is not regular.

14.2.7. Answer. If s is in A^* and L is recursive, we can answer the question "Is s in L^c?" by negating the answer to "Is s in L?"

14.2.9. Answer.

(a) List the elements of each set X_i in a sequence x_{i1}, x_{i2}, $x_{i3} \ldots$. Then draw arrows as shown below and list the elements of the union in order established by this pattern: x_{11}, x_{21}, x_{12}, x_{13}, x_{22}, x_{31}, x_{41}, x_{32}, x_{23}, x_{14}, $x_{15} \ldots$,

(b) Each of the sets A^1, A^2, $A^3, \ldots$, are countable and A^* is the union of these sets; hence A^* is countable.

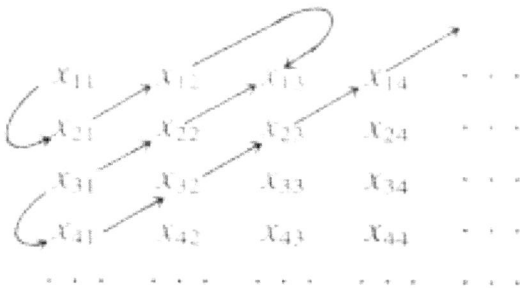

Figure B.0.3 Exercise 9

14.3 · Automata, Finite-State Machines

· **Exercises**

14.3.1. Answer.

x	s	$Z(x,s)$	$t(x,s)$
Deposit25 ¢	Locked	Nothing	Select
Deposit25 ¢	Select	Return25 ¢	Select
PressS	Locked	Nothing	Locked
PressS	Select	DispenseS	Locked
PressP	Locked	Nothing	Locked
PressP	Select	DispenseP	Locked
PressB	Locked	Nothing	Locked
PressB	Select	DispenseB	Locked

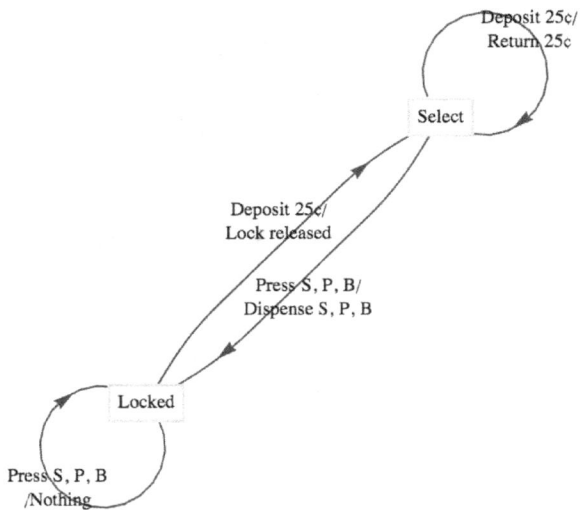

Figure B.0.4 Vending Machine Transitions

14.3.3. Answer. $\{000, 011, 101, 110, 111\}$

14.3.5. Answer.

(a)
- Input: 10110, Output: 11011 ⇒ 10110 is in position 27
- Input: 00100, Output: 00111 ⇒ 00100 is in position 7
- Input: 11111, Output: 10101 ⇒ 11111 is in position 21

(b) Let $x = x_1 x_2 \ldots x_n$ and recall that for $n \geq 1$, $G_{n+1} = \begin{pmatrix} 0 G_n \\ 1 G_n^r \end{pmatrix}$, where G_n^r is the reverse of G_n. To prove that the Gray Code Decoder always works, let $p(n)$ be the proposition "Starting in Copy state, x's output is the position of x in G_n; and starting in Complement state, x's output is the position of x in G_n^r." That p(1) is true is easy to verify for both possible values of x, 0 and 1. Now assume that for some $n \geq 1$, $p(n)$ is true and consider $x = x_1 x_2 \ldots x_n x_{n+1}$.

If $x_1 = 0$, x's output is a zero followed by the output for $(x_2 \ldots x_n x_{n+1})$ starting in Copy state. By the induction hypothesis, this is zero followed by the position of $(x_2 \ldots x_n x_{n+1})$ in G_n, which is the position of x in G_{n+1}, by the definition of G.

If $x_1 = 1$, x's output is a one followed by the output for $(x_2 \ldots x_n x_{n+1})$ starting in Complement state. By the induction hypothesis, this is one followed by the position of $(x_2 \ldots x_n x_{n+1})$ in G_n^r, which is the position of x in G_{n+1}, by the definition of G. □

14.4 · The Monoid of a Finite-State Machine

· **Exercises**

14.4.1. Hint. Where the output echoes the current state, the output can be ignored.

Answer.

(a)

Input String	a	b	c	aa	ab	ac
1	$(a,1)$	$(a,2)$	$(c,3)$	$(a,1)$	$(a,2)$	$(c,3)$
2	$(a,2)$	$(a,1)$	$(c,3)$	$(a,2)$	$(a,1)$	$(c,3)$
3	$(c,3)$	$(c,3)$	$(c,3)$	$(c,3)$	$(c,3)$	$(c,3)$

Input String	ba	bb	bc	ca	cb
1	$(a,2)$	$(a,1)$	$(c,3)$	$(c,3)$	$(c,3)$
2	$(a,1)$	$(a,2)$	$(c,3)$	$(c,3)$	$(c,3)$
3	$(c,3)$	$(c,3)$	$(c,3)$	$(c,3)$	$(c,3)$

We can see that $T_aT_a = T_{aa} = T_a$, $T_aT_b = T_{ab} = T_b$, etc. Therefore, we have the following monoid:

	T_a	T_b	T_b
T_a	T_a	T_b	T_c
T_b	T_b	T_a	T_c
T_c	T_c	T_c	T_c

Notice that T_a is the identity of this monoid.

(b)

Input String	1	2	11	12	21	22
A	C	B	A	D	D	A
B	D	A	B	C	C	B
C	A	D	C	B	B	C
D	B	C	D	A	A	D

Input String	111	112	121	122	211	212	221	222
A	C	B	B	C	B	C	C	B
B	D	A	A	D	A	D	D	A
C	B	C	C	B	C	B	B	C
D	B	C	C	B	C	B	B	C

We have the following monoid:

	T_1	T_2	T_{11}	T_{12}
T_1	T_{11}	T_{12}	T_1	T_2
T_2	T_b	T_{11}	T_2	T_1
T_{11}	T_1	T_2	T_{11}	T_{12}
T_{12}	T_2	T_1	T_{12}	T_{11}

Notice that T_{11} is the identity of this monoid.

14.4.3. Answer. Yes, just consider the unit time delay machine of Figure 14.4.4, p. 145. Its monoid is described by the table at the end of Section 14.4 where the T_λ row and T_λ column are omitted. Next consider the machine in Figure 14.5.7, p. 147. The monoid of this machine is:

	T_λ	T_0	T_1	T_{00}	T_{01}	T_{10}	T_{11}
T_λ	T_λ	T_0	T_1	T_{00}	T_{01}	T_{10}	T_{11}
T_0	T_0	T_{00}	T_{01}	T_{00}	T_{01}	T_{10}	T_{11}
T_1	T_1	T_{10}	T_{11}	T_{00}	T_{01}	T_{10}	T_{11}
T_{00}	T_{00}	T_{00}	T_{01}	T_{00}	T_{01}	T_{10}	T_{11}
T_{01}	T_{01}	T_{10}	T_{11}	T_{00}	T_{01}	T_{10}	T_{11}
T_{10}	T_{10}	T_{00}	T_{01}	T_{00}	T_{01}	T_{10}	T_{11}
T_{11}	T_{11}	T_{10}	T_{11}	T_{00}	T_{01}	T_{10}	T_{11}

Hence both of these machines have the same monoid, however, their transition diagrams are nonisomorphic since the first has two vertices and the second has seven.

14.5 · The Machine of a Monoid

· **Exercises**

14.5.1. Answer.

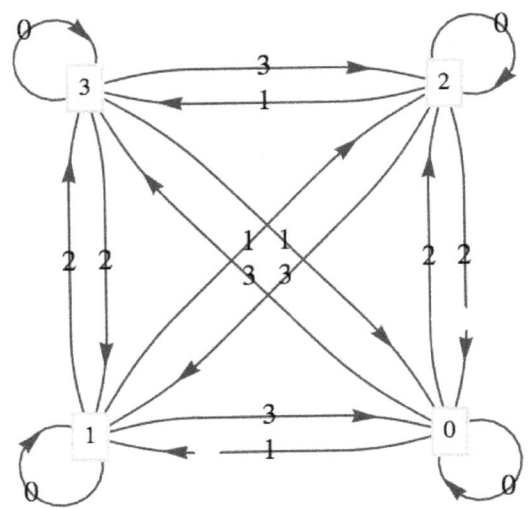

Figure B.0.5 (a)

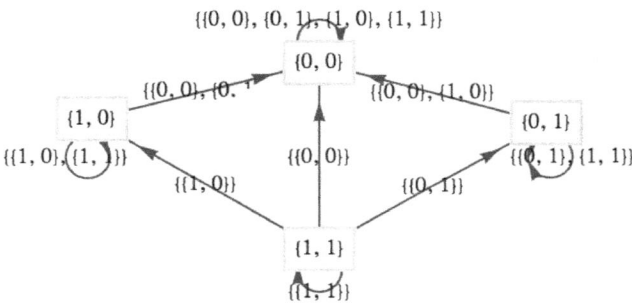

Figure B.0.6 (b)

15 · Group Theory and Applications
15.1 · Cyclic Groups

· **Exercises**

15.1.1. Answer. The only other generator is -1.

15.1.3. Answer. If $|G| = m$, $m > 2$, and $G = \langle a \rangle$, then $a, a^2, \ldots, a^{m-1}$, $a^m = e$ are distinct elements of G. Furthermore, $a^{-1} = a^{m-1} \neq a$, If $1 \leq k \leq m$, a^{-1} generates a^k:

$$\begin{aligned}(a^{-1})^{m-k} &= (a^{m-1})^{m-k} \\ &= a^{m^2 - m - mk + k} \\ &= (a^m)^{m-k-1} * a^k \\ &= e * a^k = a^k\end{aligned}$$

Similarly, if G is infinite and $G = \langle a \rangle$, then a^{-1} generates G.

15.1.5. Answer.

(a) No. Assume that $q \in \mathbb{Q}$ generates $\mathbb{Q}$. Then $\langle q \rangle = \{nq : n \in \mathbb{Z}\}$. But this gives us at most integer multiples of q, not every element in $\mathbb{Q}$.

(b) No. Similar reasoning to part a.

(c) Yes. 6 is a generator of $6\mathbb{Z}$.

(d) No.

(e) Yes, $(1,1,1)$ is a generator of the group.

15.1.7. Answer. Theorem 15.1.13, p. 152 implies that a generates $\mathbb{Z}_n$ if and only if the greatest common divisor of n and a is 1. Therefore the list of generators of $\mathbb{Z}_n$ are the integers in $\mathbb{Z}_n$ that are relatively prime to n. The generators of $\mathbb{Z}_{25}$ are all of the nonzero elements except 5, 10, 15, and 20. The generators of $\mathbb{Z}_{256}$ are the odd integers in $\mathbb{Z}_{256}$ since 256 is 2^8.

15.1.9. Answer.

(a) $\theta : \mathbb{Z}_{77} \to \mathbb{Z}_7 \times \mathbb{Z}_{11}$ maps the given integers as follows:

$$
\begin{array}{rcl}
21 & \to & (0, 10) \\
5 & \to & (5, 5) \\
7 & \to & (0, 7) \\
15 & \to & \underline{(1, 4)} \\
\text{sum} = 48 & \leftarrow & (6, 4) = \text{sum}
\end{array}
$$

The final sum, 48, is obtained by using the facts that $\theta^{-1}(1,0) = 22$ and $\theta^{-1}(0,1) = 56$

$$
\begin{aligned}
\theta^{-1}(6,4) = 6 \times_{77} \theta^{-1}(1,0) + 4 \times_{77} \theta^{-1}(0,1) \\
= 6 \times_{77} 22 +_{77} 4 \times_{77} 56 \\
= 55 +_{77} 70 \\
= 48
\end{aligned}
$$

(b) Using the same isomorphism:

$$
\begin{array}{rcl}
25 & \to & (4, 3) \\
26 & \to & (5, 4) \\
40 & \to & (5, 7) \\
& & \text{sum} = (0, 3)
\end{array}
$$

$$
\begin{aligned}
theta^{-1}(0,3) = 3 \times_{77} \theta^{-1}(0,1) \\
= 3 \times_{77} 56 \\
= 14
\end{aligned}
$$

The actual sum is 91. Our result is incorrect, since 91 is nct in $\mathbb{Z}_{77}$. Notice that 91 and 14 differ by 77. Any error that we get using this technique will be a multiple of 77.

15.2 · Cosets and Factor Groups

· **Exercises**

15.2.1. Answer. An example of a valid correct answer: Call the subsets A and B respectively. If we choose $0 \in A$ and $5 \in B$ we get $0 +_{10} 5 = 5 \in B$. On the other hand, if we choose $3 \in A$ and $8 \in B$, we get $3 +_{10} 8 = 1 \in A$. Therefore, the induced operation is not well defined on $\{A, B\}$.

15.2.3. Answer.

(a) The four distinct cosets in G/H are $H = \{(0,0),(2,0)\}$, $(1,0)+H = \{(1,0),(3,0)\}$, $(0,1)+H = \{(0,1),(2,1)\}$, and $(1,1)+H = \{(1,1),(3,1)\}$. None of these cosets generates G/H; therefore G/H is not cyclic. Hence G/H must be isomorphic to $\mathbb{Z}_2 \times \mathbb{Z}_2$.

(b) The factor group is isomorphic to $[\mathbb{R};+]$. Each coset of $\mathbb{R}$ is a line in the complex plane that is parallel to the x-axis: $\tau : \mathbb{C}/\mathbb{R} \to \mathbb{R}$, where $T(\{a+bi \mid a \in \mathbb{R}\}) = b$ is an isomorphism.

(c) $\langle 8 \rangle = \{0,4,8,12,16\} \Rightarrow |\mathbb{Z}_{20}/\langle 8 \rangle| = 4$. The four cosets are: $\bar{0}, \bar{1}, \bar{2}$, and $\bar{3}$. $\bar{1}$ generates all four cosets. The factor group is isomorphic to $[\mathbb{Z}_4;+_4]$ because $\bar{1}$ is a generator.

15.2.5. Answer.

$$a * H = b * H \Leftrightarrow a \in bH$$
$$\Leftrightarrow a = b * h \text{ for some } h \in H$$
$$\Leftrightarrow b^{-1} * a = h \text{ for some } h \in H$$
$$\Leftrightarrow b^{-1} * a \in H$$

15.3 · Permutation Groups
15.3.5 · Exercises

15.3.5.1. Answer.

(a) $\begin{pmatrix} 1 & 2 & 3 & 4 \\ 1 & 4 & 3 & 2 \end{pmatrix}$

(b) $\begin{pmatrix} 1 & 2 & 3 & 4 \\ 4 & 3 & 1 & 2 \end{pmatrix}$

(c) $\begin{pmatrix} 1 & 2 & 3 & 4 \\ 3 & 4 & 2 & 1 \end{pmatrix}$

(d) $\begin{pmatrix} 1 & 2 & 3 & 4 \\ 3 & 4 & 2 & 1 \end{pmatrix}$

(e) $\begin{pmatrix} 1 & 2 & 3 & 4 \\ 4 & 2 & 1 & 3 \end{pmatrix}$

(f) $\begin{pmatrix} 1 & 2 & 3 & 4 \\ 3 & 1 & 4 & 2 \end{pmatrix}$

(g) $\begin{pmatrix} 1 & 2 & 3 & 4 \\ 2 & 1 & 4 & 3 \end{pmatrix}$

15.3.5.3. Answer. S_3/A_3 is a group of order two. The operation on left cosets of $H = \langle f_1 \rangle$ is not well defined and so a group cannot be formed from left cosets of H.

15.3.5.5. Answer. $\mathcal{D}_4 = \{i, r, r^2, r^3, f_1, f_2, f_3, f_4\}$ Where i is the identity function, $r = \begin{pmatrix} 1 & 2 & 3 & 4 \\ 2 & 3 & 4 & 1 \end{pmatrix}$, and

$$f_1 = \begin{pmatrix} 1 & 2 & 3 & 4 \\ 4 & 3 & 2 & 1 \end{pmatrix} \quad f_2 = \begin{pmatrix} 1 & 2 & 3 & 4 \\ 2 & 1 & 4 & 3 \end{pmatrix}$$
$$f_3 = \begin{pmatrix} 1 & 2 & 3 & 4 \\ 3 & 2 & 1 & 4 \end{pmatrix} \quad f_4 = \begin{pmatrix} 1 & 2 & 3 & 4 \\ 1 & 4 & 3 & 2 \end{pmatrix}$$

The operation table for the group is

∘	i	r	r^2	r^3	f_1	f_2	f_3	f_4
i	i	r	r^2	r^3	f_1	f_2	f_3	f_4
r	r	r^2	r^3	i	f_4	f_3	f_1	f_2
r^2	r^2	r^3	i	r	f_2	f_1	f_4	f_3
r^3	r^3	i	r	r^2	f_3	f_4	f_2	f_1
f_1	f_1	f_3	f_2	f_4	i	r^2	r	r^3
f_2	f_2	f_4	f_1	f_3	r^2	i	r^3	r
f_3	f_3	f_2	f_4	f_1	r^3	r	i	r^2
f_4	f_4	f_1	f_3	f_2	r	r^3	r^2	i

A lattice diagram of its subgroups is

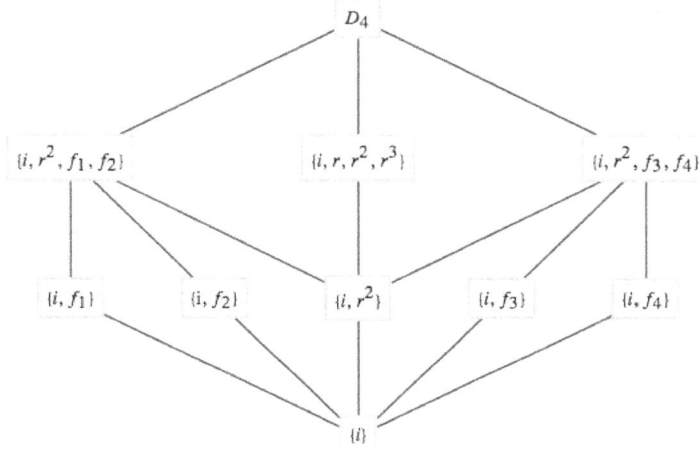

Figure B.0.7 Subgroups of $\mathcal{D}_4$

All proper subgroups are cyclic except $\{i, r^2, f_1, f_2\}$ and $\{i, r^2, f_3, f_4\}$. Each 2-element subgroup is isomorphic to $\mathbb{Z}_2$; $\{i, r, r^2, r^3\}$ is isomorphic to $\mathbb{Z}_4$; and $\{i, r^2, f_1, f_2\}$ and $\{i, r^2, f_3, f_4\}$ are isomorphic to $\mathbb{Z}_2 \times \mathbb{Z}_2$.

15.3.5.7. Answer. One solution is to cite Exercise 3 at the end of Section 11.3. It can be directly applied to this problem. An induction proof of the problem at hand would be almost identical to the proof of the more general statement. $(t_1 t_2 \cdots t_r)^{-1} = t_r^{-1} \cdots t_2^{-1} t_1^{-1}$ by Exercise 3 of Section 11.3

$= t_r \cdots t_2 t_1$ since each transposition inverts itself. ∎

15.3.5.9. Answer. Part I: That $|S_k| = k!$ follows from the Rule of Products.

Part II: Let f be the function defined on $\{1, 2, ..., n\}$ by $f(1) = 2$, $f(2) = 3$, $f(3) = 1$, and $f(j) = j$ for $4 \leq j \leq n$; and let g be defined by $g(1) = 1$, $g(2) = 3$, $g(3) = 2$, and $g(j) = j$ for $4 \leq j \leq n$. Note that f and g are elements of S_n. Next, $(f \circ g)(1) = f(g(1)) = f(1) = 2$, while $(g \circ f)(1) = g(f(1)) = g(2) = 3$, hence $f \circ g \neq g \circ f$ and S_n is non-abelian for any $n \geq 3$.

15.3.5.13. Answer.

(a) Both groups are non-abelian and of order 6; so they must be isomorphic, since only one such group exists up to isomorphism. The function θ :

$$S_3 \to R_3 \text{ defined by } \begin{array}{ll} \theta(i) = I & \theta(f_1) = F_1 \\ \theta(r_1) = R_1 & \theta(f_2) = F_2 \\ \theta(r_2) = R_2 & \theta(f_3) = F_3 \end{array} \text{ is an isomorphism,}$$

(b) Recall that since every function is a relation, it is natural to translate functions to Boolean matrices. Suppose that $f \in S_n$. We will define its image, $\theta(f)$, by

$$\theta(f)_{kj} = 1 \Leftrightarrow f(j) = k$$

That θ is a bijection follows from the existence of θ^{-1}. If A is a rook matrix,

$$\theta^{-1}(A)(j) = k \Leftrightarrow \text{The 1 in column } j \text{ of } A \text{ appears in row } k$$
$$\Leftrightarrow A_{kj} = 1$$

For $f, g \in S_n$,

$$\begin{aligned}\theta(f \circ g)_{kj} = 1 &\Leftrightarrow (f \circ g)(j) = k \\ &\Leftrightarrow \exists l \text{ such that } g(j) = l \text{ and } f(l) = k \\ &\Leftrightarrow \exists l \text{ such that } \theta(g)_{lj} = 1 \text{ and } \theta(f)_{kl} = 1 \\ &\Leftrightarrow (\theta(f)\theta(g))_{kj} = 1\end{aligned}$$

Therefore, θ is an isomorphism.

15.4 · Normal Subgroups and Group Homomorphisms
15.4.3 · Exercises
15.4.3.1. Answer.

(a) Yes, the kernel is $\{1, -1\}$

(b) No, since $\theta_2 (2 +_5 4) = \theta_2(1) = 1$, but $\theta_2(2) +_2 \theta_2(4) = 0 +_2 0 = 0$

A follow-up might be to ask what happens if 5 is replaced with some other positive integer in this part.

(c) Yes, the kernel is $\{(a, -a) | a \in \mathbb{R}\}$

(d) No. A counterexample, among many, would be to consider the two transpositions $t_1 = (1,3)$ and $t_2 = (1,2)$. Compare $\theta_4(t_1 \circ t_2)$ and $\theta_4(t_1) \circ \theta_4(t_2)$.

15.4.3.3. Answer. $\langle r \rangle = \{i, r, r^2, r^3\}$ is a normal subgroup of D_4. To see you could use the table given in the solution of Exercise 15.3.5.5, p. 170 of Section 15.3 and verify that $a^{-1}ha \in \langle r \rangle$ for all $a \in D_4$ and $h \in \langle r \rangle$. A more efficient approach is to prove the general theorem that if H is a subgroup G with exactly two distinct left cosets, than H is normal. $\langle f_1 \rangle$ is not a normal subgroup of D_4. $\langle f_1 \rangle = \{i, f_1\}$ and if we choose $a = r$ and $h = f_1$ then $a^{-1}ha = r^3 f_1 r = f_2 \notin \langle f_1 \rangle$

15.4.3.5. Answer. $(\beta \circ \alpha)(a_1, a_2, a_3) = 0$ and so $\beta \circ \alpha$ is the trivial homomorphism, but a homomorphism nevertheless.

15.4.3.7. Answer. Let $x, y \in G$.

$$\begin{aligned} q(x * y) &= (x * y)^2 \\ &= x * y * x * y \\ &= x * x * y * y \quad \text{since } G \text{ is abelian} \\ &= x^2 * y^2 \\ &= q(x) * q(y) \end{aligned}$$

Hence, q is a homomorphism. In order for q to be an isomorphism, it must be the case that no element other than the identity is its own inverse.

$$\begin{aligned} x \in \text{Ker}(q) &\Leftrightarrow q(x) = e \\ &\Leftrightarrow x * x = e \\ &\Leftrightarrow x^{-1} = x \end{aligned}$$

15.4.3.9. Answer. Proof: Recall that the inverse image of H' under θ is $\theta^{-1}(H') = \{g \in G | \theta(g) \in H'\}$.

Closure: Let $g_1, g_2 \in \theta^{-1}(H')$, then $\theta(g_1), \theta(g_2) \in H'$. Since H' is a subgroup of G',

$$\theta(g_1) \diamond \theta(g_2) = \theta(g_1 * g_2) \in H' \Rightarrow g_1 * g_2 \in \theta^{-1}(H')$$

Identity: By Theorem 15.4.14, p. 173(a), $e \in \theta^{-1}(H')$.

Inverse: Let $a \in \theta^{-1}(H')$. Then $\theta(a) \in H'$ and by Theorem 15.4.14, p. 173(b), $\theta(a)^{-1} = \theta(a^{-1}) \in H'$ and so $a^{-1} \in \theta^{-1}(H')$.

15.5 · Coding Theory, Linear Codes
15.5.4 · Exercises

15.5.4.1. Answer.

(a) Error detected, since an odd number of 1's was received; ask for retransmission.

(b) No error detected; accept this block.

(c) No error detected; accept this block.

15.5.4.3. Answer.

(a) Syndrome $= (1, 0, 1)$. Corrected coded message is $(1, 1, 0, 0, 1, 1)$ and original message was $(1, 1, 0)$.

(b) Syndrome $= (1, 1, 0)$. Corrected coded message is $(0, 0, 1, 0, 1, 1)$ and original message was $(0, 0, 1)$.

(c) Syndrome $= (0, 0, 0)$. No error, coded message is $(0, 1, 1, 1, 1, 0)$ and original message was $(0, 1, 1)$.

(d) Syndrome $= (1, 1, 0)$. Corrected coded message is $(1, 0, 0, 1, 1, 0)$ and original message was $(1, 0, 0)$.

(e) Syndrome $= (1, 1, 1)$. This syndrome occurs only if two bits have been switched. No reliable correction is possible.

(f) Syndrome $= (0, 1, 0)$. Corrected coded message is $(1, 0, 0, 1, 1, 0)$ and original message was $(1, 0, 0)$.

15.5.4.5. Answer.

(a) Blocks of two bits are encoded into code words of length 4.

(b) The code words are 0000, 1010, 0111 and 1101.

(c) Since the first two code words have a Hamming distance of 2, not all single bit errors can be corrected. For example, if 0000 is transmitted and the first bit is switch, then 1000 is received and we can't tell for sure whether this came from 0000 or 1010. To see what can be corrected, we note that $a_1 a_2$ is encoded to $a_1 a_2 (a_1 +_2 a_2) a_2$ and so if $b_1 b_2 b_3 b_4$ is recieved and no error has occurred,

$$b_1 +_2 b_2 +_2 b_3 = 0$$
$$b_2 +_2 b_4 = 0$$

We can extract the parity check matrix from this set of equations. It is

$$\begin{pmatrix} 1 & 0 \\ 1 & 1 \\ 1 & 0 \\ 0 & 1 \end{pmatrix}$$

The rows of this matrix correspond with the syndromes for errors in bits 1 through 4, which are all nonzero, so we can detect any single bit error. Notice that the syndromes for bits 1 and 3 are identical. This reflects the fact that errors in these bits can't be corrected. However, the syndromes for bits 2 and 4 are unique and so we can correct them. Therefore the second bit of the original message can be sent with more confidence than the first.

15.5.4.7. Solution. Yes, you can correct all single bit errors because the parity check matrix for the expanded code is

$$H = \begin{pmatrix} 1 & 1 & 0 \\ 1 & 0 & 0 \\ 1 & 0 & 1 \\ 0 & 1 & 0 \\ 0 & 1 & 1 \\ 0 & 0 & 1 \end{pmatrix}.$$

Since each possible syndrome of single bit errors is unique we can correct any error.

15.5.4.8. Hint. There is a parity check equation for each parity bit.

16 · An Introduction to Rings and Fields
16.1 · Rings, Basic Definitions and Concepts
16.1.6 · Exercises

16.1.6.1. Answer. All but ring d are commutative. All of the rings have a unity element. The number 1 is the unity for all of the rings except d. The unity for $M_{2\times 2}(\mathbb{R})$ is the two by two identity matrix. The units are as follows:

(a) $\{1, -1\}$

(b) $\mathbb{C}^*$

(c) $\mathbb{Q}^*$

(d) $\{A \mid A_{11} A_{22} - A_{12} A_{21} \neq 0\}$

(e) $\{1\}$

16.1.6.3. Answer.

(a) Consider commutativity

(b) Solve $x^2 = 3x$ in both rings.

16.1.6.5. Answer.

(a) We already know that $3\mathbb{Z}$ is a subgroup of the group $\mathbb{Z}$. We need only show that $3\mathbb{Z}$ is closed with respect to multiplication. Let $3m, 3n \in 3\mathbb{Z}$. $(3m)(3n) = 3(3mn) \in 3\mathbb{Z}$, since $3mn \in \mathbb{Z}$.

(b) The proper subrings are $\{0, 2, 4, 6\}$ and $\{0, 4\}$; while $\{0\}$ and $\mathbb{Z}_8$ are improper subrings.

(c) The proper subrings are $\{00, 01\}$, $\{00, 10\}$, and $\{00, 11\}$: while $\{00\}$ and $\mathbb{Z}_2 \times \mathbb{Z}_2$ are improper subrings.

16.1.6.7. Answer.

(a) The left-hand side of the equation factors into the product $(x-2)(x-3)$. Since $\mathbb{Z}$ is an integral domain, $x = 2$ and $x = 3$ are the only possible solutions.

(b) Over $\mathbb{Z}_{12}$, 2, 3, 6, and 11 are solutions. Although the equation factors into $(x-2)(x-3)$, this product can be zero without making x either 2 or 3. For example. If $x = 6$ we get $(6-2) \times_{12} (6-3) = 4 \times_{12} 3 = 0$. Notice that 4 and 3 are zero divisors.

16.1.6.9. Answer. Let R_1, R_2, and R_3 be any rings, then

(a) R_1 is isomorphic to R_1 and so "is isomorphic to" is a reflexive relation on rings.

(b) R_1 is isomorphic to $R_2 \Rightarrow R_2$ is isomorphic to R_1, and so "is isomorphic to" is a symmetric relation on rings,

(c) R_1 is isomorphic to R_2, and R_2 is isomorphic to R_3 implies that R_1 is isomorphic to R_3, and so "is isomorphic to" is a transitive relation on rings.

We haven't proven these properties here, just stated them. The combination of these observations implies that "is isomorphic to" is an equivalence relation on rings.

16.1.6.11. Answer.

(a) Commutativity is clear from examination of a multiplication table for $\mathbb{Z}_2 \times \mathbb{Z}_3$. More generally, we could prove a theorem that the direct product of two or more commutative rings is commutative. $(1, 1)$ is the unity of $\mathbb{Z}_2 \times \mathbb{Z}_3$.

(b) $\{(m, n) | m = 0 \text{ or } n = 0, (m, n) \neq (0, 0)\}$

(c) Another example is $\mathbb{Z} \times \mathbb{Z}$. You never get an integral domain in this situation. By the definition an integral domain D must contain a "zero" so we always have $(1, 0) \cdot (0, 1) = (0, 0)$ in $D \times D$.

16.1.6.13. Answer.

(a) $(a+b)(c+d) = (a+b)c + (a+b)d = ac + bc + ad + bd$

(b)
$$(a+b)(a+b) = aa + ba + ab + bb \quad \text{by part a}$$
$$= aa + ab + ab + bb \quad \text{since } R \text{ is commutative.}$$
$$= a^2 + 2ab + b^2$$

16.2 · Fields
· Exercises

16.2.5. Answer.

(a) 0 in $\mathbb{Z}_2$, 1 in $\mathbb{Z}_3$, 3 in $\mathbb{Z}_5$

(b) 2 in $\mathbb{Z}_3$, 3 in $\mathbb{Z}_5$

(c) 2 in $\mathbb{Z}_5$

16.2.7. Answer.
(a) 0 and 1   (b) 1   (c) 1   (d) none

16.3 · Polynomial Rings
· Exercises

16.3.1. Answer.

(a) $f(x) + g(x) = 2 + 2x + x^2$, $f(x) \cdot g(x) = 1 + 2x + 2x^2 + x^3$

(b) $f(x) + g(x) = x^2$, $f(x) \cdot g(x) = 1 + x^3$

(c) $1 + 3x + 4x^2 + 3x^3 + x^4$

(d) $1 + x + x^3 + x^4$

(e) $x^2 + x^3$

16.3.3. Answer.

(a) If $a, b \in \mathbb{R}$, $a - b$ and ab are in $\mathbb{R}$ since $\mathbb{R}$ is a ring in its own right. Therefore, $\mathbb{R}$ is a subring of $\mathbb{R}[x]$. The proofs of parts b and c are similar.

16.3.5. Answer.

(a) Reducible, $(x+1)(x^2+x+1)$

(b) Reducible, $x(x^2+x+1)$

(c) Irreducible. If you could factor this polynomial, one factor would be either x or $x+1$, which would give you a root of 0 or 1, respectively. By substitution of 0 and 1 into this polynomial, it clearly has no roots.

(d) Reducible, $(x+1)^4$

16.3.7. Answer. We illustrate this property of polynomials by showing that it is not true for a nonprime polynomial in $\mathbb{Z}_2[x]$. Suppose that $p(x) = x^2 + 1$, which can be reduced to $(x+1)^2$, $a(x) = x^2 + x$, and $b(x) = x^3 + x^2$. Since $a(x)b(x) = x^5 + x^3 = x^3(x^2+1)$, $p(x)|a(x)b(x)$. However, $p(x)$ is not a factor of either $a(x)$ or $b(x)$.

16.3.9. Answer. The only possible proper factors of $x^2 - 3$ are $(x - \sqrt{3})$ and $(x + \sqrt{3})$, which are not in $\mathbb{Q}[x]$ but are in $\mathbb{R}[x]$.

16.3.11. Answer. For $n \geq 0$, let $S(n)$ be the proposition: For all $g(x) \neq 0$ and $f(x)$ with $\deg f(x) = n$, there exist unique polynomials $q(x)$ and $r(x)$ such that $f(x) = g(x)q(x) + r(x)$, and either $r(x) = 0$ or $\deg r(x) < \deg g(x)$.

Basis: $S(0)$ is true, for if $f(x)$ has degree 0, it is a nonzero constant, $f(x) = c \neq 0$, and so either $f(x) = g(x) \cdot 0 + c$ if $g(x)$ is not a constant, or $f(x) = g(x)g(x)^{-1} + 0$ if $g(x)$ is also a constant.

Induction: Assume that for some $n \geq 0$, $S(k)$ is true for all $k \leq n$. If $f(x)$ has degree $n+1$, then there are two cases to consider. If $\deg g(x) > n+1$, $f(x) = g(x) \cdot 0 + f(x)$, and we are done. Otherwise, if $\deg g(x) = m \leq n+1$, we perform long division as follows, where LDT's stand for terms of lower degree than $n+1$.

$$g_m x^m + \text{LDT's} \overline{\smash{)}\begin{matrix} f_{n+1} \cdot g_m^{-1} x^{n+1-m} \\ f_{n+1} x^{n+1} + \text{LDT's} \\ \underline{f_{n+1} x^{n+1} + \text{LDT's}} \\ h(x) \end{matrix}}$$

Therefore,

$$h(x) = f(x) - \left(f_{n+1} \cdot g_m^{-1} x^{n+1-m}\right) g(x) \Rightarrow f(x) = \left(f_{n+1} \cdot g_m^{-1} x^{n+1-m}\right) g(x) + h(x)$$

Since $\deg h(x)$ is less than $n+1$, we can apply the induction hypothesis: $h(x) = g(x)q(x) + r(x)$ with $\deg r(x) < \deg g(x)$.

Therefore,

$$f(x) = g(x)\left(f_{n+1} \cdot g_m^{-1} x^{n+1-m} + q(x)\right) + r(x)$$

with $\deg r(x) < \deg g(x)$. This establishes the existence of a quotient and remainder. The uniqueness of $q(x)$ and $r(x)$ as stated in the theorem is proven as follows: if $f(x)$ is also equal to $g(x)\bar{q}(x) + \bar{r}(x)$ with $\deg \bar{r}(x) < \deg g(x)$, then

$$g(x)q(x) + r(x) = g(x)\bar{q}(x) + \bar{r}(x) \Rightarrow g(x)\left(\bar{q}(x) - q(x)\right) = r(x) - \bar{r}(x)$$

Since $\deg r(x) - \bar{r}(x) < \deg g(x)$, the degree of both sides of the last equation is less than $\deg g(x)$. Therefore, it must be that $\bar{q}(x) - q(x) = 0$, or $q(x) = \bar{q}(x)$ And so $r(x) = \bar{r}(x)$.

16.4 · Field Extensions

· Exercises

16.4.1. Answer. If $a_0 + a_1\sqrt{2} \in \mathbb{Q}\left[\sqrt{2}\right]$ is nonzero, then it has a multiplicative inverse:

$$\frac{1}{a_0 + a_1\sqrt{2}} = \frac{1}{a_0 + a_1\sqrt{2}} \frac{a_0 - a_1\sqrt{2}}{a_0 - a_1\sqrt{2}}$$

$$= \frac{a_0 - a_1\sqrt{2}}{a_0^2 - 2a_1^2}$$

$$= \frac{a_0}{a_0^2 - 2a_1^2} - \frac{a_1}{a_0^2 - 2a_1^2}\sqrt{2}$$

The denominator, $a_0^2 - 2a_1^2$, is nonzero since $\sqrt{2}$ is irrational. Since $\frac{a_0}{a_0^2 - 2a_1^2}$ and $\frac{-a_1}{a_0^2 - 2a_1^2}$ are both rational numbers, $a_0 + a_1\sqrt{2}$ is a unit of $\mathbb{Q}\left[\sqrt{2}\right]$. The field containing $\mathbb{Q}\left[\sqrt{2}\right]$ is denoted $\mathbb{Q}\left(\sqrt{2}\right)$ and so $\mathbb{Q}\left(\sqrt{2}\right) = \mathbb{Q}\left[\sqrt{2}\right]$

16.4.3. Answer. $x^4 - 5x^2 + 6 = (x^2 - 2)(x^2 - 3)$ has zeros $\pm\sqrt{2}$ and $\pm\sqrt{3}$.

$\mathbb{Q}(\sqrt{2}) = \{a + b\sqrt{2} \mid a, b \in \mathbb{Q}\}$ contains the zeros $\pm\sqrt{2}$ but does not contain $\pm\sqrt{3}$, since neither are expressible in the form $a + b\sqrt{2}$. If we consider the set $\{c + d\sqrt{3} \mid c, d \in \mathbb{Q}(\sqrt{2})\}$, then this field contains $\pm\sqrt{3}$ as well as $\pm\sqrt{2}$, and is denoted $\mathbb{Q}(\sqrt{2})(\sqrt{3}) = \mathbb{Q}(\sqrt{2}, \sqrt{3})$. Taking into account the form of c and d in the description above, we can expand to

$$\mathbb{Q}(\sqrt{2}, \sqrt{3}) = \{b_0 + b_1\sqrt{2} + b_2\sqrt{3} + b_3\sqrt{6} \mid b_i \in \mathbb{Q}\}$$

16.4.5. Answer.

(a) $f(x) = x^3 + x + 1$ is reducible if and only if it has a factor of the form $x - a$. By Theorem 16.3.14, p. 202, $x - a$ is a factor if and only if a is a zero. Neither 0 nor 1 is a zero of $f(x)$ over $\mathbb{Z}_2$.

(b) Since $f(x)$ is irreducible over $\mathbb{Z}_2$, all zeros of $f(x)$ must lie in an extension field of $\mathbb{Z}_2$. Let c be a zero of $f(x)$. $\mathbb{Z}_2(c)$ can be described several different ways. One way is to note that since $c \in \mathbb{Z}_2(c)$, $c^n \in \mathbb{Z}_2(c)$ for all n. Therefore, $\mathbb{Z}_2(c)$ includes $0, c, c^2, c^3, \ldots$. But $c^3 = c+1$ since $f(c) = 0$. Furthermore, $c^4 = c^2 + c$, $c^5 = c^2 + c + 1$, $c^6 = c^2 + 1$, and $c^7 = 1$. Higher powers of c repeat preceding powers. Therefore,

$$\mathbb{Z}_2(c) = \{0, 1, c, c^2, c+1, c^2+1, c^2+c+1, c^2+c\}$$
$$= \{a_0 + a_1 c + a_2 c^2 \mid a_i \in \mathbb{Z}_2\}$$

The three zeros of $f(x)$ are c, c^2 and $c^2 + c$.

$$f(x) = (x+c)\left(x+c^2\right)\left(x+c^2+c\right)$$

(c) Cite Theorem Theorem 16.2.10, p. 197, part 3.

16.5 · Power Series
16.5.3 · Exercises

16.5.3.5. Answer.

(a)
$$b_0 = 1$$
$$b_1 = (-1)(2 \cdot 1) = -2$$
$$b_2 = (-1)(2 \cdot (-2) + 4 \cdot 1) = 0$$
$$b_3 = (-1)(2 \cdot 0 + 4 \cdot (-2) + 8 \cdot 1) = 0$$

All other terms are zero. Hence, $f(x)^{-1} = 1 - 2x$

(b)
$$f(x) = 1 + 2x + 2^2 x^2 + 2^3 x^3 + \cdots$$
$$= (2x)^0 + (2x)^1 + (2x)^2 + (2x)^3 + \cdots$$
$$= \frac{1}{1-2x}$$

The last step follows from the formula for the sum of a geometric series.

16.5.3.7. Answer.

(a)
$$\begin{aligned}\left(x^4 - x^5\right)^{-1} &= (x^4(1-x))^{-1}\\ &= x^{-4}\frac{1}{1-x}\\ &= x^{-4}\left(\sum_{k=0}^{\infty} x^k\right).\\ &= \sum_{k=-4}^{\infty} x^k\end{aligned}$$

(b)
$$\begin{aligned}\left(x^4 - 2x^3 + x^2\right)^{-1} &= \left(x^2\left(x^2 - 2x + 1\right)\right)^{-1}\\ &= x^{-2}\left(1 - 2x + x^2\right)^{-1}\\ &= x^{-2}\left(\sum_{k=0}^{\infty}(k+1)x^k\right).\\ &= \sum_{k=-2}^{\infty}(k+2)x^k\end{aligned}$$

Appendix C

Notation

The following table defines the notation used in this book. Page numbers or references refer to the first appearance of each symbol.

Symbol	Description	Page
$a \equiv_m b$	a is congruent to b modulo m	12
Q_n	the n-cube	17
$*$	generic symbol for a binary operation	22
$string1 + string2$	The concatenation of $string1$ and $string2$	25
$[G; *]$	a group with elements G and binary operation $*$	27
$\gcd(a, b)$	the greatest common divisor of a and b	35
$a +_n b$	the mod n sum of a and b	39
$a \times_n b$	the mod n product of a and b	39
$\mathbb{Z}_n$	The Additive Group of Integer Modulo n	40
$\mathbb{U}_n$	The Multiplicative Group of Integer Modulo n	40
$W \leq V$	W is a subsystem of V	44
$\langle a \rangle$	the cyclic subgroup generated by a	47
$ord(a)$	Order of a	47
$V_1 \times V_2 \times \cdots \times V_n$	The direct product of algebraic structures $V_1, V_2, \ldots, V_n$	49
$G_1 \times G_2$	The direct product of groups G_1 and G_2	49
$G_1 \cong G_2$	G_1 is isomorphic to G_2	57
		78
$dim(V)$	The dimension of vector space V	81
$\mathbf{0}$	least element in a poset	104
$\mathbf{1}$	greatest element in a poset	104
D_n	the set of divisors of integer n	105
$a \vee b$	the join, or least upper bound of a and b	107
$a \wedge b$	the meet, or greatest lower bound of a and b	107
$[L; \vee, \wedge]$	A lattice with domain having meet and join operations	107
$\bar{a}$	The complement of lattice element a	110
$[B; \vee, \wedge, \bar{}]$	a boolean algebra with operations join, meet and complementation	110
		111

(Continued on next page)

Symbol	Description	Page
$M_{\delta_1 \delta_2 \cdots \delta_k}$	the minterm generated by $x_1, x_2, \ldots, x_k$, where $y_i = x_i$ if $\delta_i = 1$ and $y_i = \bar{x}_i$ if $\delta_i = 0$	119
A^*	The set of all strings over an alphabet A	132
A^n	The set of all strings of length n over an alphabet A	132
λ	The empty string	132
$s_1 + s_2$	The concatenation of strings s_1 and s_2	133
$L(G)$	Language created by phrase structure grammar G	135
(S, X, Z, w, t)	A finite-state machine with states S, input alphabet X, output alphabet X, and output function w and next-state function t	139
$m(M)$	The machine of monoid M	146
		149
$a * H, H * a$	the left and right cosets generated by a	156
G/H	The factor group G mod H.	160
S_A	The group of permutations of the set A	162
S_n	The group of permutations on a set with n elements	162
A_n	The Alternating Group	165
$\mathcal{D}_n$	The nth dihedral group	169
$H \triangleleft G$	H is a normal subgroup of G	172
$ker\theta$	the kernel of homomorphism θ	174
$d_H(a, b)$	Hamming distance between a and b	178
$[R; +, \cdot]$	a ring with domain R and operations $+$ and $\cdot$	187
$U(R)$	the set of units of a ring R	189
		190
		192
D	a generic integral domain	193
$\deg f(x)$	the degree of polynomial $f(x)$	199
$R[x]$	the set of all polynomials in x over R	199
$R[[x]]$	the set of all powers series in R	208
$\grave{x}, \acute{x}$	pre and post values of a variable x	217

Appendix D

Glossary

An Informal Glossary of Terms

Many of the words in this glossary are not formally defined in the book either because they are viewed as prerequisites to a course in discrete mathematics or are terms in computer science that some students may be unfamiliar with.

An. When referring to "an entity" we mean that the object can be any of the elements is some set. For example, if you say that n is an integer, it could be any integer.

Bit. The smallest unit of computer memory, normally represented as a 0 or 1.

Byte. A basic unit of computer memory containing eight Bit, p. 259s, normally modeled as a sequence of eight 0's and 1's.

Complex Number. A number of the form $a + bi$, where a and b are real numbers and $i^2 = -1$.

Composite Integer. A positive integer is composite if it is greater than one and is the product of two positive integers greater than one. For example, 10 (equal to $2 \cdot 5$) is composite. Any positive integer greater than one that is not composite is Prime, p. 260.

Constant. A numerical value that is unchanging . The value might be unknown and it still may be represented with a symbol. For example if we are discussing the process of sorting a file of N numbers, N is considered a constant with respect to the sorting algorithm. Constants can become variables, p. 260 though. If we have designed a sorting algorithm, and want to analyze its efficiency, we would consider N to be a variable.

Creative Commons. An organization which has created several open licenses for creative works such as *Applied Discrete Structures*.

Data Structure. A format for organizing, processing, retrieving and storing data.

Distinct. Two entities are distinct if they are not the same. For example, any two student ID numbers at a school should be distinct. If not, confusion could ensue. See also Unique, p. 260.

Even Integer. Any Integer, p. 259 that is equal to two times an integer. That includes 0, since $0 = 2 \cdot 0$.

Factor. If an algebraic expression is the product of several expressions, each of those expressions is a factor.

Iff. Shorthand for "if and only if"

Integer. Whole number, whether positive, negative or zero.

Irrational Number. A number that is not equal to any fraction. $\sqrt{2}$ is one we prove to be irrational in the book.

LaTeX. A markup language used for books and papers with lots of mathematics, which is built on TeX. PreTeXt uses LaTeX as an intermediate format to produce PDF and print output.

Multiples. Multiples of a number c are $\ldots, -3c, {}^-2c, -c, 0, c, 2c, 3c, \ldots$

Natural Numbers. In this book, its the numbers $0,1,2,3,4,\ldots$. There isn't 100% agreement here. Some people say its the numbers $1,2,3,4,\ldots$. We call those numbers the positive integers. The symbol we use of the natural numbers is $\mathbb{N}$. There is no consistent definition of positive complex numbers.

Nonnegative Number. A number that is either positive or zero.

Odd Integer. An integer n is odd if there exists an integer k so that $n = 2k+1$. Any integer that is not even is odd.

Positive Number. A positive number is a number that is greater than zero. Normally visualized as being to the right of zero on a conventional number line. The set of positive integers is denoted $\mathbb{P}$. The sets of positive rational and real numbers are denoted $\mathbb{Q}^+$ and $\mathbb{R}?^+$, respectively

Powers. Powers of a nonzero number c are $\ldots, c^{-3}, c^{-2}, c^{-1}, 1, c^1, c^2, c^3, \ldots$. Recall that $c^0 = 1$.

PreTeXt. An authoring and publishing system for authors of textbooks, research articles, and monographs, especially in STEM disciplines. *Applied Discrete Structures* is produced using PreTeXt.

Prime. A positive integer that is divisible by exactly two positive integers, itself and 1. One is not prime, but 2 is the oddest prime because it's even. See also Composite Integer, p. 259.

Queue. A conventional waiting line, with the first come-first serve service rule. A queue is a common Data Structure, p. 259 in computer science. See also Stack, p. 260.

Rational Number. Any real number that is equal to a quotient two integers, a/b, with $b \neq 0$.

Real Number. For the purposes of this book, think of the numbers on a standard number line. All of the points make up the set of real numbers.

SageMath. An open source computer algebra system for a wide range of symbolic and numerical mathematical computations. Originally named simply Sage.

Stack. A Data Structure, p. 259 similar to a queue, but where the last come-first serve service rule is used. This wouldn't be a fair waiting line rule, but it is a very useful data structure. See also Queue, p. 260

Subtraction. Subtraction is really addition of the negation of a number: $a - b = a + (-b)$.

Term. If an algebraic expression is the sum of several expressions, each of those expressions is a term. For example there are three terms in the expression $2y + x - (w+1)/2$. Note that subtraction is considered the same as addition here.

Unique. We say a mathematical entity is unique when there's nothing else like it. For example, the solution, $x = 3$ to the equation $2x + 1 = 7$ is unique. No other number solves the equation. See also Distinct, p. 259.

Variable. A quantity whose value that can vary within a specified set. Normally represented by an algebraic symbol. For discrete variables it is customary to use the letters in the range from i to n, but this isn't a rigid rule. Letters at the end of the alphabet are traditionally used for continuous variables.

References

Many of the references listed here were used in preparing the original 1980's version of this book. In most cases, the mathematics that they contain is still worth reading for further background. Many can be found online, in university libraries or used bookstores. A few more current references have been added.

[1] Allenby, R.B.J.T, *Rings, Fields and Groups*, Edward Arnold, 1983.

[2] Appel, K., and W. Haken, *Every Planar Map Is 4-colorable*, Bull, Am. Math. Soc. no. 82 (1976): 711–12.
This has historical significance in that it announced the first correct proof of the Four Color Theorem

[3] Austin, A. Keith, *An Elementary Approach to NP-Completeness* American Math. Monthly 90 (1983): 398-99.

[4] Beardwood, J., J. H. Halton, and J. M. Hammersley, *The Shortest Path Through Many Points* Proc. Cambridge Phil. Soc. no. 55 (1959): 299–327.

[5] Ben-Ari, M, *Principles of Concurrent Programming*, Englewood Cliffs, NJ: Prentice-Hall, 1982.

[6] Berge, C, *The Theory of Graphs and Its Applications*, New York: Wiley, 1962.

[7] Bogart, Kenneth P, *Combinatorics Through Guided Discovery*, 2005.
This book may be freely downloaded and redistributed at http://www.math.dartmouth.edu/news-resources/electronic/kpbogart/ under the terms of the GNU Free Documentation License (FDL), as published by the Free Software Foundation.

[8] Busacker, Robert G., and Thomas L. Saaty, *Finite Graphs and Networks*, New York: McGraw-Hill, 1965.

[9] Connell, Ian, *Modern Algebra, A Constructive Introduction*, New York: North-Holland, 1982.

[10] Denning, Peter J., Jack B. Dennis, and Joseph L. Qualitz, *Machines, Languages, and Computation*, Englewood Cliffs, NJ: Prentice-Hall, 1978.

[11] Denning, Peter J, *Multigrids and Hypercubes*. American Scientist 75 (1987): 234-238.

[12] Dornhoff, L. L., and F. E. Hohn, *Applied Modern Algebra*, New York: Macmillan, 1978.

[13] Ford, L. R., Jr., and D. R. Fulkerson, *Flows in Networks*, Princeton, NJ: Princeton Univesity Press, 1962.

[14] Fraleigh, John B, *A First Course in Abstract Algebra*, 3rd ed. Reading, MA: Addison-Wesley, 1982.

[15] Gallian, Joseph A, *Contemporary Abstract Algebra*, D.C. Heath, 1986.

[16] Gallian, Joseph A, *Group Theory and the Design of a Letter-Facing Machine*, American Math. Monthly 84 (1977): 285-287.

[17] Hamming, R. W, *Coding and Information Theory*, Englewood Cliffs, NJ: Prentice-Hall, 1980.

[18] Hill, F. J., and G. R. Peterson, *Switching Theory and Logical Design*, 2nd ed. New York: Wiley, 1974.

[19] Hofstadter, D. R, *Godel, Escher, Bach: An Eternal Golden Braid*, New York: Basic Books, 1979.

[20] Hohn, F. E, *Applied Boolean Algebra*, 2nd ed. New York: Macmillan, 1966.

[21] Hopcroft, J. E., and J. D. Ullman, *Formal Languages and Their Relation to Automata*, Reading, MA: Addison-Wesley, 1969.

[22] Hu, T. C, *Combinatorial Algorithms*, Reading, MA: Addison-Wesley, 1982.

[23] Knuth, D. E, *The Art of Computer Programming. Vol. 1, Fundamental Algorithms*, 2nd ed. Reading, MA: Addison-Wesley, 1973.

[24] Knuth, D. E, *The Art of Computer Programming. Vol. 2, Seminumerical Algorithms*, 2nd ed., Reading, MA: Addison-Wesley, 1981.

[25] Knuth, D. E, *The Art of Computer Programming. Vol. 3, Sorting and Searching*, Reading, MA: Addison-Wesley, 1973.

[26] Knuth, D. E, *The Art of Computer Programming. Vol. 4A, Combinatorial Algorithms, Part 1*, Upper Saddle River, New Jersey: Addison-Wesley, 2011.
https://www-cs-faculty.stanford.edu/~knuth/taocp.html

[27] Kulisch, U. W., and Miranker, W. L, *Computer Arithmetic in Theory and Practice*, New York: Academic Press, 1981.

[28] Levin, Oscar, *Discrete Mathematics: An Open Introduction*, http://discrete.openmathbooks.org.

[29] Lipson, J. D, *Elements of Algebra and Algebraic Computing*, Reading, MA: Addison-Wesley, 1981.

[30] Liu, C. L, *Elements of Discrete Mathematics*, New York: McGraw-Hill, 1977.

[31] O'Donnell, *Analysis of Boolean Functions*.
A book about Fourier analysis of boolean functions that is being developed online in a blog.

[32] *The Omnificent English Dictionary In Limerick Form* .
The source of all limericks that appear at the beginning of most chapters.
https://www.oedilf.com/

[33] Ore, O, *Graphs and Their Uses*, New York: Random House, 1963.

[34] Parry, R. T., and H. Pferrer, *The Infamous Traveling-Salesman Problem: A Practical Approach* Byte 6 (July 1981): 252-90.

[35] Pless, V, *Introduction to the Theory of Error-Correcting Codes*, New York: Wiley-Interscience, 1982.

[36] Purdom, P. W., and C. A. Brown, *The Analysis of Algorithms*, Holt, Rinehart, and Winston, 1985.

[37] Quine, W. V, *The Ways of Paradox and Other Essays*, New York: Random

House, 1966.

[38] Ralston, A, *The First Course in Computer Science Needs a Mathematics Corequisite*, Communications of the ACM 27-10 (1984): 1002-1005.

[39] Solow, Daniel, *How to Read and Do Proofs*, New York: Wiley, 1982.

[40] Sopowit, K. J., E. M. Reingold, and D. A. Plaisted *The Traveling Salesman Problem and Minimum Matching in the Unit Square.*SIAM J. Computing, 1983,**12**, 144–56.

[41] Standish, T. A, *Data Structure Techniques*, Reading, MA: Addison-Wesley, 1980.

[42] Stoll, Robert R, Sets, *Logic and Axiomatic Theories*, San Francisco: W. H. Freeman, 1961.

[43] Strang, G, *Linear Algebra and Its Applications*, 2nd ed. New York: Academic Press, 1980.

[44] Tucker, Alan C, *Applied Combinatorics*, 2nd ed. New York: John Wiley and Sons, 1984.

[45] Wand, Mitchell, *Induction, Recursion, and Programming*, New York: North-Holland, 1980.

[46] Warshall, S, *A Theorem on Boolean Matrices* Journal of the Association of Computing Machinery, 1962, 11-12.

[47] Weisstein, Eric W. *Strassen Formulas*, MathWorld--A Wolfram Web Resource, http://mathworld.wolfram.com/StrassenFormulas.html.

[48] Wilf, Herbert S, *Some Examples of Combinatorial Averaging*, American Math. Monthly 92 (1985).

[49] Wilf, Herbert S. *generatingfunctionology*, A K Peters/CRC Press, 2005 The 1990 edition of this book is available at https://www.math.upenn.edu/~wilf/DownldGF.html

[50] Winograd, S, *On the Time Required to Perform Addition*, J. Assoc. Comp. Mach. 12 (1965): 277-85.

[51] Wilson, R., *Four Colors Suffice - How the Map Problem Was Solved*Princeton, NJ: Princeton U. Press, 2013.

Index

Abelian Group, 28
    a Limerick, 21
Algebraic Systems, 25
algorithm
    a Limerick, 213
Alternating Group, 165
Alternating group
    a Limerick, 149
An, 259
Associative Property, 23
Atom of a Boolean Algebra, 113
augmented matrix
    a Limerick, 63
Automata, 138
Automorphism
    Inner, 62

Basis, 80
Binary Conversion Algorithm, 1
Binary Operation., 22
Bit, 259
Boolean Algebra, 110
Boolean Algebras, 109
Boolean Expression, 117
Boolean Expressions, 117
Bounded Lattice, 109
Byte, 259
Bézout's lemma, 37

Cancellation in Groups, 31
Chinese Remainder Theorem, 152
Closure Property, 23
Code
    Polynomial, 206
Codes
    Linear, 177
Coding Theory, 177
Commutative Property, 22
Complement of a Lattice Element, 109
    as an operation, 110

Complemented Lattice, 109
Complex Number, 259
Composite Integer, 259
Concatenation, 133
Congruence Modulo m, 12
Constant, 259
Coset, 156
Coset Counting Formula, 157
Coset Representative, 156
Cosets
    Operation on, 158
Cosets and Factor Groups, 155
covering relation, 114
Creative Commons, 259
Cycle Notation, 163
Cyclic Group, 47, 149
Cyclic Subgroup, 47

Data Structure, 259
Degree Sequence of a Graph, 17
Diagonalizable Matrix, 85
Diagonalization Process, The, 83
Dihedral Group, 166
    Definition, 169
Dimension of a Vector Space, 81
Direct Product, 49
    of Two Groups, 49
Direct Products, 49
Disjoint Cycles, 163
Distinct, 259
Distributive Lattice, 107
Distributive Property, 23
Division Property for Integers, 35
Division Property for Polynomials, 202
Divisors of an Integer, 105
Doyle, Chris, 149
Duality for Boolean Algebras, 111

Eigenvalue, 83
Eigenvector, 83

Elementary Operations on
    Equations, 64
Elementary Row Operations, 66
Euclidean Algorithm, The, 36
Even Integer, 259
Exponentiation in Groups, 32

Factor, 259
Factor Group, 160
Factor Theorem, 202
Fibonacci Sequence
    Matrix Representation, 92
Field, 195
field extension
    a Limerick, 187
Finite-State Machine, 139
Finite-State Machines, 138
Formal Language, 133
Frankovich, Jesse, 213
Free Monoids and Languages, 132
Fundamental Theorem of Group
    Homomorphisms, 175

Gauss-Jordan Algorithm, 69
Generate, 78
Generator, 47
George Boole
    a Limerick, 103
Glossary, 259
Gray Code Decoder, 143
Greatest Common Divisor (gcd), 35
Greatest Element, 104
Greatest Lower Bound, 104
Group, 27

Hamming Distance, 178
Homomorphism, 173
    Group, 170

Idempotent Property, 23
Identity Property, 23
Iff, 259
Integer, 259
Integers Modulo n
    Additive Group, 40
    Multiplicative Group, 40
Integral Domain, 193
Inverse Property, 23
Involution Property, 23
Irrational Number, 260
Irreducibility of a Polynomial, 203
Isomorphism
    Group, 57
Isomorphisms, 55

Join, 107

Karnaugh map, 120
Kernel, 174

Lagrange's Theorem, 158
LaTeX, 260
Lattice, 107
Lattices, 107
Least Element, 104
Least Upper Bound, 104
Left Distributive Property, 23
Levels of Abstraction, 26
Linear Code, 181
Linear Combination., 78
Linear Dependence, 80
Linear Equations
    over the Integers Mod 2, 97
Linear Equations in a Group, 32
Linear Independence, 80
Logic Design, 121
Lower Bound, 103

Machine of a Monoid, 146
Matrix Inversion, 72
Meet, 107
Minset, 7
Minset Normal Form, 7
Minterm, 119
Minterm Normal Form, 119
Modular Addition, 39
Modular Arithmetic, 38
    Properties, 40
Modular Multiplication, 39
Monoid, 129
    of a Finite-State Machine, 143
Monoids, 129
Multiples, 260
Multiplicative Inverses, 189

N-cube, 17
Natural Homomorphism, 174
Natural Numbers, 260
Ngai, Steve, 63
Nonnegative Number, 260
Normal Subgroup., 172
Normal Subgroups, 170

Odd Integer, 260
Operation Tables, 24
Operations, 21
Order
    of elements of a finite cyclic
        group, 152
Order of a Group Element, 47
Order Sequence, 59

Oslan, Steven, ix

Partial Ordering, 103
Partition
    of a group by cosets, 157
Perfect Codes, 186
Permutation Groups, 161
Permutations
    Composition, 163
Phrase Structure Grammar, 135
Polynomial
    Irreducible, 203
Polynomial Addition, 199
Polynomial Code, 206
Polynomial Multiplication, 200
Polynomial over a Ring, 199
Polynomial Units, 210
Posets Revisited, 103
Positive Number, 260
Power Series, 208
Power Series Units, 210
Powers, 260
PreTeXt, 260
Prime, 260
Properties of Operations, 22
Propp, Jim, 59

Queue, 260

Rational Number, 260
Real Number, 260
Rectangular codes, 185
Recursive Language, 134
Reducible Polynomial, 203
References, 263
Regular Grammar, 137
Relatively Prime, 36
Right Distributive Property, 23
Ring, 187
    Commutative, 188
Ring Isomorphism., 190
Ring with unity, 188
Robinson, Andrew, 103
Row Equivalent Matrices, 66

SageMath, 260

SageMath Note
    Matrix Diagonalization, 89
    Matrix Exponential, 95
    Matrix Reduction, 70
    Modular Arithmetic, 41
Solution Set, 63
Some General Properties of
    Groups, 30
Span, 78
Spindel, Howard, 21
Stack, 260
Strings over an Alphabet, 132
STT, *See* Sun Tzu's Theorem
Subgroup, 44
    Conditions, 45
Submonoid
    Generated by a Set, 130
Subsystem, 44
Subsystems, 44
Subtraction, 260
Sun Tzu's Theorem, 152
Switching Theory, 121
Symmetric Group, 162
Syndrome, 182
Systems of Linear Equations, 63

Term, 260
Transposition, 164

Unary Operation., 22
Unique, 260
Units
    of a ring, 189
    of Polynomial Rings, 210
    of Power Series Rings, 210
Unity of a Ring, 188
Upper Bound, 103

Variable, 260
Vector Space, 76
Vector Spaces, 76

XOR linked list, 53

Zero Divisor, 192
zqms, 187